D0078129

WORKBOOK FOR GENERAL CHEMISTRY

Third Edition

Bassam Z. Shakhashiri
Rodney Schreiner

Department of Chemistry
University of Wisconsin-Madison
Madison, Wisconsin

Stipes Publishing Company
Champaign, Illinois

ISBN 1-58874-431-0

Copyright © by Bassam Z. Shakhashiri and Rodney Schreiner. All rights reserved.

No part of this publication may be reproduced or transmitted in any form or by any means, electronic or mechanical, including photocopying and recording, or by any information storage or retrieval system, without prior written permission of the publisher, Stipes Publishing L.L.C., 204 W. University Ave., Champaign, IL 61820

Second Printing

PREFACE

This Workbook contains a series of lessons that deal with many of the skills needed for success in a general chemistry course. These lessons help students to develop, on their own and at their own pace, the ability to solve various sorts of problems encountered in college-level and advanced high-school courses. This book is intended as a complement to a textbook that provides a context for the exercises and problems presented here.

At the beginning of each lesson is a list describing the material presented in the lesson, and with a concept map that represents relationships between the important concepts in the lesson. A special notation is used in these maps:

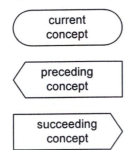

Concepts that are introduced in the current lesson are represented in ovals.

Relevant concepts from previous lessons are represented in boxes with arrows pointing left.

Relevant concepts that appear in following lessons are represented in boxes with arrows pointing right.

The lesson proceeds with a concise outline of steps needed to master the material. The steps are presented in brief descriptions and completed examples. The lessons are subdivided into sections by exercises that provide an opportunity to check understanding and mastery of the material. Students should complete the exercises and check their results with the step-by-step solutions provided in the answer section at the back of the Workbook before proceeding to the following sections of the lesson. If the answers are not correct, they should return to the section and attempt the exercise again. When the answers are correct, they may proceed to the next section. This approach allows students to study and learn at their own pace. Many lessons also conclude with a set of problems whose answers are also provided.

This Workbook is based on a series of lessons that have been developed and used at the University of Wisconsin-Madison and at many other schools for over 25 years. Over the years, we have benefitted from the comments and suggestions of many students and colleagues. We express deep gratitude to our first-year students, their graduate teaching assistants, and the staff of the Initiative for Science Literacy for much helpful feedback and suggestions.

Bassam Z. Shakhashiri
Rodney Schreiner

Madison, Wisconsin
July, 2004

CONTENTS

CHEMICAL SYMBOLS AND FORMULAS: MOLAR MASS CALCULATIONS

This lesson deals with:

1. Writing and interpreting symbols of the elements and their isotopes.
2. Writing and interpreting chemical formulas of pure substances.
3. Calculating the molar mass of a substance from its formula and the atomic weights of the elements.
4. Calculating the percent composition by weight of a substance from its formula.
5. Calculating the mass of an element contained in a given mass of a compound.
6. Determining the empirical formula of a compound from its percent composition by mass.
7. Determining the molecular formula of a compound from its empirical formula and its molar mass.
8. Calculating the molar mass of an element from percent abundance and masses of its isotopes.

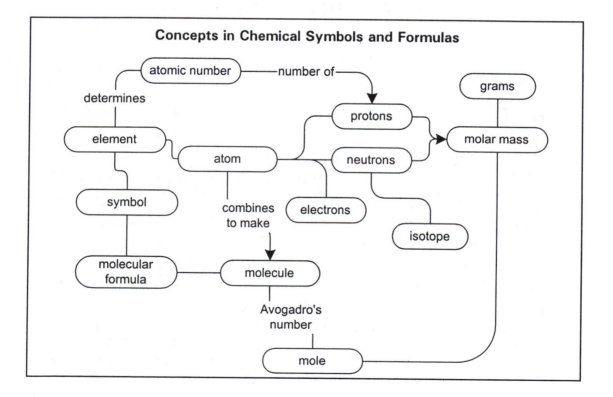

Concepts in Chemical Symbols and Formulas

ELEMENTS

An **element** is a substance whose atoms are all alike. There are over 100 different kinds of atoms, so there are over 100 different elements.

The elements have names and symbols.

    The symbols of the elements are listed on the Periodic Table of the Elements.

        Above each symbol is the atomic number of the element.

        Below each symbol is the molar mass of the element (also called its atomic weight).

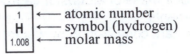

1	⟵ atomic number
H	⟵ symbol (hydrogen)
1.008	⟵ molar mass

Some elements are more common than others.

    Table 1 lists the elements most commonly encountered in General Chemistry.

Memorize the names and symbols of the elements listed in Table 1.

    A good way to do this is to make a set of flash cards with the name of an element on one side and the corresponding symbol on the other.

    It is not necessary to memorize the atomic numbers; the atomic numbers are listed on the Periodic Table.

PERIODIC TABLE OF THE ELEMENTS

IA	IIA											IIIA	IVA	VA	VIA	VIIA	VIIIA
						1 **H** 1.008											2 **He** 4.003
3 **Li** 6.941	4 **Be** 9.012											5 **B** 10.81	6 **C** 12.01	7 **N** 14.01	8 **O** 16.00	9 **F** 19.00	10 **Ne** 20.18
11 **Na** 22.99	12 **Mg** 24.30	IIIB	IVB	VB	VIB	VIIB		VIIIB		IB	IIB	13 **Al** 26.98	14 **Si** 28.09	15 **P** 30.97	16 **S** 32.06	17 **Cl** 35.45	18 **Ar** 39.95
19 **K** 39.10	20 **Ca** 40.08	21 **Sc** 44.96	22 **Ti** 47.87	23 **V** 50.94	24 **Cr** 52.00	25 **Mn** 54.94	26 **Fe** 55.85	27 **Co** 58.93	28 **Ni** 58.69	29 **Cu** 63.55	30 **Zn** 65.39	31 **Ga** 69.72	32 **Ge** 72.61	33 **As** 74.92	34 **Se** 78.96	35 **Br** 79.90	36 **Kr** 83.80
37 **Rb** 85.47	38 **Sr** 87.62	39 **Y** 88.90	40 **Zr** 91.22	41 **Nb** 92.91	42 **Mo** 95.94	43 **Tc** (98)	44 **Ru** 101.1	45 **Rh** 102.9	46 **Pd** 106.4	47 **Ag** 107.9	48 **Cd** 112.4	49 **In** 114.8	50 **Sn** 118.7	51 **Sb** 121.8	52 **Te** 127.6	53 **I** 126.9	54 **Xe** 131.3
55 **Cs** 132.9	56 **Ba** 137.3	57 **La** 138.9	72 **Hf** 178.5	73 **Ta** 180.9	74 **W** 183.8	75 **Re** 186.2	76 **Os** 190.2	77 **Ir** 192.2	78 **Pt** 195.1	79 **Au** 197.0	80 **Hg** 200.6	81 **Tl** 204.4	82 **Pb** 207.2	83 **Bi** 209.0	84 **Po** (209)	85 **At** (210)	86 **Rn** (222)
87 **Fr** (223)	88 **Ra** (226)	89 **Ac** (227)	104 **Rf** (261)	105 **Db** (262)	106 **Sg** (263)	107 **Bh** (262)	108 **Hs** (265)	109 **Mt** (266)	110 **Ds** (271)	111 **Uuu** (272)							

58 **Ce** 140.1	59 **Pr** 140.9	60 **Nd** 144.2	61 **Pm** (145)	62 **Sm** 150.4	63 **Eu** 152.0	64 **Gd** 157.2	65 **Tb** 158.9	66 **Dy** 162.5	67 **Ho** 164.9	68 **Er** 167.3	69 **Tm** 168.9	70 **Yb** 173.0	71 **Lu** 175.0
90 **Th** 232.0	91 **Pa** 231.0	92 **U** 238.0	93 **Np** (237)	94 **Pu** (244)	95 **Am** (243)	96 **Cm** (247)	97 **Bk** (247)	98 **Cf** (251)	99 **Es** (252)	100 **Fm** (257)	101 **Md** (258)	102 **No** (259)	103 **Lr** (262)

The symbol of an element may represent:
   (a) a sample of the element or
   (b) an atom of the element.

Atoms are composed of **protons, neutrons, and electrons**.

   The **atomic number**:   the number of protons in an atom.
   The **mass number**:   the sum of the number of protons and the number of neutrons in an atom.

   In a neutral atom, the number of electrons is the same as the number of protons.

Indicating the composition of an atom:

mass number
$$^{24}_{12}\text{Mg} \leftarrow \text{symbol of element}$$
atomic number

Atoms that have the same atomic number but different mass numbers are called **isotopes**.

   Three isotopes of magnesium: ^{24}Mg ^{25}Mg ^{26}Mg
       number of neutrons = mass number – atomic number

      ^{24}Mg: 24 – 12 = 12 neutrons
      ^{25}Mg: 25 – 12 = 13 neutrons
      ^{26}Mg: 26 – 12 = 14 neutrons

The mass number is used in a symbol only when a particular isotope needs to be specified.

Example 1

A. What is the symbol for the atom containing 16 protons, 17 neutrons, and 16 electrons?

   The atomic number is 16, therefore the element is sulfur: S

   The mass number is the number of protons plus the number of neutrons:

         16 protons + 17 neutrons = 33 particles (mass number)

   The symbol is ^{33}S. The name is written as sulfur-33.

B. What particles are contained in a neutral atom of ^{60}Co?

   Co is the symbol for cobalt, element with atomic number 27.
      Therefore the atom contains 27 protons and 27 electrons.
   The mass number is 60, which is the sum of the number of protons and the number of neutrons. Therefore, the number of neutrons is 60 – 27, namely 33.

   The particles in a neutral atom of ^{60}Co are: 27 protons, 33 neutrons, and 27 electrons.

Table 1. Names and Symbols of Selected Elements

Atomic Number	Name	Symbol	Atomic Number	Name	Symbol
1	hydrogen	H	26	iron	Fe
2	helium	He	27	cobalt	Co
3	lithium	Li	28	nickel	Ni
4	beryllium	Be	29	copper	Cu
5	boron	B	30	zinc	Zn
6	carbon	C	31	gallium	Ga
7	nitrogen	N	32	germanium	Ge
8	oxygen	O	33	arsenic	As
9	fluorine	F	34	selenium	Se
10	neon	Ne	35	bromine	Br
11	sodium	Na	36	krypton	Kr
12	magnesium	Mg	38	strontium	Sr
13	aluminum	Al	47	silver	Ag
14	silicon	Si	50	tin	Sn
15	phosphorus	P	51	antimony	Sb
16	sulfur	S	53	iodine	I
17	chlorine	Cl	56	barium	Ba
18	argon	Ar	74	tungsten	W
19	potassium	K	78	platinum	Pt
20	calcium	Ca	79	gold	Au
21	scandium	Sc	80	mercury	Hg
22	titanium	Ti	82	lead	Pb
23	vanadium	V	83	bismuth	Bi
24	chromium	Cr	92	uranium	U
25	manganese	Mn	94	plutonium	Pu

Exercise One

A. Complete the following (refer to the Periodic Table and to Table 1 if necessary):

name of atom	symbol	number of protons	number of neutrons
chlorine-35	^{35}Cl	_____	_____
carbon-14	_____	_____	_____
_____	^{90}Sr	_____	_____
_____	_____	8	10
cobalt-60	_____	_____	_____
_____	^{235}U	_____	_____

B. Complete the following (refer to the Periodic Table and to Table 1 if necessary):

name of element	atomic number	symbol	name of element	atomic number	symbol
oxygen	8	O	antimony	51	___
carbon	___	___	sulfur	___	___
phosphorus	___	___	___	___	Mg
___	___	Mn	bromine	___	___
calcium	___	___	___	___	Ar
strontium	38	___	copper	___	___
lead	___	___	tin	___	___
___	___	Ni	bismuth	___	___
sodium	___	___	cobalt	___	___
potassium	___	___	barium	___	___
plutonium	94	___	gold	___	___
mercury	___	___	iodine	___	___

MOLAR MASS RELATIONSHIPS

The protons and neutrons contribute most of the mass of an atom.

mass of proton $\approx$ mass of neutron

The electron has a mass only 1/1800 times that of a proton or neutron. Its mass is negligible.

One atom of ^{24}Mg weighs twice as much as one atom of ^{12}C.

Five atoms of ^{24}Mg weigh twice as much as five atoms of ^{12}C.

Any number of atoms of ^{24}Mg weigh twice as much as the same number of atoms of ^{12}C.

A **mole** is defined as the *number* of atoms in exactly 12 grams of ^{12}C.

A mole of ^{12}C atoms has a mass of exactly 12 grams.

A mole of ^{24}Mg atoms has a mass that is twice the mass of a mole of ^{12}C atoms, because each ^{24}Mg atom has a mass twice that of a ^{12}C atom.

The mass of one mole of ^{24}Mg is

2×12 grams $= 24$ grams

The **molar mass** is the mass of one mole.
   The molar mass of ^{24}Mg is 24 grams/mole.

The mole is a number of items, just as a dozen is a number of items.
   A mole is the number of atoms in exactly 12 grams of ^{12}C.
   The number of atoms in 12 grams of ^{12}C is 6.022×10^{23}.
   Therefore, the number of items in a mole is 6.022×10^{23}.

The value of the number of items in a mole is called **Avogadro's number**.
   Therefore, Avogadro's number is 6.022×10^{23}.

The molar masses of the elements are listed on the periodic table.
   (They are not whole numbers because they are the weighted average mass of the various
   isotopes of the elements. See the end of this lesson for an example of how the molar
   masses are related to the masses of the isotopes.)

Example 2

What mass of aluminum contains twice as many atoms as 35.00 grams of copper?

There are two ways of finding the answer to this question.

1. Find the number of atoms in 35.00 grams of copper, double it, then find the mass of
    aluminum that contains that number of atoms.

$$35.00 \text{ g Cu} \left(\frac{1 \text{ mol Cu}}{63.55 \text{ g Cu}} \right) = 0.5507 \text{ mol Cu}$$

$$0.5507 \text{ mol Cu} \left(\frac{6.022 \times 10^{23} \text{ atoms}}{1 \text{ mol}} \right) = 3.316 \times 10^{23} \text{ atoms Cu}$$

$$6.632 \times 10^{23} \text{ atoms Al} \left(\frac{1 \text{ mol}}{6.022 \times 10^{23} \text{ atoms}} \right) = 1.101 \text{ mol Al}$$

$$1.101 \text{ mol Al} \left(\frac{26.98 \text{ g Al}}{1 \text{ mol Al}} \right) = 29.70 \text{ g Al}$$

2. The second way is simpler. Because a mole is a number, twice as many atoms
    corresponds to twice as many moles. Therefore, we can find the moles of copper,
    double it, and find the mass of that many moles of aluminum.

$$35.00 \text{ g Cu} \left(\frac{1 \text{ mol Cu}}{63.55 \text{ g Cu}} \right) = 0.5507 \text{ mol Cu}$$

$$1.101 \text{ mol Al} \left(\frac{26.98 \text{ g Al}}{1 \text{ mol Al}} \right) = 29.70 \text{ g Al}$$

Working with moles is the same as working with numbers of items. This is one of the
fundamental concepts of chemistry.

Exercise Two

What mass of sulfur contains half the number of atoms as 46.00 grams of oxygen?

CHEMICAL FORMULAS

A **chemical formula** consists of symbols of the elements followed by subscripts which indicate the ratio of atoms in a substance.

The chemical formula indicates the kind and number of atoms in the smallest unit of a substance, called a **formula unit**.

A **molecule** is a distinct group of atoms whose composition is indicated by the chemical formula. Some substances are composed of molecules, other substances are large aggregates of atoms.

Table 2 gives the chemical formulas for a variety of common substances.

Example 3

Indicate the kind and number of atoms represented by one formula unit of each of the following.

A. ammonium chloride NH_4Cl

one atom of chlorine
four atoms of hydrogen } in one formula unit
one atom of nitrogen

B. oxygen gas O_2
one molecule contains two atoms of oxygen

C. calcium nitrate $Ca(NO_3)_2$

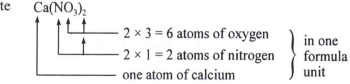

$2 \times 3 = 6$ atoms of oxygen
$2 \times 1 = 2$ atoms of nitrogen } in one formula unit
one atom of calcium

Table 2. Names and Formulas of Some Common Substances

A. *Elements that occur as polyatomic molecules*:

		normal state			normal state
hydrogen	H_2	gas	nitrogen	N_2	gas
fluorine	F_2	gas	oxygen	O_2	gas
chlorine	Cl_2	gas	ozone	O_3	gas
bromine	Br_2	liquid	sulfur	S_8	solid
iodine	I_2	solid	phosphorus	P_4	solid

B. *Compounds that are solids at room temperature*:

ammonium chloride	NH_4Cl	silver nitrate	$AgNO_3$
calcium chloride	$CaCl_2$	sodium acetate	$NaC_2H_3O_2$
calcium nitrate	$Ca(NO_3)_2$	sodium carbonate	Na_2CO_3
magnesium chloride	$MgCl_2$	sodium chloride	$NaCl$
nickel sulfate	$NiSO_4$	sodium sulfate	Na_2SO_4
potassium bromide	KBr	sodium	
potassium cyanide	KCN	hydrogen carbonate †	$NaHCO_3$

C. *Compounds that are liquids at room temperature*:

benzene	C_6H_6	ethanol (ethyl alcohol)	C_2H_5OH
carbon tetrachloride	CCl_4	methanol (methyl alcohol)	CH_3OH
chloroform	$CHCl_3$	hydrogen peroxide	H_2O_2
water	H_2O	octane	C_8H_{18}

D. *Compounds that are gases at room temperature*:

ammonia	NH_3	methane	CH_4
carbon dioxide	CO_2	ethane	C_2H_6
carbon monoxide	CO	propane	C_3H_8
hydrogen chloride	HCl	butane	C_4H_{10}
hydrogen cyanide	HCN	nitric oxide	NO
hydrogen sulfide	H_2S	nitrogen dioxide	NO_2
sulfur dioxide	SO_2	sulfur trioxide	SO_3

E. *Acids*

acetic acid	$HC_2H_3O_2$	perchloric acid	$HClO_4$
sulfuric acid	H_2SO_4	nitric acid	HNO_3
sulfurous acid	H_2SO_3	nitrous acid	HNO_2
hydrochloric acid	HCl (aqueous solution of hydrogen chloride)		

F. *Bases*

ammonia	NH_3	potassium hydroxide	KOH
calcium hydroxide	$Ca(OH)_2$	sodium hydroxide	$NaOH$

† Sodium hydrogen carbonate is often called sodium bicarbonate. It is also baking soda.

Exercise Three

A. Write the name of each of the following:

HCl   _____ NH_3 _____

H_2SO_4 _____ $NaHCO_3$ _____

B. Write the chemical formula for each of the following:

sodium chloride   _____   sulfur trioxide   _____

gaseous hydrogen   _____   methane   _____

carbon dioxide   _____   nitric acid   _____

C. Indicate the number of oxygen atoms in one formula unit of:

O_3 _____ $Ba_3(PO_4)_2$ _____ Na_3PO_4 _____ $Al_2(SO_4)_3$ _____

D. Indicate the total number of atoms in one formula unit of:

H_2O _____ $Ca(OH)_2$ _____ $CHCl_3$ _____ $NaAl(SO_4)_2$ _____

MOLAR MASS

The **molar mass** of a substance is the mass that contains one mole of molecules (or formula units).

$$1 \text{ molar mass} = 1 \text{ } mole$$

Example 4

What is the mass of one mole of water?

H_2O

one atom of oxygen
two atoms of hydrogen $\}$ in one molecule

1 mole of H_2O molecules contain:

1 mole of O atoms   and   2 moles of H atoms
16.00 grams   plus $2 \times (1.008 \text{ grams})$

equals 18.02 grams

$$1 \text{ mole } H_2O = 18.02 \text{ grams } H_2O$$

Example 5

What is the mass, in grams, of one mole of calcium nitrate?

Formula: $Ca(NO_3)_2$

one mole of Ca	=	1 mole × 40.08 g/mol	=	40.08 grams
two moles of N	=	2 moles × 14.01 g/mol	=	28.02 grams
six moles of O	=	6 moles × 16.00 g/mol	=	96.00 grams
		1 mole $Ca(NO_3)_2$	=	164.10 grams

PERCENT COMPOSITION BY MASS

The **percent composition by mass** (or weight) of a substance is the mass percent of all of the elements in the substance.

Example 6

What is the percent composition by mass of each of the elements in calcium nitrate?

$$1 \text{ mole } Ca(NO_3)_2 \quad = 40.08 \text{ grams Ca} + 28.02 \text{ grams N} + 96.00 \text{ grams O}$$
$$= 164.10 \text{ grams}$$

$$\% \text{ Ca} = \frac{\text{mass Ca}}{\text{mass } Ca(NO_3)_2} \times 100 = \frac{40.08 \text{ g}}{164.10 \text{ g}} \times 100 = 24.42 \% \text{ Ca}$$

$$\% \text{ N} = \frac{\text{mass N}}{\text{mass } Ca(NO_3)_2} \times 100 = \frac{28.02 \text{ g}}{164.10 \text{ g}} \times 100 = 17.07 \% \text{ N}$$

$$\% \text{ O} = \frac{\text{mass O}}{\text{mass } Ca(NO_3)_2} \times 100 = \frac{96.00 \text{ g}}{164.10 \text{ g}} \times 100 = 58.50 \% \text{ O}$$

Note: the sum of the mass percents of the elements in a compound is 100%

24.42% + 17.07% + 58.50% = 99.99% = 100% (3 significant figures).

The percent composition of a substance can be used to calculate the mass of an element in a given mass of the substance.

Example 7

How many grams of nitrogen are in 25.0 grams of calcium nitrate?

$$\text{mass nitrogen} = 17.07\% \text{ of 25.0 grams}$$
$$= (0.1707)(25.0 \text{ grams}) = 4.27 \text{ grams}$$

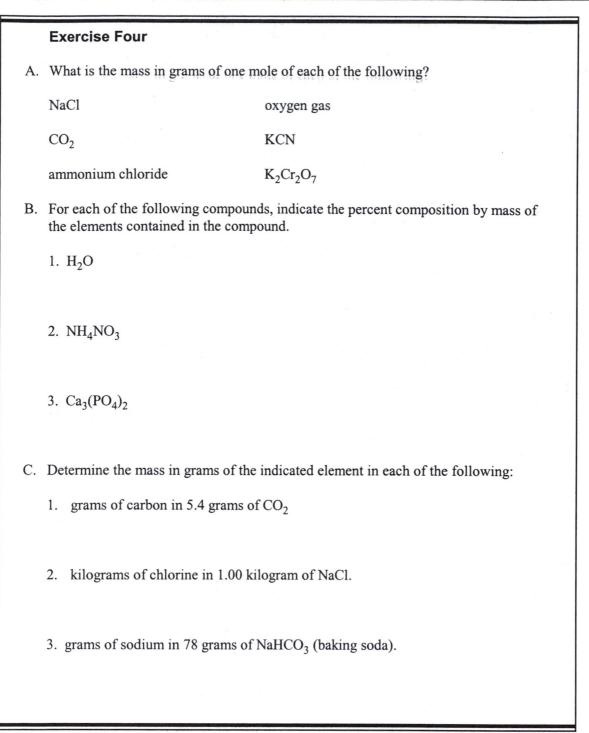

Exercise Four

A. What is the mass in grams of one mole of each of the following?

NaCl   oxygen gas

CO_2 KCN

ammonium chloride $K_2Cr_2O_7$

B. For each of the following compounds, indicate the percent composition by mass of the elements contained in the compound.

1. H_2O

2. NH_4NO_3

3. $Ca_3(PO_4)_2$

C. Determine the mass in grams of the indicated element in each of the following:

1. grams of carbon in 5.4 grams of CO_2

2. kilograms of chlorine in 1.00 kilogram of NaCl.

3. grams of sodium in 78 grams of $NaHCO_3$ (baking soda).

DETERMINING THE FORMULA OF A COMPOUND

The **empirical formula** indicates the simplest whole number ratio between the numbers of atoms of the elements in a substance.

The **molecular formula** indicates the number of atoms of each element in a molecule of substance.

Consider hydrogen peroxide –   molecular formula: H_2O_2
empirical formula: HO

To determine the empirical formula from percent composition:
1. find the number of grams of each element in a certain mass of the compound (such as 100 grams).
2. calculate the number of moles of each element in this mass.
3. find the ratios of these numbers of moles.
4. convert these ratios to the smallest whole numbers.

To determine the molecular formula:
1. divide the molar mass by the empirical formula mass; the result will be an integer.
2. multiply the subscripts of the empirical formula by this integer.

Example 8

One type of mothballs, para-dichlorobenzene, contains 49.0% carbon, 48.3% chlorine, and 2.72% hydrogen by mass. Its molar mass is 147.0 grams per mole. What is the molecular formula of para-dichlorobenzene?

1. The percent composition indicates that 100 grams of para-dichlorobenzene contains
    49.0 grams carbon (C)
    48.3 grams chlorine (Cl)
    2.72 grams hydrogen (H)

2. Find moles of each element in 100 grams

$$49.0 \text{ g C} \left(\frac{1 \text{ mol C}}{12.01 \text{ g C}} \right) = 4.08 \text{ mol C}$$

$$48.3 \text{ g Cl} \left(\frac{1 \text{ mol Cl}}{35.45 \text{ g Cl}} \right) = 1.36 \text{ mol Cl}$$

$$2.72 \text{ g H} \left(\frac{1 \text{ mol H}}{1.008 \text{ g H}} \right) = 2.70 \text{ mol H}$$

3. Find the ratio of moles of the elements by dividing by the smallest number of moles. The moles of chlorine is the smallest, namely 1.36 moles.

Ratio of moles of C to moles of Cl:

$$\frac{4.08 \text{ mol C}}{1.36 \text{ mol Cl}} = 3.00 \text{ mol C/mol Cl}$$

Ratio of moles of Cl to moles of Cl:

$$\frac{1.36 \text{ mol Cl}}{1.36 \text{ mol Cl}} = 1.00 \text{ mol Cl/mol Cl}$$

Ratio of moles of H to moles of Cl:

$$\frac{2.70 \text{ mol H}}{1.36 \text{ mol Cl}} = 1.98 \text{ mol H/mol Cl}$$

1.98 is very close to 2 – round off to 2.

4.   Convert the ratios to simple whole numbers.

   3.00 moles C/mole Cl $\longrightarrow$ 3 C
   1.00 mole Cl/mole Cl $\longrightarrow$ 1 Cl
   1.98 moles H/mole Cl $\longrightarrow$ rounded off to 2 H

The empirical formula of para-dichlorobenzene is C_3ClH_2.

Determine the molecular formula:

1.   divide the molar mass by the empirical formula mass; the result will be an integer

Empirical formula mass of C_3ClH_2:
$$3(12.01) + (35.45) + 2(1.008) = 73.50 \text{ g}$$

$$\frac{\text{molar mass}}{\text{empirical formula mass}} = \frac{147.0 \text{ g}}{73.50 \text{ g}} = 2$$

2.   multiply the subscripts of the empirical formula by this integer.

The molecular formula of para-dichlorobenzene is $C_6Cl_2H_4$.

Exercise Five

A. Determine the empirical formula of a compound that contains 40.0% carbon, 53.3% oxygen, and 6.67% hydrogen by mass.

B. Determine the molecular formula for the compound that contains 85.66% carbon and 14.34% hydrogen by mass. Its molar mass is 70.13 grams per mole.

ANOTHER EXAMPLE OF DETERMINING AN EMPIRICAL FORMULA

Example 9

The percent composition by mass of glycerol is 39.12% carbon, 8.756% hydrogen, and 52.11% oxygen. What is the empirical formula of glycerol?

1. In 100 grams of glycerol there are
    39.12 grams carbon
    8.756 grams hydrogen
    52.11 grams oxygen

2. Find moles of each element

$$39.12 \text{ g C} \left(\frac{1 \text{ mol C}}{12.01 \text{ g C}} \right) = 3.257 \text{ mol C}$$

$$8.756 \text{ g H} \left(\frac{1 \text{ mol H}}{1.008 \text{ g H}} \right) = 8.686 \text{ mol H}$$

$$52.11 \text{ g O} \left(\frac{1 \text{ mol O}}{16.00 \text{ g O}} \right) = 3.257 \text{ mol O}$$

3. Find ratios:

$$\frac{3.257 \text{ mol C}}{3.257 \text{ mol C}} = 1.00 \text{ mol C/mol C}$$

$$\frac{8.686 \text{ mol H}}{3.257 \text{ mol C}} = 2.667 \text{ mol H/mol C}$$

$$\frac{3.257 \text{ mol O}}{3.257 \text{ mol C}} = 1.00 \text{ mol O/mol C}$$

4. Convert ratios to whole numbers
    either

    (a) round off — if change is less than 3%.
         If 2.667 is rounded to 3, the change is $3 - 2.667 = 0.333$, and the percent change is

$$\frac{0.333}{2.667} \times 100 = 12.5\%$$

    Here, rounding off produces a change of 12%, which is too large.

or

    (b) multiply all ratios by same integer to convert all to integers.
         Multiplying by three,
         C — 3
         H — 8.001, round off to 8
         O — 3

The empirical formula of glycerol is $C_3H_8O_3$.

If the ratio of moles is	then multiply by
1.5, 2.5, 3.5, etc.....	2
0.33, 1.33, etc....or 0.67, 1.67, etc....	3
0.25, 1.25, etc....or 0.75, 1.75, etc....	4

Exercise Six

Determine the empirical formula of a compound that contains 26.58% potassium, 35.35% chromium, and 38.07% oxygen by mass.

DETERMINING ELEMENTAL MOLAR MASS FROM ISOTOPE DATA

The **molar mass** of an element is the average molar mass of all of its isotopes.

Example 10

Use the information in the table below to calculate the molar mass of naturally-occurring magnesium.

The molar mass of an element is the average mass of its isotopes.

Percent Abundance of Magnesium Isotopes

Isotope	% Abundance	Mass on ^{12}C Scale
^{24}Mg	78.70%	23.98 grams/mole
^{25}Mg	10.13%	24.98 grams/mole
^{26}Mg	11.17%	25.98 grams/mole

78.70% of 23.98 g/mol = (0.7870)(23.98 g/mol) = 18.87 g/mol

10.13% of 24.98 g/mol = (0.1013)(24.98 g/mol) = 2.530 g/mol

11.17% of 25.98 g/mol = (0.1117)(25.98 g/mol) = 2.902 g/mol

Average Molar Mass = 24.30 g/mol

One molar mass of magnesium is 24.30 grams and contains 6.022×10^{23} atoms.

Exercise Seven

Calculate the molar mass of boron. The natural abundance and molar masses on the carbon-12 scale of boron isotopes are:

^{10}B	19.78%	10.01 g/mol
^{11}B	80.22%	11.01 g/mol

PROBLEMS

1. Write the chemical formula for each of the following:

   (a) nitric acid
   (b) solid iodine
   (c) ozone
   (d) sodium hydroxide
   (e) methane

   (f) sulfur dioxide
   (g) hydrogen cyanide
   (h) acetic acid
   (i) hydrogen peroxide
   (j) sodium carbonate

2. Give the mass in grams (to 4 significant figures) of one mole of each of the following.

   (a) H_2SO_4
   (b) $(NH_2)_2CO$
   (c) $[Ag(NH_3)_2]_3PO_4$

   (d) liquid bromine
   (e) sodium hydrogen carbonate
   (f) ammonia

3. Emerald gemstones are composed mainly of the mineral beryl, which has the chemical formula $Be_3Al_2Si_6O_{18}$. What is the percent composition by mass of beryl?

4. A. What mass of iron metal can be obtained from one metric ton (1000.0 kg) of hematite, an iron ore which has the chemical formula Fe_2O_3?

   B. What mass of hematite is needed to produce one metric ton of iron metal?

5. Vitamin C has a molar mass of about 176 g/mol, and it is 40.9% carbon, 4.55% hydrogen, and 54.5% oxygen by mass. What is the molecular formula of vitamin C?

WRITING AND BALANCING CHEMICAL EQUATIONS

This lesson deals with:

1. Writing a chemical equation to express a verbal description of a chemical reaction.
2. Determining whether a chemical equation is balanced.
3. Balancing chemical equations by inspection.

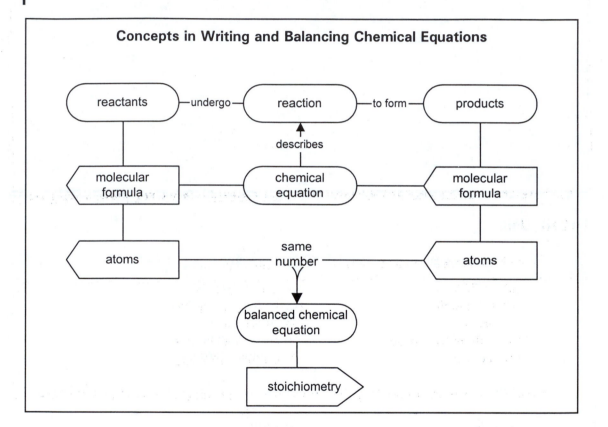

CHEMICAL EQUATIONS

A **chemical equation** represents a chemical reaction. It is a statement describing in chemical symbols what occurs in a chemical reaction.

Example 1

A chemical reaction described in words:

> When methane gas burns in gaseous oxygen, the products are carbon dioxide gas and water vapor.

The same chemical reaction described by a chemical equation:

$$CH_4(g) \quad + \quad O_2(g) \quad \longrightarrow \quad CO_2(g) \quad + \quad H_2O(g)$$

methane gas	reacts with	oxygen gas	to give, yield, or form	carbon dioxide gas	and	water vapor

reactants   products

States of reactants and products are indicated parenthetically after their symbols.

(s)	—	solid
(l)	—	liquid
(g)	—	gas (vapor)
(aq)	—	aqueous (dissolved in water)

Example 2

Solid sodium reacts with liquid water to form aqueous sodium hydroxide and hydrogen gas. Write a chemical equation to represent this reaction.

sodium solid	reacts with	water liquid	to form	sodium hydroxide aqueous	and	hydrogen gas

$$Na(s) \quad + \quad H_2O(l) \quad \longrightarrow \quad NaOH(aq) \quad + \quad H_2(g)$$

Exercise One

Write a representation of each of the following chemical changes:

A. Ammonia gas reacts with oxygen gas to form gaseous nitric oxide and water vapor.

B. Aqueous sodium carbonate and hydrochloric acid react and form aqueous sodium chloride, liquid water, and carbon dioxide gas.

C. When heated, solid potassium chlorate ($KClO_3$) decomposes to form potassium chloride, a solid, and oxygen gas.

D. Sulfur dioxide gas reacts with liquid water to form aqueous sulfurous acid.

BALANCED CHEMICAL EQUATIONS

A *balanced* chemical equation indicates

(a) that the number of atoms of each element in the reactants is the same as the number of atoms of each element in the products.

(b) the relationship between the numbers of moles of the reactants and products in a reaction.

Example 3

When methane gas burns in gaseous oxygen, the products are carbon dioxide gas and water vapor.

$$CH_4(g) + O_2(g) \longrightarrow CO_2(g) + H_2O(g)$$

reactants	products
4 H	2 H
1 C	1 C
2 O	3 O

The number of atoms of H and of O are not the same in the reactants as in the products. Therefore, this equation is not balanced.

The balanced equation is

$$CH_4(g) + 2 O_2(g) \longrightarrow CO_2(g) + 2 H_2O(g)$$

reactants	products
4 H	4 H
1 C	1 C
4 O	4 O

What the balanced chemical equation indicates:

one molecule CH_4 + two molecules $O_2 \longrightarrow$

one molecule CO_2 + two molecules H_2O

12 molecules $CH_4 + 2 \times 12$ molecules $O_2 \longrightarrow$

12 molecules $CO_2 + 2 \times 12$ molecules H_2O

6.022×10^{23} molecules $CH_4 + 2 \times (6.022 \times 10^{23})$ molecules $O_2 \longrightarrow$

6.022×10^{23} molecules $CO_2 + 2 \times (6.022 \times 10^{23})$ molecules H_2O

1 mole CH_4 + 2 moles $O_2 \longrightarrow$ 1 mole CO_2 + 2 moles H_2O

Exercise Two

Indicate whether each of the following equations is balanced:

A. $AgNO_3(aq) + NaCl(aq) \longrightarrow NaNO_3(aq) + AgCl(s)$

B. $C_6H_{12}O_6(s) + O_2(g) \longrightarrow 6\,CO_2(g) + 5\,H_2O(l)$

C. $2\,NaHCO_3(s) + H_2SO_4(aq) \longrightarrow Na_2SO_4(aq) + 2\,CO_2(g) + 2\,H_2O(l)$

D. $3\,CuO(s) + 2\,NH_3(g) \longrightarrow 3\,Cu(s) + N_2(g) + 3\,H_2O(g)$

E. $3\,Ca(NO_3)_2(aq) + Na_3PO_4(aq) \longrightarrow Ca_3(PO_4)_2(s) + 3\,NaNO_3(aq)$

F. $3\,H_2SO_4(l) + H_2S(g) \longrightarrow 4\,SO_2(g) + 4\,H_2O(l)$

BALANCING CHEMICAL EQUATIONS

Balance a chemical equation by placing numbers before the chemical formulas. *Never* change a subscript in a formula.

Example 4

The decomposition of liquid water by electrolysis:

unbalanced: $H_2O(l) \longrightarrow H_2(g) + O_2(g)$

balanced: $2\,H_2O(l) \longrightarrow 2\,H_2(g) + O_2(g)$

WRONG: $H_2O_2(l) \longrightarrow H_2(g) + O_2(g)$

This is wrong, because changing a subscript changes the identity of the substance. This equation no longer involves water, but hydrogen peroxide instead.

Example 5

Balance this chemical equation.

$$Mg_3N_2(s) + H_2O(l) \longrightarrow Mg(OH)_2(aq) + NH_3(g)$$

1. Count the number of atoms of each element in the reactants and in the products.

reactants	products
3 Mg	1 Mg
2 N	1 N
2 H	5 H
1 O	2 O

    The reaction is not balanced.

2. Begin by identifying the elements that occur in only two substances (compounds or elements) in the equation.

    Hydrogen occurs in three compounds: H_2O, $Mg(OH)_2$, and NH_3.

    Magnesium, nitrogen, and oxygen occur in only two compounds.

3. Among the elements that occur in only two compounds identify the one that has the largest subscript.

    Magnesium has a subscript of 3.

    Balance magnesium first.

$$Mg_3N_2(s) + H_2O(l) \longrightarrow 3 Mg(OH)_2(aq) + NH_3(g)$$

reactants	products
3 Mg	3 Mg
2 N	1 N
2 H	9 H
1 O	6 O

4. Balance elements other than hydrogen and oxygen.
    Balance nitrogen:

$$Mg_3N_2(s) + H_2O(l) \longrightarrow 3 Mg(OH)_2(aq) + 2 NH_3(g)$$

5. Balance hydrogen and oxygen.
    Balance oxygen:

$$Mg_3N_2(s) + 6 H_2O(l) \longrightarrow 3 Mg(OH)_2(aq) + 2 NH_3(g)$$

reactants	products
3 Mg	3 Mg
2 N	2 N
12 H	12 H
6 O	6 O

6. Check to see if the equation is balanced. It is balanced!

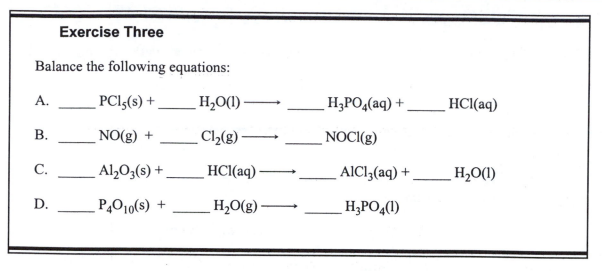

Exercise Three

Balance the following equations:

A. _____ $PCl_5(s)$ + _____ $H_2O(l) \longrightarrow$ _____ $H_3PO_4(aq)$ + _____ $HCl(aq)$

B. _____ $NO(g)$ + _____ $Cl_2(g) \longrightarrow$ _____ $NOCl(g)$

C. _____ $Al_2O_3(s)$ + _____ $HCl(aq) \longrightarrow$ _____ $AlCl_3(aq)$ + _____ $H_2O(l)$

D. _____ $P_4O_{10}(s)$ + _____ $H_2O(g) \longrightarrow$ _____ $H_3PO_4(l)$

MORE EXAMPLES OF BALANCING CHEMICAL EQUATIONS

Example 6

Balance this chemical equation.

$$CuO(s) + NH_3(g) \longrightarrow Cu(s) + N_2(g) + H_2O(l)$$

Note: reactants: 3 H   products: 2 H

The number of atoms of an element may be even on one side and odd on the other side of the equation.

If the element occurs in only two substances in the equation, balance this element first, by using two coefficients that increase the number of atoms on each side of the equation to the least common multiple.

Least common multiple for balancing H is 3 × 2 or 6

$$CuO(s) + 2 NH_3(g) \longrightarrow Cu(s) + N_2(g) + 3 H_2O(l)$$

Balance oxygen:

$$3 CuO(s) + 2 NH_3(g) \longrightarrow Cu(s) + N_2(g) + 3 H_2O(l)$$

Balance copper:

$$3 CuO(s) + 2 NH_3(g) \longrightarrow 3 Cu(s) + N_2(g) + 3 H_2O(l)$$

Exercise Four

Balance the following equations:

A. _____ $AsH_3(g)$ + _____ $KClO_3(s) \longrightarrow$ _____ $KCl(s)$ + _____ $H_3AsO_4(s)$

B. _____ $C_3H_8(g) +$ _____ $O_2(g) \longrightarrow$ _____ $CO_2(g) +$ _____ $H_2O(l)$

C. _____ $SiF_4(g) +$ _____ $H_2O(l) \longrightarrow$ _____ $H_4SiO_4(aq) +$ _____ $H_2SiF_6(aq)$

D. _____ $Sb_2S_3(s) +$ _____ $Fe(s) \longrightarrow$ _____ $Sb(s) +$ _____ $FeS(s)$

Example 7

Balance this chemical equation:

$$Ce(s) + H_2O(l) \longrightarrow Ce(OH)_3(aq) + H_2(g)$$

Oxygen occurs in only two substances. Balance oxygen:

$$Ce(s) + 3 H_2O(l) \longrightarrow Ce(OH)_3(aq) + H_2(g)$$

Make the number of hydrogens at the right an even number:

$$Ce(s) + 3 H_2O(l) \longrightarrow 2 Ce(OH)_3(aq) + H_2(g)$$

Rebalance oxygen:

$$Ce(s) + 6 H_2O(l) \longrightarrow 2 Ce(OH)_3(aq) + H_2(g)$$

Balance hydrogen:

$$Ce(s) + 6 H_2O(l) \longrightarrow 2 Ce(OH)_3(aq) + 3 H_2(g)$$

Balance cerium:

$$2 Ce(s) + 6 H_2O(l) \longrightarrow 2 Ce(OH)_3(aq) + 3 H_2(g)$$

Exercise Five

Balance the following equations:

A. _____ $C_2H_6(g) +$ _____ $O_2(g) \longrightarrow$ _____ $CO_2(g) +$ _____ $H_2O(g)$

B. _____ $H_2O(l) +$ _____ $F_2(g) \longrightarrow$ _____ $HF(aq) +$ _____ $O_2(g)$

C. _____ $I_2(s) +$ _____ $Cl_2(g) \longrightarrow$ _____ $ICl_3(s)$

D. _____ $Al_2O_3(s) +$ _____ $C(s) +$ _____ $Cl_2(g) \longrightarrow$ _____ $AlCl_3(s) +$ _____ $CO(g)$

E. _____ $Ca(s) +$ _____ $H_2O(l) \longrightarrow$ _____ $Ca(OH)_2(aq) +$ _____ $H_2(g)$

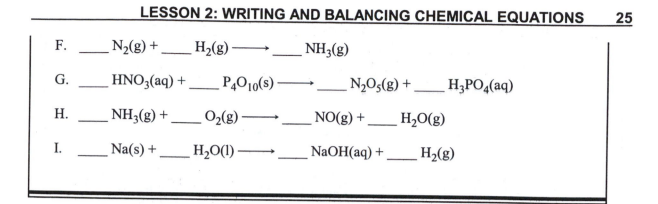

F. ____ $N_2(g)$ + ____ $H_2(g)$ ⟶ ____ $NH_3(g)$

G. ____ $HNO_3(aq)$ + ____ $P_4O_{10}(s)$ ⟶ ____ $N_2O_5(g)$ + ____ $H_3PO_4(aq)$

H. ____ $NH_3(g)$ + ____ $O_2(g)$ ⟶ ____ $NO(g)$ + ____ $H_2O(g)$

I. ____ $Na(s)$ + ____ $H_2O(l)$ ⟶ ____ $NaOH(aq)$ + ____ $H_2(g)$

FRACTIONAL COEFFICIENTS

A chemical equation can always be balanced using integer coefficients. However, in some circumstances, it may be desirable to use fractional coefficients.

Example 8

Write the equation for the formation of *one mole* of liquid water from its elements, hydrogen gas and oxygen gas.

$$H_2(g) + O_2(g) \longrightarrow H_2O(l)$$

Balanced:

$$2 H_2(g) + O_2(g) \longrightarrow 2 H_2O(l)$$

Divide all coefficients by the coefficient of H_2O, namely 2:

$$H_2(g) + \tfrac{1}{2} O_2(g) \longrightarrow H_2O(l)$$

Exercise Six

A. 1. Write the balanced equation for the formation of *one mole* of sulfur trioxide gas from sulfur dioxide gas and gaseous oxygen.

2. Write the balanced equation for the decomposition of *one mole* of ammonia gas into its elements, nitrogen gas and hydrogen gas.

3. When methane gas is burned in gaseous oxygen, carbon dioxide gas and liquid water are formed. Write the balanced equation for the formation of *one mole* of water by this reaction.

4. Write the balanced equation for the reaction of *one mole* of solid aluminum with hydrochloric acid to form aqueous aluminum chloride ($AlCl_3$) and hydrogen gas.

B. 1. State what is represented by the chemical equation:

$$16\ H_2S(g)\ +\ 8\ SO_2(g) \longrightarrow 3\ S_8(s)\ +\ 16\ H_2O(l)$$

2. State what is represented by the chemical equation:

$$NH_3(g)\ +\ ^5/_4\ O_2(g) \longrightarrow NO(g)\ +\ ^3/_2\ H_2O(l)$$

PROBLEMS

1. Write a balanced chemical equation for each of the following reactions.

(a) When hydrochloric acid is mixed with sodium hydrogen carbonate, the mixture fizzes. The fizzing occurs because the reaction produces a gas, namely carbon dioxide. The other products are a solution of sodium chloride and liquid water.

(b) Carbon tetrachloride is an industrial liquid solvent. It is made by combining methane gas with elemental chlorine. These react to form carbon tetrachloride and hydrogen chloride gas.

(c) Silicon carbide, SiC, is one of the hardest solids known. It is made by heating sand, which is SiO_2, with elemental carbon to a temperature over 2000°C. The other product of the reaction is carbon monoxide gas.

2. Balance the following chemical equations.

(a) $P_2O_5(s) + H_2O(l) \longrightarrow H_3PO_4(l)$

(b) $KClO_3(s) \longrightarrow KClO_4(s) + KCl(s)$

(c) $KClO_3(s) \longrightarrow KCl(s) + O_2(g)$

(d) $Al(s) + H_2SO_4(aq) \longrightarrow Al_2(SO_4)_3(s) + H_2(g)$

(e) $UF_4(g) + Mg(s) \longrightarrow U(s) + MgF_2(s)$

(f) $N_2H_4(g) + N_2O_4(g) \longrightarrow N_2(g) + H_2O(g)$

USING THE MOLE CONCEPT

This lesson deals with:

1. The relationship between the number of grams and the number of moles.
2. The relationship between grams and numbers of molecules.
3. Calculating how many moles of a substance will react with a given number of moles of another substance, using the balanced chemical equation for the reaction.
4. Calculating how many grams of a substance will be produced by a given number of grams of another substance, using the balanced chemical equation for the reaction.
5. Relating grams of one substance to molecules of another substance in a chemical reaction.

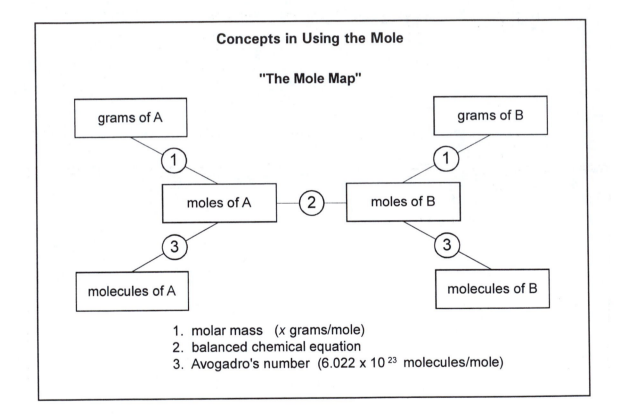

Concepts in Using the Mole

"The Mole Map"

1. molar mass   (*x* grams/mole)
2. balanced chemical equation
3. Avogadro's number (6.022×10^{23} molecules/mole)

THE MOLE MAP

The Mole Map is a diagram of the relationships between the various quantities involved in chemical reactions. Chemists call these relationships the **stoichiometry** of a reaction.

On the map, each rectangle represents one of these quantities: grams, moles, and molecules. The numbered lines are like highways linking the rectangles. Each line represents a factor that converts one quantity into the other.

For example, grams of A are related to moles of A by the molar mass of A.

GRAMS AND MOLES

Example I

What is the mass in grams of 2.65 moles of NaOH?

Step 1. What is given? 2.65 moles of NaOH

Step 2. What are we looking for?   grams of NaOH

Step 3. Look at the map to find where we are and where we're going. Then see how to get there.

We're at moles of A. We want to get to grams of A. We need the molar mass of NaOH.

40.0 g NaOH = 1 mole NaOH.

Step 4. Write the relation as a conversion factor so the units cancel. The units in the denominator (bottom) of the factor cancel those in the given amount.

$$2.65 \; \cancel{\text{moles NaOH}} \times \frac{40.0 \text{ grams NaOH}}{1 \; \cancel{\text{mole NaOH}}}$$

Step 5. Do the arithmetic.

$$2.65 \times 40.0 \text{ g NaOH} = 106 \text{ g NaOH}$$

Exercise One

How many moles is 50.0 grams of $CaBr_2$?

NUMBERS OF MOLECULES

Example 2

What is the mass in grams of 3.0×10^{24} molecules of O_2?

Step 1.   What's given? 3.0×10^{24} molecules of O_2

Step 2.   What are we looking for?   grams of O_2

Step 3.   Where are we?   molecules of A

         Where are we going?   grams of A

         How do we get there?

            First use Avogadro's number

               $(6.022 \times 10^{23}$ molecules $O_2 = 1$ mole $O_2)$,

            then use the molar mass of oxygen

               $(32.00$ g $O_2 = 1$ mole $O_2)$

Step 4.   Set up the problem so that the units cancel.

$$3.0 \times 10^{24} \text{ molecules } O_2 \left(\frac{1 \text{ mole } O_2}{6.022 \times 10^{23} \text{ molecules } O_2} \right) \left(\frac{32.00 \text{ g } O_2}{1 \text{ mole } O_2} \right)$$

Step 5.   Do the arithmetic.

$$\frac{(3.0 \times 10^{24}) \times 32.00}{6.022 \times 10^{23}} \text{ g } O_2 = 160 \text{ g } O_2$$

Exercise Two

A.   How many molecules are in 160 grams of water (H_2O)?

B.   How many atoms are in 160 grams of water?

MOLES AND THE CHEMICAL EQUATION

Example 3

How many moles of O_2 will react with 0.50 mole of Al?

$$4 \, Al(s) + 3 \, O_2(g) \longrightarrow 2 \, Al_2O_3(s)$$

Step 1.   What is given?   0.50 mole of Al
Step 2.   What is to be found?   moles of O_2
Step 3.   Use the map.

         Moles of A —③— Moles of B

Moles of A are related to Moles of B by the balanced chemical equation.

$$4 \, Al(s) \quad + \quad 3 \, O_2(g) \quad \longrightarrow \quad 2 \, Al_2O_3(s)$$
4 moles of Al react with 3 moles of O_2

Step 4.   Set up the problem so that the units cancel.

$$0.50 \; \cancel{\text{moles Al}} \times \frac{3 \text{ moles } O_2}{4 \; \cancel{\text{moles Al}}}$$

Note how the coefficients from the balanced chemical equation are used here.

Step 5.   Do the arithmetic.

$$\frac{0.50 \times 3}{4} \text{ moles } O_2 = 0.38 \text{ moles } O_2$$

Exercise Three

We just found that 0.50 mole of Al will react with 0.38 mole of O_2.

$$4 \, Al(s) + 3 \, O_2(g) \longrightarrow 2 \, Al_2O_3(s)$$

A.   How many moles of Al_2O_3 will be formed by 0.50 mole of Al?

B.   How many moles of Al_2O_3 will be formed by 0.38 mole of O_2?

C.   How does the answer to part A compare to the answer from part B?

GRAMS AND THE CHEMICAL EQUATION

Example 4

Photosynthesis is the name of the process by which plants combine CO_2 with H_2O to form sugar, $C_{12}H_{22}O_{11}$ (342.3 g/mol), and oxygen. The chemical equation for the overall process is

$$12\ CO_2\ +\ 11\ H_2O\ \longrightarrow\ C_{12}H_{22}O_{11}\ +\ 12\ O_2$$

How many grams of CO_2 does a plant use to make 150 grams of sugar?

Step 1.   What is given?   150 g sugar
Step 2.   What is to be found?   grams of CO_2
Step 3.   Use the map.
          To get from grams of sugar to grams of CO_2, we use:
          1.   the molar mass of sugar
               342.3 g $C_{12}H_{22}O_{11}$ = 1 mole $C_{12}H_{22}O_{11}$
          2.   the balanced chemical equation
               12 moles CO_2 produce 1 mole $C_{12}H_{22}O_{11}$
          3.   the molar mass of CO_2
               44.0 g CO_2 = 1 mole CO_2

Step 4.   Set up the problem so the units cancel.

$$150\ \cancel{g\ C_{12}H_{22}O_{11}}\left(\frac{1\ \cancel{mol\ C_{12}H_{22}O_{11}}}{342.3\ \cancel{g\ C_{12}H_{22}O_{11}}}\right)\left(\frac{12\ \cancel{mol\ CO_2}}{1\ \cancel{mol\ C_{12}H_{22}O_{11}}}\right)\left(\frac{44.0\ g\ CO_2}{1\ \cancel{mol\ CO_2}}\right)$$

Step 5. Do the arithmetic.

$$\frac{150\ \times\ 12\ \times\ 44.0}{342.3}\ g\ CO_2\ =\ 230\ g\ CO_2$$

Exercise Four

When ammonium dichromate, $(NH_4)_2Cr_2O_7$ (252.1 g/mol), is heated, it decomposes to nitrogen gas, chromium oxide, and water.

$$(NH_4)_2Cr_2O_7(s) \longrightarrow N_2(g) + Cr_2O_3(s) + 4\,H_2O(l)$$

How many molecules of H_2O will be produced when 1.0 gram of ammonium dichromate is heated?

PROBLEMS

1. What is the mass in grams of 4.25 moles of CH_4?

2. How many molecules are in 32.4 grams of HBr?

3. How many moles of O_2 are produced by the decomposition of 0.30 mole of $KClO_3$?

$$2\,KClO_3(s) \longrightarrow 2\,KCl(s) + 3\,O_2(g)$$

4. Natural gas, which is used in gas stoves, is mostly methane, CH_4. When methane burns, it forms carbon dioxide, CO_2, and water, H_2O.

$$CH_4(g) + 2\,O_2(g) \longrightarrow CO_2(g) + 2\,H_2O(l)$$

Suppose you burn 20.0 grams of CH_4.
   a. How many grams of O_2 would you use?
   b. How many grams of CO_2 would be formed?
   c. How many grams of H_2O would be formed?
   d. Add the masses of the substances on the left of the equation (CH_4 and O_2).
   e. Add the masses of the substances on the right side of the equation (CO_2 and H_2O).
   f. How does the mass from part d compare to the mass from part e? What chemical law does this illustrate?

LIMITING REACTANTS AND PERCENT YIELD

This lesson deals with:

1. Identifying the limiting reactant in a chemical reaction.
2. Calculating the amount of product obtained when a limiting reactant is involved.
3. Calculating the amount of excess reactant.
4. Calculating the percent yield in a chemical reaction.

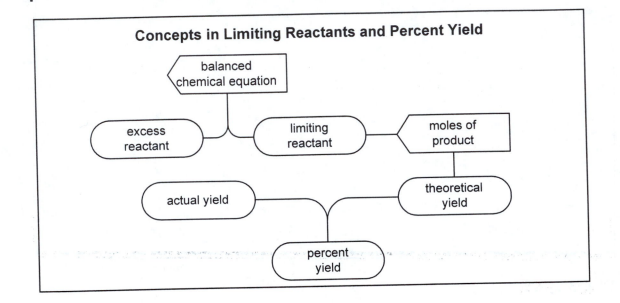

Concepts in Limiting Reactants and Percent Yield

LIMITING REACTANT IN CHEMICAL REACTIONS

The amount of product obtained in a chemical reaction is limited by the amounts of the reactants.

If all but one reactant are in excess, the amount of product is determined by the one reactant not in excess. This is the **limiting reactant**.

> How many grams of $MgCl_2$ will be produced when 15.8 grams of Mg react with excess HCl?
> This type of question was dealt with in the previous lesson.

If it is not known or indicated which reactant is limiting, then the limiting reactant must be determined.

> How many grams of $MgCl_2$ will be produced when 15.8 grams of Mg are mixed with 14.6 grams of HCl?
> This type of question is dealt with in this lesson.

Example 1

How many moles of water can be made from 3.0 moles of H_2 and 3.0 moles of O_2? The balanced equation for the reaction is

$$2 H_2(g) + O_2(g) \longrightarrow 2 H_2O(l)$$

Given: 3.0 moles H_2
         3.0 moles O_2
Find: moles H_2O

When the amounts of more than one reactant are given, then which one is the limiting reactant must be determined.

We find which is in excess and which is the limiting reactant. We use the limiting reactant to calculate the number of moles of H_2O produced.

To determine which is the limiting reactant, select one of the reactants and calculate how much of the other reactant is needed to react completely with it.

How many moles of O_2 will react with 3.0 moles of H_2?

$$3.0 \text{ moles } H_2 \left(\frac{1 \text{ mole } O_2}{2 \text{ moles } H_2} \right) = 1.5 \text{ moles } O_2$$

Either there will be enough of the other reactant and the other reactant is in excess, or there will not be enough of the other reactant and the other reactant is limiting.

We were given 3.0 moles of O_2, which is more than the 1.5 moles of O_2 needed to react with 3.0 moles of H_2. Therefore, O_2 is in excess. H_2 is the limiting reactant.

Use the limiting reactant to calculate the amount of product.

Use 3.0 moles H_2 to calculate the number of moles of H_2O produced.

$$3.0 \text{ moles } H_2 \left(\frac{2 \text{ moles } H_2O}{2 \text{ moles } H_2} \right) = 3.0 \text{ moles } H_2O$$

Example 2

What is the limiting reactant in the reaction between 5.0 moles of NH_3 and 7.0 moles of CuO? The balanced equation for the reaction is

$$2 NH_3(g) + 3 CuO(s) \longrightarrow N_2(g) + 3 Cu(s) + 3 H_2O(l)$$

Given: 5.0 moles of NH_3
         7.0 moles of CuO

To determine the limiting reactant, we must answer one of two questions:

How many moles of NH_3 will react with 7.0 moles of CuO?

or

How many moles of CuO will react with 5.0 moles of NH_3?

The answer to the second question is obtained by the calculation

$$5.0 \text{ moles } NH_3 \left(\frac{3 \text{ moles CuO}}{2 \text{ moles } NH_3} \right) = 7.5 \text{ moles CuO}$$

We do not have 7.5 moles of CuO, only 7.0 moles. Therefore, CuO is the limiting reactant.

Exercise One

What is the limiting reactant for the reaction between 5.0 moles of C_3H_6 and 12.0 moles of O_2? The balanced equation for the reaction is:

$$2 \, C_3H_6(g) + 9 \, O_2(g) \longrightarrow 6 \, CO_2(g) + 6 \, H_2O(l)$$

MASS OF LIMITING REACTANT

Example 3

How many grams of $MgCl_2$ will be produced when 15.8 grams of Mg are mixed with 14.6 grams of HCl? The balanced equation is

$$Mg(s) + 2 \, HCl(aq) \longrightarrow MgCl_2(aq) + H_2(g)$$

Given: 15.8 grams Mg   Find: grams $MgCl_2$
          14.6 grams HCl

The amounts of two reactants are given. Only one determines the amount of product, namely the limiting reactant.

We need to find the limiting reactant.

    First, convert the masses of reactants to moles of reactants.

$$(1) \quad 15.8 \text{ g Mg} \left(\frac{1 \text{ mol Mg}}{24.30 \text{ g Mg}} \right) = 0.650 \text{ mol Mg}$$

$$(2) \quad 14.6 \text{ g HCl} \left(\frac{1 \text{ mol HCl}}{36.46 \text{ g HCl}} \right) = 0.400 \text{ mol HCl}$$

How many moles of HCl are needed to react with 0.650 mole of Mg?

$$0.650 \text{ mol Mg} \left(\frac{2 \text{ mol HCl}}{1 \text{ mol Mg}} \right) = 1.30 \text{ mol HCl}$$

We do not have 1.30 moles of HCl, so HCl is the limiting reactant. Use the amount of HCl (0.400 mole) to find the grams of $MgCl_2$.

$$\text{moles HCl} \longrightarrow \text{moles MgCl}_2 \longrightarrow \text{grams MgCl}_2$$

$$0.400 \text{ mol HCl} \left(\frac{1 \text{ mol MgCl}_2}{2 \text{ mol HCl}} \right) \left(\frac{95.20 \text{ g MgCl}_2}{1 \text{ mol MgCl}_2} \right) = 19.0 \text{ g MgCl}_2$$

Exercise Two

What is the maximum mass (in grams) of ammonia, NH_3, that can be obtained from 5.0 grams of H_2 and 30.0 grams of N_2? The balanced equation is

$$3 \, H_2(g) + N_2(g) \longrightarrow 2 \, NH_3(g)$$

EXCESS REACTANT

The limiting reactant in a chemical reaction is completely consumed. When it is gone, the reactions stops. When the reaction stops, some of the other reactant(s) remain. This is the **excess reactant**.

Example 4

Phosphorus reacts with chlorine, producing phosphorus trichloride. What is the excess reactant and how many grams of it are left, when 31.3 g of chlorine are combined with 45.0 g of phosphorus?

$$P_4(s) + 6\,Cl_2(g) \longrightarrow 4\,PCl_3(g)$$

Given: 45.0 g P_4
        31.3 g Cl_2
Find: excess reactant

First, determine the limiting reactant.
    Find the moles of reactants:

$$45.0 \text{ g } P_4 \left(\frac{1 \text{ mol } P_4}{123.88 \text{ g } P_4} \right) = 0.363 \text{ mol } P_4$$

$$31.3 \text{ g } Cl_2 \left(\frac{1 \text{ mol } Cl_2}{70.90 \text{ g } Cl_2} \right) = 0.441 \text{ mol } Cl_2$$

How many moles of Cl_2 are required to react with 0.363 moles P_4?

$$0.363 \text{ mol } P_4 \left(\frac{6 \text{ mol } Cl_2}{1 \text{ mol } P_4} \right) = 2.18 \text{ mol } Cl_2$$

We do not have this much Cl_2, so Cl_2 is the limiting reactant. Therefore, P_4 is the excess reactant. To find how much is left after the reaction, first find how much reacts with the limiting reactant.

$$0.441 \text{ mol } Cl_2 \left(\frac{1 \text{ mol } P_4}{6 \text{ mol } Cl_2} \right) = 0.0735 \text{ mol } P_4 \text{ reacts}$$

The amount remaining after the reaction is the initial amount minus the amount that reacts.

$$0.363 \text{ mol } P_4 - 0.0735 \text{ mol } P_4 = 0.290 \text{ mol } P_4 \text{ remains}$$

Find the mass of the excess P_4.

$$0.290 \text{ mol } P_4 \left(\frac{123.88 \text{ g } P_4}{1 \text{ mol } P_4} \right) = 35.9 \text{ g } P_4 \text{ remain}$$

Exercise Three

Aluminum metal reacts with nickel(II) oxide, forming nickel metal and aluminum oxide.

$$2 \text{ Al(s)} + 3 \text{ NiO(s)} \longrightarrow 3 \text{ Ni(s)} + Al_2O_3(s)$$

What is the excess reactant and how many grams of it remain, after a reaction between 4.55 g of aluminum and 9.62 g of nickel(II) oxide?

PERCENT YIELD

The **percent yield** of a reaction is the fraction of the maximum possible yield that is actually obtained, expressed as a percent.

$$\text{percent yield} = \frac{\text{actual yield}}{\text{maximum yield}} \times 100\%$$

The yield of a reaction may be less than maximum when there are competing reactions. For example, carbon burns in oxygen to form both carbon dioxide and carbon monoxide.

$$C(s) + O_2(g) \longrightarrow CO_2(g)$$

$$2 \, C(s) \, + \, O_2(g) \longrightarrow 2 \, CO(g)$$

When 1.0 mole of carbon is burned in excess oxygen, the maximum amount of carbon dioxide that can be produced is 1.0 mole. However, some of the carbon may combine with oxygen to form carbon monoxide instead. When this happens, less than 1.0 mole of carbon dioxide is formed.

Suppose only 0.83 mole of carbon dioxide is formed. Then the percent yield of carbon dioxide is

$$\text{percent yield} = \frac{\text{actual yield}}{\text{maximum yield}} \times 100\% = \frac{0.83 \text{ mole}}{1.0 \text{ mole}} \times 100\% = 83\%$$

Exercise Four

A. What is the percent yield if, in example 3, instead of getting the calculated 19.0 grams of $MgCl_2$, only 15.0 grams were obtained?

B. What is the percent yield if 4.0 moles of NaCl are obtained when 5.0 moles of NaOH react with 6.0 moles of HCl? The balanced equation is

$$\text{NaOH(aq)} \, + \, \text{HCl(aq)} \longrightarrow \text{NaCl(aq)} \, + \, H_2O(l)$$

c

PROBLEMS

1. Wine turns into vinegar when the ethanol in the wine, C_2H_5OH, reacts with oxygen to form acetic acid, CH_3CO_2H.

    $$C_2H_5OH(aq) + O_2(g) \longrightarrow CH_3CO_2H(aq) + H_2O(l)$$

    Suppose 1.12 g of oxygen were sealed in a wine bottle that contains 2.28 g of ethanol. Which reactant is limiting, oxygen or ethanol?

2. Sulfur dioxide gas reacts with hydrogen sulfide gas, producing solid sulfur and water vapor.

    $$8\ SO_2(g) + 16\ H_2S(g) \longrightarrow 3\ S_8(s) + 16\ H_2O(g)$$

    Suppose 32.4 g of SO_2 are combined with 28.7 g of H_2S. Which reactant is limiting and which is in excess? How many grams of sulfur can be produced? How many grams of the excess reactant remain when the reaction is complete?

3. The combustion of N_2 gas yields the pollutant NO_2. Calculate the percent yield of NO_2 if 60.0 grams of NO_2 form when 56 grams of nitrogen gas burn.

4. The reaction between $SO_2(g)$ and molecular oxygen yields $SO_3(g)$. Calculate the percent yield of SO_3 if 40 grams of SO_3 form when 32 grams of SO_2 react with an excess of oxygen.

5. In the reaction of 4.0 moles of N_2 with 6.0 moles of H_2, 1.6 moles of NH_3 were obtained. What is the percent yield?

WRITING NET IONIC EQUATIONS

This lesson deals with:

1. Identifying formulas of ionic compounds.
2. Identifying the ions in a compound from its formula.
3. Predicting the solubility of ionic compounds.
4. Predicting whether a compound is a strong electrolyte, weak electrolyte, or non-electrolyte in aqueous solution.
5. Determining the mole number of a solute.
6. Writing a net ionic equation, given an complete equation or a verbal description of a reaction.
7. Predicting what precipitation reaction occurs when two aqueous solutions of ionic compounds are mixed.

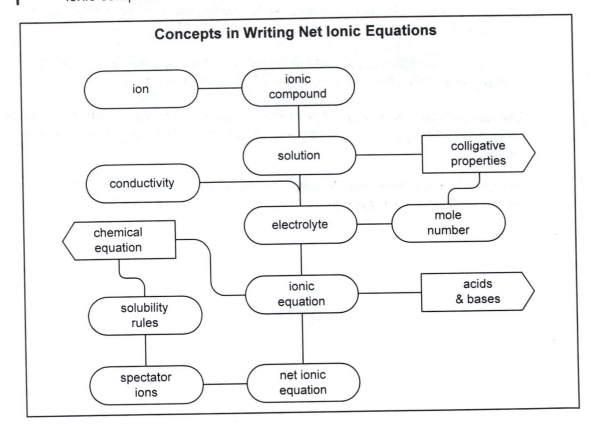

IONS AND IONIC COMPOUNDS

Ionic compounds are composed of ions.
Ions are atoms or groups of atoms that have an electrical charge.
A **cation** is an ion with a positive (+) charge.
An **anion** is an ion with a negative (−) charge.

Ionic compounds contain both cations and anions. The cation is frequently a metal ion or the ammonium ion (NH_4^+). Compounds that contain metals are usually ionic. The anion is often

Table 1. Common Anions and Cations

A. Anions

F^-	fluoride ion	ClO_4^-	perchlorate ion
Cl^-	chloride ion	ClO_3^-	chlorate ion
Br^-	bromide ion	ClO_2^-	chlorite ion
I^-	iodide ion	ClO^-	hypochlorite ion
O^{2-}	oxide ion	MnO_4^-	permanganate ion
S^{2-}	sulfide ion	NO_3^-	nitrate ion
Se^{2-}	selenide ion	NO_2^-	nitrite ion
O_2^{2-}	peroxide ion	PO_4^{3-}	phosphate ion
OH^-	hydroxide ion	HPO_4^{2-}	hydrogen phosphate ion
CO_3^{2-}	carbonate ion	$H_2PO_4^-$	dihydrogen phosphate ion
HCO_3^-	hydrogen carbonate ion[†]	SO_4^{2-}	sulfate ion
$C_2H_3O_2^-$	acetate ion	HSO_4^-	hydrogen sulfate ion
CrO_4^{2-}	chromate ion	SO_3^{2-}	sulfite ion
$Cr_2O_7^{2-}$	dichromate ion	HSO_3^-	hydrogen sulfite ion

B. Cations

Li^+	lithium ion	Co^{2+}	cobalt(II) ion
Na^+	sodium ion	Co^{3+}	cobalt(III) ion
K^+	potassium ion	Fe^{2+}	iron(II) ion
Mg^{2+}	magnesium ion	Fe^{3+}	iron(III) ion
Ca^{2+}	calcium ion	Pb^{2+}	lead(II) ion
Sr^{2+}	strontium ion	Mn^{2+}	manganese(II) ion
Ba^{2+}	barium ion	Hg_2^{2+}	mercury(I) ion
Al^{3+}	aluminum ion	Hg^{2+}	mercury(II) ion
NH_4^+	ammonium ion	Ni^{2+}	nickel(II) ion
Cr^{3+}	chromium(III) ion	Ag^+	silver ion
Cu^{2+}	copper(II) ion	Zn^{2+}	zinc ion

[†] The hydrogen carbonate ion is often call the bicarbonate ion.

one of those listed in Table 1. Ionic compounds have high melting points and are good conductors of electricity in the molten state.

Memorize the names and formulas of the ions listed in Table 1.

An effective method of memorizing the names and formulas of these ions is by making and using flash cards (cards with the name on one side and the formula on the other).

In the formula of an ionic compound, the cation is usually indicated first, followed by the anion. The total charge of all of the cations in a formula must balance the total charge of all of the anions.

Example 1

Which of the following formulas represent ionic compounds? For each ionic compound, indicate the ions of which it is composed.

A.   calcium chloride: $CaCl_2$
> Contains calcium, a metal. Therefore, it is ionic.
> $CaCl_2$ contains one Ca^{2+} and two Cl^-

B. $(NH_4)_2CO_3$
> It contains NH_4^+ ions. Therefore, it is ionic.
> $(NH_4)_2CO_3$ contains two NH_4^+ and one CO_3^{2-}

C. $Fe_2(SO_4)_3$
> It contains iron, a metal. Therefore, it is ionic.
> What is the charge of the Fe ion?
> The total charge of the anions must balance the total charge of the cations.
> three SO_4^{2-} total charge $= 3 \times (-2) = -6$
> two Fe ions   total charge $= +6 = 2 \times (?) = 2 \times (+3)$
> $Fe_2(SO_4)_3$ contains two Fe^{3+} and three SO_4^{2-}

D.   hydrogen chloride gas: $HCl(g)$
> Contains neither a metal ion nor NH_4^+. Therefore, it is not ionic.
> $HCl(g)$ contains *no* ions.

Exercise One

Which of the following are ionic compounds? For each ionic compound, indicate the ions of which it is composed.

compound	ionic?	ions (if any)
A. KOH		
B. HBr		
C. $CoCl_2$		
D. $NaHCO_3$		
E. NH_3BF_3		
F. $AgNO_3$		
G. ClF		
H. $(NH_4)_2Cr_2O_7$		

BEHAVIOR OF IONIC COMPOUNDS IN WATER

When placed in water, an ionic compound will dissolve to some extent to form a solution. The extent to which a compound dissolves is called its solubility. Those that dissolve to a significant extent are called soluble, those that dissolve to only a very small extent are called insoluble.

Solubility guidelines help predict the extent to which an ionic compound will dissolve. These guidelines are often called "solubility rules."

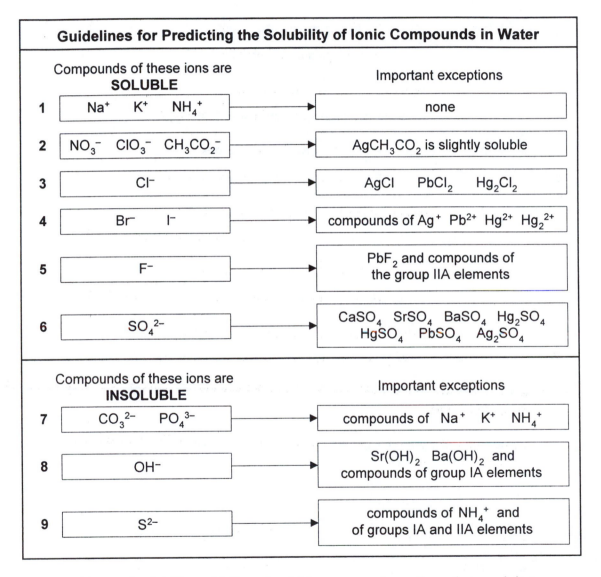

Guidelines for Predicting the Solubility of Ionic Compounds in Water

	Compounds of these ions are **SOLUBLE**	Important exceptions
1	Na^+ K^+ NH_4^+	none
2	NO_3^- ClO_3^- $CH_3CO_2^-$	$AgCH_3CO_2$ is slightly soluble
3	Cl^-	$AgCl$ $PbCl_2$ Hg_2Cl_2
4	Br^- I^-	compounds of Ag^+ Pb^{2+} Hg^{2+} Hg_2^{2+}
5	F^-	PbF_2 and compounds of the group IIA elements
6	SO_4^{2-}	$CaSO_4$ $SrSO_4$ $BaSO_4$ Hg_2SO_4 $HgSO_4$ $PbSO_4$ Ag_2SO_4

	Compounds of these ions are **INSOLUBLE**	Important exceptions
7	CO_3^{2-} PO_4^{3-}	compounds of Na^+ K^+ NH_4^+
8	OH^-	$Sr(OH)_2$ $Ba(OH)_2$ and compounds of group IA elements
9	S^{2-}	compounds of NH_4^+ and of groups IA and IIA elements

(For this set of solubility guidelines, insoluble compounds are those that precipitate upon mixing equal volumes of 0.1 M solutions of the respective ions. Precipitation reactions are discussed later in this lesson.)

Example 2

Apply the solubility guidelines to predict whether each of the following ionic compounds is soluble in water.

First, determine the ions the compound contains. Then, go down the list of solubility guidelines to find the one that applies.

A. Na_3PO_4 contains sodium ions and phosphate ions

                 A sodium compound, by guideline 1, *soluble.*

B. $Pb(NO_3)_2$ contains lead(II) ions and nitrate ions

                 A nitrate, by guideline 2, *soluble.*

C. AgBr   contains silver(I) ions and bromide ions

                 Bromide of silver, by guideline 4, *insoluble.*

D. $NiSO_4$ contains nickel(II) ions and sulfate ions

                 A sulfate, not an exception in guideline 6, *soluble.*

E. $(NH_4)_2S$ contains ammonium ions and sulfide ions

                 A compound of ammonium, by guideline 1, *soluble.*

F. $CaCO_3$ contains calcium ions and carbonate ions

                 A carbonate, not an exception in guideline 7, *insoluble.*

G. $Fe(OH)_3$ contains iron(III) ions and hydroxide ions

                 A hydroxide, not an exception in guideline 8, *insoluble.*

Exercise Two

Apply the solubility guidelines to predict whether each of the following compounds is soluble or insoluble in water.

A. Na_2S _____   F. $LiC_2H_3O_2$ _____

B. $Ni(OH)_2$ _____   G. $(NH_4)_2Cr_2O_7$ _____

C. $PbSO_4$ _____   H. $Ca_3(PO_4)_2$ _____

D. FeI_2 _____   I. Ag_2SO_4 _____

E. $Ca(ClO_3)_2$ _____   J. CoS   _____

ELECTRICAL CONDUCTIVITY OF AQUEOUS SOLUTIONS

If the solution being tested conducts electricity, the light bulb will glow. The more it conducts, the brighter the glow.

Solutes whose solutions conduct electricity are called electrolytes.

Strong electrolytes: their solutions conduct electricity well.
Weak electrolytes: their solutions conduct electricity weakly.
Non-electrolytes: their solutions do not conduct electricity.

Pure water is a non-electrolyte.

Electrical conductivity of a solution can be attributed to the presence of ions in solution.

Ions carry electrical charge in a solution.

Strong electrolytes form a large number of mobile ions.
Weak electrolytes form relatively few ions.
Non-electrolytes form no ions.

Strong Electrolytes

All solute particles are ions; there are *no* neutral solute particles (molecules).

A.   Soluble ionic compounds.
  $Na_2SO_4(aq)$ is more descriptively written as $2\,Na^+(aq) + SO_4^{2-}(aq)$

B.   Strong acids.

Table 2. The Seven Strong Acids			
$HCl(aq)$	hydrochloric acid	$HClO_3(aq)$	chloric acid
$HBr(aq)$	hydrobromic acid	$HClO_4(aq)$	perchloric acid
$HI(aq)$	hydroiodic acid	$H_2SO_4(aq)$	sulfuric acid
$HNO_3(aq)$	nitric acid		

Note: As pure compounds, the strong acids are not ionic compounds. They form ions when dissolved in water.

HCl(aq) is more descriptively written as $H^+(aq) + Cl^-(aq)$.

$H_2SO_4(aq)$ is more descriptively written as $H^+(aq) + HSO_4^-(aq)$.

Weak Electrolytes

When a weak electrolyte dissolves, only some molecules break into ions; most remain as molecules.

Table 3. Some Common Weak Acids and Bases

A. Weak Acids

$HF(aq)$	hydrofluoric acid	$H_2SO_3(aq)$	sulfurous acid
$HNO_2(aq)$	nitrous acid	$H_2CO_3(aq)$	carbonic acid
$HClO_2(aq)$	chlorous acid	$H_3PO_4(aq)$	phosphoric acid
$HClO(aq)$	hypochlorous acid	$H_2S(aq)$	hydrosulfuric acid
$HC_2H_3O_2(aq)$	acetic acid		

B. Weak Bases

$NH_3(aq)$	ammonia	$C_5H_5N(aq)$	pyridine

Only a few $HF(aq)$ molecules break into $H^+(aq) + F^-(aq)$. Most of the $HF(aq)$ remains as molecules. Therefore, hydrofluoric acid is written as $HF(aq)$, and not as ions.

Non-Electrolytes

These are *not* ionic compounds, strong acids, weak acids, or weak bases.

Examples: $C_{12}H_{22}O_{11}$, sugar and C_2H_5OH, ethyl alcohol

Non-electrolytes form *no* ions and are written as aqueous molecules when dissolved in water.

$$C_{12}H_{22}O_{11}(aq) \text{ and } C_2H_5OH(aq)$$

Mole number (*i*)

i = number of moles of particles per mole of compound dissolved.

$Na_2SO_4(aq)$ is $2\,Na^+(aq) + SO_4^{2-}(aq)$

3 moles of particles $i = 3$

$HCl(aq)$ is $H^+(aq) + Cl^-(aq)$

2 moles of particles $i = 2$

$HF(aq)$ is mostly $HF(aq)$ plus some $H^+(aq)$ and $F^-(aq)$

more than 1 but less than 2 moles of particles $1 < i < 2$

$C_{12}H_{22}O_{11}(aq)$ is all $C_{12}H_{22}O_{11}(aq)$

1 mole of particles $i = 1$

Summary:

Strong Electrolytes: $i \geq 2$
Weak Electrolytes: $1 < i < 2$
Non-Electrolytes: $i = 1$

Example 3

State whether each compound is a strong, weak, or non-electrolyte. Also, give a descriptive representation of the way it exists in aqueous solution and give its mole number.

A. K_2CrO_4
A compound of potassium is ionic and soluble in water.
Therefore, it is a strong electrolyte.
Descriptive representation: $2 K^+(aq) + CrO_4^{2-}(aq)$ $i = 3$

B. C_3H_6O (acetone, soluble in water)
It is not ionic, not a strong acid, nor a weak acid or weak base.
Therefore, it is a non-electrolyte.
Descriptive representation: $C_3H_6O(aq)$ $i = 1$

C. $HC_2H_3O_2$
It is neither an ionic compound nor a strong acid.
It is the weak acid acetic acid.
Descriptive representation: $HC_2H_3O_2(aq)$ $1 < i < 2$

Exercise Three

Indicate whether each of the following compounds is a strong, weak, or non-electrolyte. Also give a descriptive representation of the way it exists in aqueous solution, and give its mole number.

compound	strong, weak, or non-electrolyte	descriptive representation	mole number
A. $HNO_3(aq)$			
B. $NH_3(aq)$			
C. $Na_3PO_4(aq)$			
D. $CH_3OH(aq)$			
E. $H_3PO_4(aq)$			
F. $H_2SO_4(aq)$			

NET IONIC EQUATIONS

A complete chemical equation for a reaction that occurs in aqueous solution shows all of the substances (molecules or ions) that are present in the solution. Some of the ions may remain unchanged during the course of the reaction. These ions are called **spectator ions**. A **net ionic equation** shows only the substances that are changed in the reaction; it does not include spectator ions.

Characteristics of net ionic equations:
1. Soluble strong electrolytes are represented as aqueous ions.
2. Weak electrolytes and non-electrolytes are represented as aqueous molecules.
3. Insoluble solids, liquids, and gases are represented as solids, liquids, or gases.
4. Ions not directly involved in the reaction (spectator ions) are omitted.

Example 4

A.   When $Ba(NO_3)_2(aq)$ and $Na_2SO_4(aq)$ are mixed, a white precipitate of $BaSO_4(s)$ forms, while $NaNO_3(aq)$ remains in solution.

Complete Equation:

$$Ba(NO_3)_2(aq) \; + \; Na_2SO_4(aq) \longrightarrow BaSO_4(s) \; + \; 2\,NaNO_3(aq)$$

1.   Write a complete ionic equation, in which all strong electrolytes are written as aqueous ions.

$$Ba(NO_3)_2(aq) \qquad + \qquad Na_2SO_4(aq)$$

$$Ba^{2+}(aq) + 2\,NO_3^{-}(aq) \quad + \quad 2\,Na^{+}(aq) + SO_4^{2-}(aq) \longrightarrow$$

$$BaSO_4(s) \qquad\qquad NaNO_3(aq)$$

$$BaSO_4(s) \quad + \quad 2\,Na^{+}(aq) + 2\,NO_3^{-}(aq)$$

This is the complete ionic equation:

$$Ba^{2+}(aq) + 2\,NO_3^{-}(aq) + 2\,Na^{+}(aq) + SO_4^{2-}(aq) \longrightarrow BaSO_4(s) + 2\,Na^{+}(aq) + 2\,NO_3^{-}(aq)$$

2.   There are no weak or non-electrolytes in this equation, so no aqueous molecules are written.

3.   Insoluble $BaSO_4$ is written as a solid

4.   Omit the spectator ions, the ions that appear as both reactants and products.

$$Ba^{2+}(aq) + \cancel{2\,NO_3^{-}(aq)} + \cancel{2\,Na^{+}(aq)} + SO_4^{2-}(aq) \longrightarrow BaSO_4(s) + \cancel{2\,Na^{+}(aq)} + \cancel{2\,NO_3^{-}(aq)}$$

This is the net ionic equation:

$$Ba^{2+}(aq) \; + \; SO_4^{2-}(aq) \longrightarrow BaSO_4(s)$$

B. When NiF_2(aq) is mixed with HCl(aq), they react to form ( ) and $NiCl_2$(aq).

Complete Equation:

$$NiF_2(aq) + 2\ HCl(aq) \longrightarrow 2\ HF(aq) + NiCl_2(aq)$$

1. Write the complete ionic equation, in which all strong electrolytes are written as aqueous ions.

$$Ni^{2+}(aq) + 2\ F^-(aq) + 2\ H^+(aq) + 2\ Cl^-(aq) \longrightarrow 2\ HF(aq) + Ni^{2+}(aq) + 2Cl^-(aq)$$

2. HF(aq) is a weak electrolyte, so it is written as aqueous molecules.

3. There are no solids or gases in this reaction.

4. Omit the spectator ions, those that appear as both reactants and products.

$$2\ F^-(aq) + 2\ H^+(aq) \longrightarrow 2\ HF(aq)$$

or $\qquad F^-(aq) + H^+(aq) \longrightarrow HF(aq)$

This is the net ionic equation.

C. Ba(s) reacts with H_2SO_4(aq) to form H_2(g) and $BaSO_4$(s).

Complete Equation:

$$Ba(s) + H_2SO_4(aq) \longrightarrow H_2(g) + BaSO_4(s)$$

1. Write the complete ionic equation.

$$Ba(s) + H^+(aq) + HSO_4^-(aq) \longrightarrow H_2(g) + BaSO_4(s)$$

2. No weak electrolytes nor non-electrolytes.

3. Ba(s) and $BaSO_4$(s) are written as solids, H_2(g) as a gas.

4. There are no spectator ions, so no ions may be deleted.
    The net ionic equation is

$$Ba(s) + H^+(aq) + HSO_4^-(aq) \longrightarrow H_2(g) + BaSO_4(s)$$

Exercise Four

A. Write the net ionic equation for each of the following complete equations.

1. HCl(aq) + NaOH(aq) $\longrightarrow$ NaCl(aq) + H_2O(l)

2. $AgNO_3(aq) + NaCl(aq) \longrightarrow NaNO_3(aq) + AgCl(s)$

3. $H_2S(aq) + 2\,NH_3(aq) \longrightarrow (NH_4)_2S(aq)$

4. $Na_3PO_4(aq) + H_2O(l) \longrightarrow Na_2HPO_4(aq) + NaOH(aq)$

5. $2\,HCl(aq) + CaCO_3(s) \longrightarrow CaCl_2(aq) + CO_2(g) + H_2O(l)$

B.   Write a net ionic equation for each of the following reactions.

1.   When aqueous sodium chloride is added to a solution of $Pb(NO_3)_2$, a white precipitate of $PbCl_2$ forms, and sodium nitrate remains in solution.

2.   When aqueous sodium hydroxide is added to aqueous ammonium chloride, ammonia gas is given off as sodium chloride and water are formed.

PREDICTING PRECIPITATION REACTIONS

Example 5

Identify the precipitates, if any, that form when each of the following pairs of 0.1 M solutions is mixed?

A. $Na_2CO_3(aq)$ and $CaCl_2(aq)$

　　1. Identify the ions in each solution.

$$Na_2CO_3(aq) \qquad\qquad \text{and} \qquad\qquad CaCl_2(aq)$$

$$\overbrace{2\,Na^+(aq) + CO_3{}^{2-}(aq)} \qquad\qquad \overbrace{Ca^{2+}(aq)\ +\ 2\,Cl^-(aq)}$$

　　2. Combine cations from one solution with anions of the other.

　　3. Check solubilities of new compounds.

　　　　NaCl is soluble and remains dissolved.

　　　　$CaCO_3$ is insoluble and forms a precipitate.

　　4.　Write net ionic equation for the formation of any insoluble compounds.

$$Ca^{2+}(aq) + CO_3{}^{2-}(aq) \longrightarrow CaCO_3(s)$$

B. $AgNO_3(aq)$ and $NH_4F(aq)$

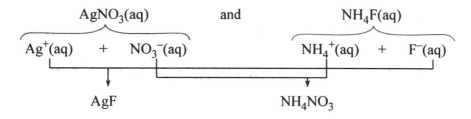

Both AgF and NH_4NO_3 are soluble; no precipitate forms, and no net reaction occurs.

Exercise Five

Predict what precipitate, if any, will form when each of the following pairs of solutions is mixed. Write a net ionic equation for the formation of all precipitates.

A. $(NH_4)_2S$(aq) and $MnCl_2$(aq)

B. NH_4Br(aq) and $Fe(NO_3)_3$(aq)

C. KOH(aq) and $NiSO_4$(aq)

D. K_3PO_4(aq) and $MgSO_4$(aq)

PROBLEMS

1. Write a complete ionic equation and a net ionic equation for each of the following.

   (a) $Pb(ClO_3)_2$(aq) + 2 NaBr(aq) $\longrightarrow$ $PbBr_2$(s) + 2 $NaClO_3$(aq)

   (b) 2 NaOH(aq) + $CuSO_4$(aq) $\longrightarrow$ Na_2SO_4(aq) + $Cu(OH)_2$(s)

(c) $3 H_2S(g) + 2 CrCl_3(aq) \longrightarrow Cr_2S_3(s) + 6 HCl(aq)$

(d) $CaCO_3(s) + 2 HCH_3CO_2(aq) \longrightarrow Ca(CH_3CO_2)_2(aq) + H_2O(l) + CO_2(g)$

2. A precipitation reaction occurs when each of the following pairs of solutions is mixed. For each, write the complete equation, the complete ionic equation, and the net ionic equation.

   (a) $FeCl_3(aq)$ and $NaOH(aq)$

   (b) $Pb(NO_3)_2(aq)$ and $K_2SO_4(aq)$

   (c) $Hg(NO_3)_2(aq)$ and $K_2CO_3(aq)$

   (d) $H_3PO_4(aq)$ and $Sr(OH)_2(aq)$

3. Indicate if a reaction occurs when each of the following pairs of solutions is mixed. If a reaction occurs, write the net ionic equation and identify the spectator ions.

   (a) $H_3PO_4(aq)$ and $CuCl_2(aq)$

   (b) $K_2S(aq)$ and $AgNO_3(aq)$

   (c) $Mn(NO_3)_2(aq)$ and $KBr(aq)$

   (d) $NiSO_4(aq)$ and $Ba(OH)_2(aq)$

4. Suggest two solutions of soluble compounds that could be mixed together to produce the reaction described by each of the following net ionic equations.

   (a) $Ag^+(aq) + Cl^-(aq) \longrightarrow AgCl(s)$

   (b) $Ca^{2+}(aq) + 2 F^-(aq) \longrightarrow CaF_2(s)$

   (c) $2 H^+(aq) + CO_3^{2-}(aq) \longrightarrow H_2O(l) + CO_2(g)$

   (d) $HF(aq) + OH^-(aq) \longrightarrow F^-(aq) + H_2O(l)$

MOLARITY CALCULATIONS

This lesson deals with:

1. Calculating the molarity of a solution.
2. Calculating the amount of solute needed to prepare a solution of given molarity.
3. Describing the preparation of a solution of given molarity by dilution of a more concentrated solution.
4. Calculating the molarity of a solution obtained by mixing two other solutions of known molarity.
5. Calculating the volume of solution required to react with a given volume of another solution, when both solution molarities are known.
6. Determining the molarity of a solution from titration data.

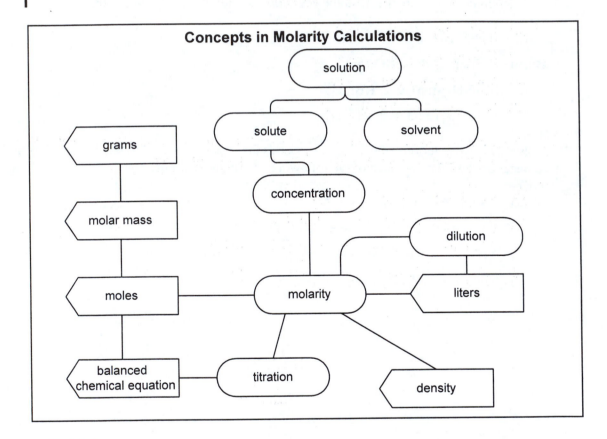

Concepts in Molarity Calculations

SOLUTIONS

A **solution** is a homogeneous mixture of two or more components.

The components are often described using these terms:

Solvent: the component present in the largest amount

Solute: the component that is considered to be dissolved in the solvent

Concentration expresses the amounts of the components in a solution.

Molarity

The **molarity** or **molar concentration** of a solute in a solution is defined as the number of moles of solute per liter of solution.

$$\text{Molarity} \;=\; \frac{\text{moles of solute}}{\text{liters of solution}}$$

Molarity is abbreviated by a capital M.

For example, a 2.0 M solution of HCl in water contains

       2.0 moles of HCl in 1.0 liter of solution,

or   4.0 moles of HCl in 2.0 liters of solution,

or   1.0 mole of HCl in 0.50 liter of solution,

etc.

Molarity is a commonly used unit of concentration, because it allows the moles of a substance, the solute, to be measured by measuring the volume of its solution, which is easy to do.

Example 1

Calculate the molarity of a solution made by adding 141 g of acetic acid, $HC_2H_3O_2$ (60.05 g/mol), to enough water to make 750.0 mL of solution.

$$\text{Molarity} \;=\; \frac{\text{moles of solute}}{\text{liters of solution}}$$

Calculations:

$$\text{moles of solute} = 141 \text{ g } HC_2H_3O_2 \left(\frac{1 \text{ mol } HC_2H_3O_2}{60.05 \text{ g } HC_2H_3O_2} \right) = 2.35 \text{ mol } HC_2H_3O_2$$

$$\text{liters of solution} \;=\; 750.0 \text{ mL} \left(\frac{1 \text{ L}}{1000 \text{ mL}} \right) = 0.7500 \text{ L}$$

$$\text{molarity} = \frac{2.35 \text{ mol } HC_2H_3O_2}{0.7500 \text{ L}} = 3.13 \text{ M } HC_2H_3O_2$$

Exercise One

Calculate the molarity of a solution made by dissolving 2.42 g of rubidium chloride, RbCl (120.92 g/mol), in enough water to make 250.0 mL of solution.

PREPARING A SOLUTION OF SPECIFIED MOLARITY

One way to prepare a solution of specified molarity is to dissolve the solid solute in the desired volume.

Example 2

A chemist wishes to prepare 0.500 liter of an 1.5 M aqueous solution of sodium chloride, NaCl (58.44 g/mol). How many grams of NaCl should be used?

$$\text{Molarity} = \frac{\text{moles of solute}}{\text{liters of solution}} = \frac{\text{mol NaCl}}{\text{L solution}}$$

$$\text{mol NaCl} = (\text{molarity}) \times (\text{L solution})$$

$$= (1.5 \text{ mol NaCl/L}) \times (0.500 \text{ L})$$

$$= 0.75 \text{ mol NaCl}$$

$$\text{grams of NaCl} = 0.75 \text{ mol NaCl} \left(\frac{58.44 \text{ g NaCl}}{1 \text{ mol NaCl}} \right) = 44 \text{ g NaCl}$$

To prepare the solution, dissolve 44 g NaCl in an amount of water less than 0.5 L, then dilute the resulting solution to 0.500 L.

Exercise Two

A chemist wishes to prepare 250.0 mL of a 2.00 M aqueous solution of sodium hydroxide, NaOH (40.00 g/mol). How many grams of NaOH should be used?

DILUTING A SOLUTION

Another way to prepare a solution of desired molarity is to dilute a solution that has a concentration higher than the needed concentration.

Example 3

Commercial "concentrated" ammonia solution is 15.0 M aqueous NH_3. How can this solution be used to prepare 500.0 mL of 6.0 M aqueous ammonia?

$$\text{Molarity} = \frac{\text{moles of solute}}{\text{liters of solution}}$$

1. How many moles of NH_3 are needed to prepare 500.0 mL of 6.0 M NH_3?

$$\text{desired molarity} = \frac{\text{required moles of solute}}{\text{required liters of solution}}$$

$$\text{required moles } NH_3 = (\text{desired Molarity } NH_3) \times (\text{liters required})$$

$$= (\text{ 6.0 mol } NH_3/L) \times (0.5000 \text{ L })$$

$$= 3.0 \text{ mol } NH_3$$

2. What volume of 15.0 M NH_3 contains 3.0 moles NH_3?

$$\text{given Molarity} = \frac{\text{required moles of solute}}{\text{liters of solution to dilute}}$$

$$\text{liters of solution to dilute} = \frac{\text{required moles of solute}}{\text{given molarity}}$$

$$= \frac{3.0 \text{ mol NH}_3}{15.0 \text{ mol NH}_3/\text{L}} = 0.20 \text{ L}$$

To prepare 500 mL of 6.0 M NH_3, dilute 0.20 L (200 mL) of 15.0 M NH_3 to a volume of 500.0 mL.

Exercise Three

The laboratory acid called "concentrated" sulfuric acid is 18 M H_2SO_4. How can 500.0 mL of 3.0 M aqueous H_2SO_4 be prepared from "concentrated" sulfuric acid?

ANOTHER EXAMPLE OF DILUTING A SOLUTION

When a solution is diluted by adding solvent, there is a simple relationship between the starting concentration and the concentration after dilution. In Example 3, we calculated that 0.20 L of 15.0 M NH_3 solution should be diluted to 0.50 L to make a 6.0 M NH_3 solution. The initial solution had a volume of 0.20 L and a concentration of 15.0 M. The diluted solution has a volume of 0.50 L and a concentration of 6.0 M. Here is the relationship:

$$(0.20 \text{ L}) \times (15.0 \text{ M}) = 3.0 \text{ mol} = (0.50 \text{ L}) \times (6.0 \text{ M})$$

This relationship is often represented as

$$M_1 V_1 = M_2 V_2$$

and it works because the product of molarity and liters is moles of solute, which remains constant when a solution is diluted. In fact, the equation works with any units of volume, as long as they are the same, because molarity expresses moles per volume.

Remember, this equation applies only when a solution is diluted with pure solvent.

Example 4

What is the concentration of a solution made by diluting 15.0 mL of a 0.468 M HCl solution to a volume of 100.0 mL?

$$M_1 V_1 = M_2 V_2$$

$$M_1 = 0.468 \text{ M} \qquad M_2 = ?$$
$$V_1 = 15.0 \text{ mL} \qquad V_2 = 100.0 \text{ mL}$$

$$(0.468 \text{ M}) (15.0 \text{ mL}) = (M_2) (100.0 \text{ mL})$$

$$M_2 = \frac{(0.468 \text{ M})(15.0 \text{ mL})}{100.0 \text{ mL}} = 0.0702 \text{ M}$$

Exercise Four

What volume of 0.800 M $CuSO_4$ solution must be diluted to make 250.0 mL of 0.150 M $CuSO_4$ solution?

MIXING TWO SOLUTIONS

In the previous examples, we've seen what happens to the concentration when a solution is diluted with solvent. What happens to the concentration when two different solutions are mixed?

Example 5

What is the concentration of a solution prepared by mixing 200.0 mL of 0.150 M $CaCl_2$ with 50.0 mL of a 0.200 M $CaCl_2$ solution? Assume that the volume of the mixture is the sum of the volumes of the separate solutions.

$$\text{molarity} = \frac{\text{moles of solute}}{\text{liters of solution}}$$

To calculate the molarity of the mixture, we need the volume of the mixture and the moles of $CaCl_2$ in the mixture.

The volume of the mixture is the sum of the volumes of the separate solutions.

$$200.0 \text{ mL} + 50.0 \text{ mL} = 250.0 \text{ mL}$$

The moles of $CaCl_2$ in the mixture is the sum of the moles of $CaCl_2$ in each of the separate solutions.

For one solution,

$$(0.2000 \text{ L}) (0.150 \text{ mol } CaCl_2 / \text{L}) = 0.0300 \text{ mol } CaCl_2$$

For the other solution,

$$(0.0500 \text{ L}) (0.200 \text{ mol } CaCl_2 / \text{L}) = 0.0100 \text{ mol } CaCl_2$$

The total moles of $CaCl_2$ in the mixture,

$$0.0300 \text{ mol } CaCl_2 + 0.0100 \text{ mol } CaCl_2 = 0.0400 \text{ mol } CaCl_2$$

Then, the molarity of the mixture is

$$\text{molarity} = \frac{0.0400 \text{ mol } CaCl_2}{0.2500 \text{ L}} = 0.160 \text{ M } CaCl_2$$

We could not use the equation $M_1V_1 = M_2V_2$ in this situation, because it applies only when a solution is diluted with pure solvent. Here we diluted one solution with another solution.

Exercise Five

What is the molarity of a solution prepared by mixing 100.0 mL of 0.200 M $CuSO_4$ with 500.0 mL of 0.150 M $CuSO_4$? Assume that the volume of the mixture is the sum of the volumes of the separate solutions.

MIXING DIFFERENT SOLUTIONS

In the previous example we saw how to calculate the concentration of a solution prepared by mixing two solutions of the same solute. What happens to the concentration of the solutes when they are different?

Example 6

A solution is prepared by adding 0.200 L of 1.00 M $ZnCl_2$(aq) to 0.300 L of 0.500 M NaCl(aq). What are the molar concentrations of the ions (Na^+, Zn^{2+}, and Cl^-) in the mixture?

For each of the ions, the molarity is equal to the number of moles of the ion divided by the total liters of the solution.

The volume of the mixture is the sum of the volumes of the two solutions.
   liters of solution = 0.200 L + 0.300 L = 0.500 L

The moles of Zn^{2+} ions is the same as the moles of $ZnCl_2$ because there is one mole of Zn^{2+} ions in one mole of $ZnCl_2$.

$$\text{mol } Zn^{2+} = (0.200 \text{ L})(1.00 \text{ mol } ZnCl_2 / \text{L}) = 0.200 \text{ mol } Zn^{2+}$$

$$\text{molarity of } Zn^{2+} = \frac{0.200 \text{ mol } Zn^{2+}}{0.500 \text{ L}} = 0.400 \text{ M } Zn^{2+}$$

The moles of Na^+ ions is the same as the moles of NaCl because there is one mole of Na^+ ions in one mole of NaCl.

$$\text{mol } Na^+ = (0.300 \text{ L})(0.500 \text{ mol NaCl} / \text{L}) = 0.150 \text{ mol } Na^+$$

$$\text{molarity of } Na^+ = \frac{0.150 \text{ mol } Na^+}{0.500 \text{ L}} = 0.300 \text{ M } Na^+$$

The chloride ions in the mixture come from both the $ZnCl_2$ and the NaCl solutions. The moles of chloride ions is the sum of the moles from $ZnCl_2$ and the moles from NaCl.

The moles of Cl^- from $ZnCl_2$ is twice the number of moles of Zn^{2+} because there are two moles of Cl^- ions for every Zn^{2+} ion in $ZnCl_2$

$$\text{moles } Cl^- \text{ from } ZnCl_2 = 0.200 \text{ mol } Zn^{2+} \left(\frac{2 \text{ mol } Cl^-}{1 \text{ mol } Zn^{2+}} \right) = 0.400 \text{ mol } Cl^-$$

The moles of Cl^- from NaCl is the same as the moles of Na^+.

$$\text{moles } Cl^- \text{ from NaCl} = 0.150 \text{ mol } Na^+ \left(\frac{1 \text{ mol } Cl^-}{1 \text{ mol } Na^+} \right) = 0.150 \text{ mol } Cl^-$$

The total moles of Cl^- is

$$0.400 \text{ mol } Cl^- + 0.150 \text{ mol } Cl^- = 0.550 \text{ mol } Cl^-$$

$$\text{molarity of } Cl^- = \frac{0.550 \text{ mol } Cl^-}{0.500 \text{ L}} = 1.10 \text{ M } Cl^-$$

Exercise Six

A solution is prepared by mixing 500.0 mL of a 1.00 M solution of $Mg(NO_3)_2$, with 250.0 mL of a 2.00 M solution of $AgNO_3$. What are the molar concentrations of the Mg^{2+} ions, the Ag^+ ions, and the NO_3^- ions in the final mixture?

TITRATIONS

One of the most important uses for molarity is in the experimental procedure called a **titration**. A titration involves two different solutions. The solute in one solution reacts with the solute in the other solution. In a titration, one solution is added to a carefully measured volume of the other solution until the reaction between them is complete, and there is no excess of either reactant. At this point, the volume of the added solution is determined. Thus, the volumes of the two solutions are determined. If the concentration of one of the solutions is known, the concentration of the other can be calculated.

The relationships in a titration can be summarized by a map, an extension to the mole map from Lesson 3.

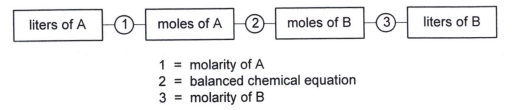

1 = molarity of A
2 = balanced chemical equation
3 = molarity of B

The solutes in the two solutions in a titration are called A and B. The balanced chemical equation relates moles of A to moles of B. The molarity of solution A relates the liters of solution A in the titration to the moles of A in the reaction. The molarity of solution B relates the liters of solution B to the moles of B in the reaction.

Example 7

What volume of 0.320 M I_2(aq) is required for a complete reaction with 50.00 mL of 0.226 M $FeSO_4$? The reaction is represented by the equation below.

$$3\ I_2(aq)\ +\ 6\ FeSO_4(aq)\ \longrightarrow\ 2\ Fe_2(SO_4)_3(aq)\ +\ 2\ FeI_3(aq)$$

We're given liters of $FeSO_4$(aq), and we need to find liters of I_2(aq).

Following the map above, where A = $FeSO_4$ and B = I_2,
    to get from liters of $FeSO_4$(aq) to moles of $FeSO_4$, we use the molarity of $FeSO_4$
    to get from moles of $FeSO_4$ to moles of I_2 , we use the balanced chemical equation
    to get from moles of I_2 to liters of I_2, we use the molarity of I_2

Then,

$$0.05000\ \text{L FeSO}_4 \left(\frac{0.226\ \text{mol FeSO}_4}{1\ \text{L FeSO}_4} \right) \left(\frac{3\ \text{mol I}_2}{6\ \text{mol FeSO}_4} \right) \left(\frac{1\ \text{L I}_2}{0.320\ \text{mol I}_2} \right) = 0.0176\ \text{L I}_2$$

Therefore, we need 17.6 mL of 0.320 M solution of I_2 to react to completion with 50.00 mL of 0.266 M solution of $FeSO_4$.

Actually, this example does not describe a titration experiment because both solution concentrations are known. The example following the next Exercise will present a titration calculation. The calculation shown in this example is one a chemist might do before performing a titration experiment: estimating the amount of one solution needed for the titration based on a guess of the unknown concentration of the second solution.

Exercise Seven

What volume of 1.024 M H_2SO_4(aq) is required for a complete reaction with 100.0 mL of 0.884 M NaOH? The reaction is represented by the equation below.

$$2\,NaOH(aq) + H_2SO_4(aq) \longrightarrow Na_2SO_4(aq) + 2\,H_2O(l)$$

A TITRATION CALCULATION

Example 8

In the titration of 75.00 mL of a 0.1434 M solution of HCl with a solution of $KMnO_4$, 38.24 mL of the $KMnO_4$ solution was required for a complete reaction with the HCl solution. The equation for the reaction is

$$2\,KMnO_4(aq) + 10\,HCl(aq) + 3\,H_2SO_4(aq)$$
$$\longrightarrow 2\,MnSO_4(aq) + 5\,Cl_2(aq) + K_2SO_4(aq) + 8\,H_2O(l)$$

What is the molarity of the $KMnO_4$ solution?

The molarity of the $KMnO_4$ solution is given by

$$\text{molarity of } KMnO_4 = \frac{\text{mol } KMnO_4}{\text{liters of solution}}$$

The volume of the $KMnO_4$ solution used is 38.24 mL, or 0.03824 L.

The moles of $KMnO_4$ in this volume is related to the volume of and concentration of HCl, as represented in the mole map diagram.

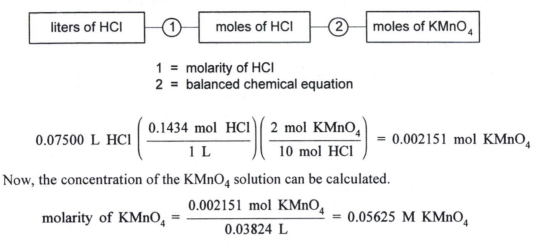

1 = molarity of HCl
2 = balanced chemical equation

$$0.07500 \text{ L HCl} \left(\frac{0.1434 \text{ mol HCl}}{1 \text{ L}} \right) \left(\frac{2 \text{ mol KMnO}_4}{10 \text{ mol HCl}} \right) = 0.002151 \text{ mol KMnO}_4$$

Now, the concentration of the $KMnO_4$ solution can be calculated.

$$\text{molarity of KMnO}_4 = \frac{0.002151 \text{ mol KMnO}_4}{0.03824 \text{ L}} = 0.05625 \text{ M KMnO}_4$$

Exercise Eight

A titration is carried out using 50.00 mL of a solution of oxalic acid, $H_2C_2O_4$, whose concentration is unknown. A 0.3242 M solution of $K_2Cr_2O_7$ is added until the reaction below is complete.

$$3 \text{ H}_2\text{C}_2\text{O}_4(aq) + \text{K}_2\text{Cr}_2\text{O}_7(aq) + 4 \text{ H}_2\text{SO}_4(aq)$$

$$\longrightarrow 6 \text{ CO}_2(g) + \text{Cr}_2(\text{SO}_4)_3(aq) + \text{K}_2\text{SO}_4(aq) + 7 \text{ H}_2\text{O}(l)$$

The titration required 28.44 mL of the $K_2Cr_2O_7$ solution. What is the concentration of the oxalic acid solution?

MOLARITY AND DENSITY OF SOLUTION

Molarity expresses concentration in terms of volume of solution, not of solvent. If the volume of solvent, not of solution, is measured in preparing a solution, then the density of the solution must be known to determine the molarity of the solution.

Example 9

A solution prepared by dissolving 50.0 grams of hydrogen chloride in 97.0 grams of water has a density of 1.171 g/mL. What is the concentration of the solution?

$$\text{molarity} = \frac{\text{moles of solute}}{\text{liters of solution}}$$

To determine the molarity of the solution, we need to know both the moles of HCl and the volume of the solution.

$$\text{moles HCl} = 50.0 \text{ g HCl}\left(\frac{1 \text{ mol HCl}}{36.46 \text{ g HCl}}\right) = 1.37 \text{ mol HCl}$$

The volume of the solution can be determined from its density:

$$\text{density} = \frac{\text{mass}}{\text{volume}}$$

So,

$$\text{volume} = \frac{\text{mass}}{\text{density}}$$

The mass of the solution is the sum of the masses of all its components, namely, 50.0 g HCl and 97.0 g H_2O. Therefore, the mass of the solution is 147.0 g.

$$\text{volume} = \frac{147.0 \text{ g}}{1.171 \text{ g/mL}} = 125.5 \text{ mL}$$

Then, the volume of the solution is 0.1255 L, so its molarity can be calculated.

$$\text{molarity} = \frac{1.37 \text{ mol HCl}}{0.1255 \text{ L}} = 10.9 \text{ M HCl}$$

Exercise Nine

What is the molarity of a solution made by dissolving 32.0 g of silver nitrate, $AgNO_3$ (169.9 g/mol), in 68.0 g of water, H_2O (18.02 g/mol)? The density of water is 0.990 g/mL, and the density of the solution is 1.35 g/mL.

PROBLEMS

1. How many grams of $Ca(OH)_2$ are contained in 25.20 mL of a 0.0930 M aqueous solution?

2. When 46.2 g of NaOH are dissolved in water to make 350 mL of solution, what is the molarity of the solution?

3. What volume of 0.77 M H_2SO_4 contains 0.50 mole of solute?

4. How many milliliters of 0.54 M $AgNO_3$ are needed to obtain 0.34 grams of solute?

5. A solution is prepared by mixing 25.0 mL of 0.200 M HCl(aq) and 50.0 mL of 0.150 M HCl(aq). What is the molarity of HCl in the resulting mixture?

6. When 0.015 g of a pure enzyme is dissolved in water to make 250.0 mL of solution, the molarity of the solution is found to be 8.8×10^{-7} M. What is the molar mass of the enzyme?

7. The concentration of a solution of I_3^- can be determined by titration with sodium thiosulfate solution.

$$2 \, S_2O_3^{2-}(aq) + I_3^-(aq) \longrightarrow S_4O_6^{2-}(aq) + 3 \, I^-(aq)$$

What is the molar concentration of I_3^- if 28.67 mL of 0.107 M $Na_2S_2O_3$ is needed for complete reaction with 10.00 mL of the I_3^- solution?

8. In a titration of oxalic acid with sodium hydroxide, 37.83 mL of the NaOH solution were needed for complete reaction with 50.00 mL of a 0.1376 M solution of $H_2C_2O_4$. What is the molarity of the NaOH solution?

$$NaOH(aq) + H_2C_2O_4(aq) \longrightarrow Na_2C_2O_4(aq) + H_2O(l)$$

CONCENTRATION UNITS OF SOLUTIONS: MOLE FRACTION AND MOLALITY

This lesson deals with:

1. Calculating the mole fractions of the components in a mixture from their masses.
2. Calculating the mole fractions of the components in a mixture from their volumes.
3. Determining the amounts of components needed to prepare a mixture of specified mole fraction.
4. Calculating the molality of a solution.
5. Calculating the amount of solute needed to prepare a solution of given molality.
6. Calculating the molar mass of an unknown from the molality of its solution.

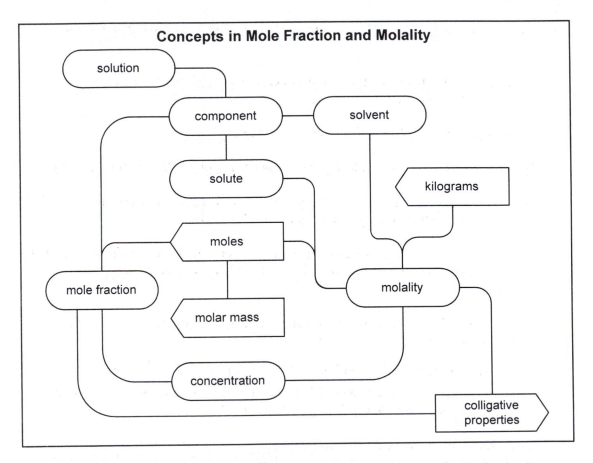

Concepts in Mole Fraction and Molality

The concentration of a solution can be expressed in a number of ways. In this lesson, two methods are described, **mole fraction** and **molality**. Both of these concentration units have the property of being independent of temperature.

MOLE FRACTION

The **mole fraction** of a component in a mixture is equal to the number of moles of that component divided by the total number of moles of all components of the mixture.

$$\text{mole fraction of A} = \frac{\text{moles of A}}{\text{total moles of all components}}$$

Mole fraction of component A in a mixture is represented by X_A.

$$X_A = \frac{\text{moles of A}}{\text{total moles in mixture}}$$

Example 1

What are the mole fractions of butyl alcohol, C_4H_9OH (74.12 g/mol), and water, H_2O (18.02 g/mol), in a mixture made by combining 100.0 grams of butyl alcohol with 50.0 grams of water?

$$X_{C_4H_9OH} = \frac{\text{mol } C_4H_9OH}{\text{mol } C_4H_9OH + \text{mol } H_2O}$$

$$X_{H_2O} = \frac{\text{mol } H_2O}{\text{mol } C_4H_9OH + \text{mol } H_2O}$$

To calculate the values of these mole fractions, we need the moles of both components.

$$\text{mol } C_4H_9OH = 100.0 \text{ g } C_4H_9OH \left(\frac{1 \text{ mol } C_4H_9OH}{74.12 \text{ g } C_4H_9OH} \right) = 1.349 \text{ mol } C_4H_9OH$$

$$\text{mol } H_2O = 50.0 \text{ g } H_2O \left(\frac{1 \text{ mol } H_2O}{18.02 \text{ g } H_2O} \right) = 2.77 \text{ mol } H_2O$$

Now the mole fractions can be calculated.

$$X_{C_4H_9OH} = \frac{1.349 \text{ mol } C_4H_9OH}{1.349 \text{ mol } C_4H_9OH + 2.77 \text{ mol } H_2O} = \frac{1.349 \text{ mol}}{4.12 \text{ mol}} = 0.327$$

$$X_{H_2O} = \frac{2.77 \text{ mol } H_2O}{1.349 \text{ mol } C_4H_9OH + 2.77 \text{ mol } H_2O} = \frac{2.77 \text{ mol}}{4.12 \text{ mol}} = 0.672$$

The mole fraction does not have units.

The value of a mole fraction is always between zero and one, and the sum of the mole fractions of all components in a mixture is one. In this example, $0.327 + 0.672 = 0.999$, which is one to the accuracy of the data provided.

Exercise One

Calculate the mole fractions of ethanol, C_2H_5OH (46.07 g/mol), and water, H_2O (18.02 g/mol), in a solution made by adding 100.0 g of ethanol to 100.0 g of water.

MOLE FRACTION FROM VOLUME

Example 2

What are the mole fractions of butyl alcohol, C_4H_9OH (74.12 g/mol), and water, H_2O (18.02 g/mol), in a mixture made by combining 100.0 mL of butyl alcohol with 50.0 mL of water? The density of butyl alcohol is 0.790 g/mL, and the density of water is 0.990 g/mL.

In this example, the volumes of the components are given. However, to determine the mole fractions, the amounts in moles must be known. Moles can be determined from grams, which are related to volume by density.

So, $$\text{mass} = \text{volume} \times \text{density}$$

$$\text{mass of } C_4H_9OH = (100.0 \text{ mL})(0.790 \text{ g/mL}) = 79.0 \text{ g } C_4H_9OH$$

$$\text{mass of } H_2O = (50.0 \text{ mL})(0.990 \text{ g/mL}) = 49.5 \text{ g } H_2O$$

From the masses, the moles are determined.

$$\text{mol } C_4H_9OH = 79.0 \text{ g } C_4H_9OH \left(\frac{1 \text{ mol } C_4H_9OH}{74.12 \text{ g } C_4H_9OH} \right) = 1.06 \text{ mol } C_4H_9OH$$

$$\text{mol } H_2O = 49.5 \text{ g } H_2O \left(\frac{1 \text{ mol } H_2O}{18.02 \text{ g } H_2O} \right) = 2.75 \text{ mol } H_2O$$

Now the mole fractions can be calculated.

$$X_{C_4H_9OH} = \frac{1.06 \text{ mol } C_4H_9OH}{1.06 \text{ mol } C_4H_9OH + 2.75 \text{ mol } H_2O} = \frac{1.06 \text{ mol}}{3.81 \text{ mol}} = 0.278$$

$$X_{H_2O} = \frac{2.75 \text{ mol } H_2O}{1.06 \text{ mol } C_4H_9OH + 2.75 \text{ mol } H_2O} = \frac{2.75 \text{ mol}}{3.81 \text{ mol}} = 0.722$$

Check that the sum of the mole fractions is 1.

$$0.278 + 0.722 = 1.000$$

Exercise Two

Calculate the mole fractions of ethanol, C_2H_5OH (46.07 g/mol), and water, H_2O (18.02 g/mol), in a solution made by adding 100.0 mL of ethanol to 100.0 mL of water. The density of ethanol is 0.710 g/mL, and the density of water is 0.990 g/mL.

PREPARING A SOLUTION OF SPECIFIED MOLE FRACTION

Example 3

How many grams of methanol, CH_3OH, must be added to 250.0 grams of water to make a solution in which the mole fraction of methanol is 0.350?

This equation represents the situation we want:

$$X_{CH_3OH} = \frac{\text{mol } CH_3OH}{\text{mol } CH_3OH + \text{mol } H_2O} = 0.350$$

We can determine the moles of water from the mass of water. Then, we will have an equation with one unknown, the moles of methanol.

$$\text{mol } H_2O = 250.0 \text{ g } H_2O \left(\frac{1 \text{ mol } H_2O}{18.02 \text{ g } H_2O} \right) = 13.87 \text{ mol } H_2O$$

Then, if we let n represent the moles of methanol,

$$\frac{n}{n + 13.87 \text{ mol}} = 0.350$$

Solving this for n,

$$n = 0.350 \, (n + 13.87 \text{ mol}) = 0.350 \, n + 4.85 \text{ mol}$$

$$0.650 \, n = 4.85 \text{ mol}$$

$$n = 7.46 \text{ mol } CH_3OH$$

From the moles of methanol, we can calculate the grams required.

$$7.46 \text{ mol } CH_3OH \left(\frac{32.04 \text{ g } CH_3OH}{1 \text{ mol } CH_3OH} \right) = 239 \text{ g } CH_3OH$$

Exercise Three

How many grams of water must be added to 150.0 g of ethanol, C_2H_5OH, to make a solution in which the mole fraction of ethanol is 0.200?

MOLALITY

The molality of a solute in a solution is equal to the moles of solute divided by the number of kilograms of solvent.

$$\text{molality} = \frac{\text{moles of solute}}{\text{kilograms of solvent}}$$

Molality is abbreviated as a lower-case m.

For example, a 2.0 m solution of HCl in water contains
   2.0 moles of HCl dissolved in 1.0 kg of water,
or   1.0 mole of HCl in 0.5 kg of water,
or   4.0 moles of HCl in 2.0 kg of water,
   etc.

Example 4

Calculate the molality of a solution made by adding 5.00 g of ethanol, C_2H_5OH, to 100.0 g of water.

$$molality = \frac{moles \ of \ solute}{kilograms \ of \ solvent}$$

To calculate the molality, we need the moles of solute and the mass, in kg, of solvent. The solvent is water, and the solution contains 100.0 g of water.

$$100.0 \ g \ H_2O \left(\frac{1 \ kg}{1000 \ g} \right) = 0.1000 \ kg \ H_2O$$

The solute is ethanol, and the moles of ethanol can be determined from the mass of ethanol in the solution.

$$5.00 \ g \ C_2H_5OH \left(\frac{1 \ mol \ C_2H_5OH}{46.07 \ g \ C_2H_5OH} \right) = 0.108 \ mol \ C_2H_5OH$$

Now, the molality of the solution can be calculated.

$$molality = \frac{0.108 \ mol \ C_2H_5OH}{0.1000 \ kg \ solvent} = 1.08 \ m \ C_2H_5OH$$

Exercise Four

Calculate the molality of a solution made by adding 25.3 g of butyl alcohol, C_4H_9OH (74.12 g/mol), to 752.6 g of water.

PREPARING A SOLUTION OF SPECIFIED MOLALITY

Example 5

How many grams of water should be added to 5.82 g of sucrose, $C_{12}H_{22}O_{11}$, to prepare a 1.00 m aqueous solution of sucrose?

$$\text{molality} = \frac{\text{moles of solute}}{\text{kilograms of solvent}}$$

From the definition of molality, we can find the mass of water needed to make the solution.

$$\text{kilograms of solvent} = \frac{\text{moles of solute}}{\text{molality}}$$

To do this, however, we must also know the moles of solute in the solution.

$$\text{mol } C_{12}H_{22}O_{11} = 5.82 \text{ g } C_{12}H_{22}O_{11} \left(\frac{1 \text{ mol } C_{12}H_{22}O_{11}}{342.30 \text{ g } C_{12}H_{22}O_{11}} \right) = 0.0170 \text{ mol } C_{12}H_{22}O_{11}$$

Now, we can calculate the kg of water needed.

$$\text{kg } H_2O = \frac{0.0170 \text{ mol } C_{11}H_{22}O_{11}}{1.00 \text{ mol } C_{11}H_{22}O_{11} / \text{kg } H_2O} = 0.0170 \text{ kg } H_2O$$

We need to add 0.0170 kg (17.0 g) of water.

Exercise Five

How many grams of water should be added to 12.64 g of sodium chloride, NaCl, to prepare a 0.500 m aqueous solution of NaCl?

MOLAR MASS FROM MOLALITY

There are experimental methods, such as freezing-point depression, by which to determine the molality of an existing solution. If the mass of solute and solvent in the solution is known, and the molality is determined experimentally, then the molar mass of the solute can be calculated.

Example 6

A solution was prepared by dissolving 42.1 grams of urea in 500.0 g of water. The molality of the solution was determined to be is 1.40 m. Calculate the molar mass of urea.

The molar mass of urea is equal to the mass of urea divided by the corresponding number of moles.

$$\text{molar mass} = \frac{\text{grams}}{\text{moles}}$$

The mass of urea is known, 42.1 g. The corresponding number of moles can be determined from the molality of the solution.

$$\text{molality urea} = \frac{\text{mol urea}}{\text{kg water}}$$

$$
\begin{aligned}
\text{mol urea} &= (\text{molality urea})(\text{kg water}) \\
&= (1.40 \text{ mol urea} / \text{kg water})(0.5000 \text{ kg water}) \\
&= 0.700 \text{ mol urea}
\end{aligned}
$$

Now, the molar mass can be calculated.

$$\text{molar mass urea} = \frac{42.1 \text{ g urea}}{0.700 \text{ mol urea}} = 60.1 \text{ g/mol urea}$$

Exercise Six

A solution was made by dissolving 1.08 g of an unknown substance in 38.96 g of carbon tetrachloride, CCl_4. The molality of the solution was determined to be 0.0856 m. Calculate the molar mass of the unknown substance.

PROBLEMS

1. What mass in grams of KOH must be dissolved in 250.0 grams of water to prepare a 0.110 m KOH solution?

2. A solid sample weighing 0.524 g is dissolved in 20.0 mL of ethanol, C_2H_5OH. The molality of the resulting solution is 0.24 m. What is the molar mass of the solid? (The density of ethanol is 0.810 g/mL.)

3. What is the mole fraction of bromine in a 0.40 molal solution of Br_2 in carbon tetrachloride (CCl_4) solvent?

4. A solution is prepared by mixing 25.0 g of ethanol, C_2H_5OH, with 150.0 g of water.
   (a) What is the mole fraction of ethanol in this solution?
   (b) What is the molality of ethanol in this solution?

5. What is the mole fraction of pentane, C_5H_{12}, in a mixture that contains 25.0 g of pentane, 35.0 g of hexane, C_6H_{14}, and 50.0 grams of octane, C_8H_{18}?

6. When 20.0 grams of sugar, $C_{12}H_{22}O_{11}$, is dissolved in 100.0 grams of water, what is the molality of the solution?

7. What is the mole fraction of ammonium chloride, NH_4Cl, in a 1.80 m aqueous solution?

8. How many grams of urea, $CO(NH_2)_2$ are needed to prepare a 4.00 m solution in 80.0 grams of water?

9. What is the mole fraction of acetic acid, $HC_2H_3O_2$, prepared by dissolving 12 grams of acetic acid in 150.0 grams of water?

10. A solution is prepared by mixing equal masses of methanol, CH_3OH, and ethanol, CH_3CH_2OH.
    (a) What is the mole fraction of each component in the mixture?
    (b) If ethanol is considered the solvent, what is the molality of methanol?

THERMAL ENERGY IN CHEMICAL REACTIONS

This lesson deals with:

1. Calculating the amount of thermal energy absorbed by a sample of water when its temperature changes.
2. Calculating the molar enthalpy of a chemical reaction.
3. Calculating the enthalpy of reaction from calorimetric data.
4. Calculating the standard enthalpy of formation from calorimetric data.
5. Calculating the standard enthalpy of reaction using Hess's Law.
6. Calculating the standard enthalpy of reaction using a table of standard enthalpies of formation.

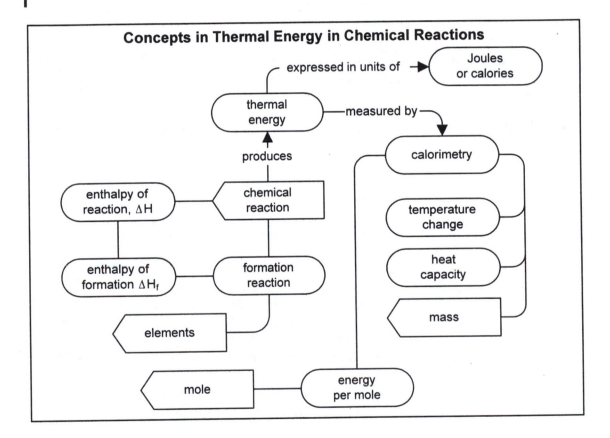

MEASUREMENT OF THERMAL ENERGY

Thermal energy is the form of energy that is transferred from a hot object to a cold object when they are brought together.

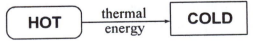

Temperature is a measure of thermal energy. The higher the temperature of an object, the more thermal energy it has.

The absolute temperature (Kelvin temperature) of an object is proportional to its thermal energy.

Definition of a unit of thermal energy:

1 **calorie** = the amount of thermal energy required to raise the temperature of one gram of water by one Kelvin

The dietician's Calorie is 1000 of these calories.

1 dietician's Calorie = 1000 calories = 1 kilocalorie

The standard unit of energy is the **Joule** (abbreviated J).

$$1 \text{ J} = 1 \text{ kg-m}^2/\text{sec}^2$$

The amount of energy required to raise the temperature of 1 gram of water by 1 Kelvin is 4.184 Joules.

$$1 \text{ cal} = 4.184 \text{ J}$$

The amount of thermal energy involved in a process is often determined by measuring the temperature change the process causes in some standard object. This measurement is called **calorimetry**.

The amount of thermal energy gained by an object when its temperature changes is given by the equation

$$q = c \, m \, \Delta T$$

where:

ΔT = change in Kelvin temperature = $T_f - T_i$
m = mass of the object, in grams
c = heat capacity of the object, in J/g K
q = thermal energy absorbed, in Joules

The sign of q calculated using this equation indicates if thermal energy is absorbed or released by the object. When an object absorbs thermal energy, its final temperature is higher than its initial temperature. Then, ΔT is positive. Because both m and c are positive numbers, q is positive, when an object absorbs thermal energy. Conversely, when an object gives off thermal energy, its temperature goes down, ΔT is negative, and q is also negative.

$q > 0$ means   thermal energy absorbed
$q < 0$ means   thermal energy given off

Example 1

How much thermal energy is absorbed by 80.0 grams of water when its temperature increases from 22.7°C to 28.3°C? (The heat capacity of water is 4.184 J/g K.)

The amount of thermal energy absorbed is given by the equation

$$q = c \, m \, \Delta T$$

For water, the heat capacity is 4.184 J/g K.

$$c = 4.184 \text{ J/g K}$$

In this case, we have 80.0 grams of water.
$$m = 80.0 \text{ g}$$

The change in Kelvin temperature can be calculated from the Celsius temperatures.

$$\Delta T = T_f - T_i$$
$$T_f = 28.3 + 273.15 = 301.45 \text{ K}$$
$$T_i = 22.7 + 273.15 = 295.85 \text{ K}$$

Now, substitute the values into the expression for thermal energy absorbed.

$$
\begin{aligned}
q &= (4.184 \text{ J/g K})(80.0 \text{ g})(301.4 \text{ K} - 295.8 \text{ K}) \\
&= (4.184 \text{ J/g K})(80.0 \text{ g})(5.6 \text{ K}) \\
&= 1900 \text{ J}
\end{aligned}
$$

Because q is positive, the water absorbed thermal energy.

Exercise One

A. How much thermal energy is absorbed by 750 grams of water when its temperature increases from 15.4°C to 86.3°C?

B. How much thermal energy is absorbed by 150 grams of water when its temperature *decreases* from 24.36°C to 4.40°C? (Be careful about the sign of q.)

C. Suppose 3260 joules are absorbed by 135 grams of water. If the initial temperature of the water is 21.4°C, what is its final temperature?

TERMINOLOGY OF THERMAL ENERGY IN CHEMICAL REACTIONS

An **exothermic** reaction gives off thermal energy to its surroundings.

Example: the burning of natural gas

$$CH_4(g) + 2 O_2(g) \longrightarrow CO_2(g) + 2 H_2O(g) + \text{thermal energy}$$

An **endothermic** reaction absorbs thermal energy from its surroundings.

Example: the roasting of limestone to form lime and carbon dioxide

$$CaCO_3(s) + \text{thermal energy} \longrightarrow CaO(s) + CO_2(g)$$

The amount of thermal energy involved in a chemical reaction is called the **change in enthalpy (ΔH)** of the reaction.

For a reaction carried out at constant pressure (e.g., in an open container) $\Delta H_{reaction}$ (or ΔH_{rxn}) is the thermal energy *absorbed* in the reaction.

For an exothermic reaction, $\Delta H_{rxn} < 0$ (ΔH is negative)

For an endothermic reaction, $\Delta H_{rxn} > 0$ (ΔH is positive)

The quantity of thermal energy involved in a reaction
(a) is directly proportional to the quantity of reactants, and
(b) depends upon the conditions under which the reaction occurs.

Therefore,
(a) the ΔH is always associated with a balanced chemical equation interpreted in terms of moles of reactants and products, and
(b) the temperature and pressure of the reaction are specified.

A commonly used set of temperature and pressure are the **Standard State Conditions**, which are

Temperature	= 25°C
Pressure	= 1 atmosphere
Solute concentration	= 1 molar

When the ΔH_{rxn} is measured at standard state conditions, then the ΔH_{rxn} is called a standard enthalpy of reaction, and it is represented with a superscript zero.
$$\Delta H^\circ_{rxn} = \text{standard enthalpy of reaction}$$

Example 2

The standard enthalpy of reaction in which two moles of hydrogen chloride gas react with one mole of fluorine gas is –352 kJ.

$$2 HCl(g) + F_2(g) \longrightarrow 2 HF(g) + Cl_2(g) \qquad \Delta H^\circ_{rxn} = -352 \text{ kJ}$$

Use this information to determine the standard enthalpies for the reactions represented by the following equations.

(A) $HCl(g) + \frac{1}{2} F_2(g) \longrightarrow HF(g) + \frac{1}{2} Cl_2(g)$

This equation represents half the amounts of reactants and products as the equation above. Because ΔH_{rxn} is proportional to the amounts of reactants, $\Delta H°_{rxn}$ for this equation is half that above.

$$\Delta H°_{rxn} = -176 \text{ kJ}$$

(B) $HF(g) + \frac{1}{2} Cl_2(g) \longrightarrow HCl(g) + \frac{1}{2} F_2(g)$

This equation is the reverse of that in (A) above. Because ΔH_{rxn} for (A) is negative, reaction (A) is exothermic, and it releases thermal energy. When the equation is reversed, the chemical reaction it represents absorbs thermal energy rather than releasing it. Therefore, ΔH_{rxn} for (B) is positive. Because the amounts of materials are the same, the amount of thermal energy is the same, so the magnitude of ΔH_{rxn} for (B) is the same.

$$\Delta H°_{rxn} = +176 \text{ kJ}$$

Exercise Two

A.   Indicate whether each reaction below is exothermic or endothermic:

    1. $N_2(g) + O_2(g) + \text{thermal energy} \longrightarrow 2 NO(g)$

    2. $S(s) + O_2(g) \longrightarrow SO_2(g)$ $\Delta H°_{rxn} = -297 \text{ kJ}$

    3. $PCl_3(g) + Cl_2(g) \longrightarrow PCl_5(g) + \text{thermal energy}$

    4. $H_2(g) + I_2(s) \longrightarrow 2 HI(g)$ $\Delta H°_{rxn} = +51.8 \text{ kJ}$

B.   What are the standard state conditions?

C.   The standard enthalpy of reaction, $\Delta H°_{rxn}$, for the equation below is -349 kJ.

     $NH_3(g) + \frac{7}{4} O_2(g) \longrightarrow NO_2(g) + \frac{3}{2} H_2O(l)$

Give the standard enthalpy of reaction for each equation below:

    1. $\frac{3}{2} H_2O(l) + NO_2(g) \longrightarrow NH_3(g) + \frac{7}{4} O_2(g)$ $\Delta H°_{rxn} =$

    2. $2 NH_3(g) + \frac{7}{2} O_2(g) \longrightarrow 2 NO_2(g) + 3 H_2O(l)$ $\Delta H°_{rxn} =$

    3. $4 NH_3(g) + 7 O_2(g) \longrightarrow 4 NO_2(g) + 6 H_2O(l)$ $\Delta H°_{rxn} =$

    4. $2 NO_2(g) + 3 H_2O(l) \longrightarrow \frac{7}{2} O_2(g) + 2 NH_3(g)$ $\Delta H°_{rxn} =$

CALCULATION OF $\Delta H°_{rxn}$

Example 3

When 6.0 grams of NO(g) are mixed with excess O_2(g), they react to form 9.2 grams of NO_2(g). At standard state conditions, this reaction releases 22.6 kilojoules of thermal energy. Find $\Delta H°_{rxn}$ for the reaction represented by the equation

$$2\ NO(g)\ +\ O_2(g) \longrightarrow 2\ NO_2(g)$$

$\Delta H°_{rxn}$ is the amount of thermal energy **absorbed** in a reaction involving the molar amounts of substance represented by the chemical equation.

The chemical equation represents the reaction of 2 moles of NO. Therefore, the $\Delta H°_{rxn}$ is the thermal energy absorbed when 2 moles of NO react.

We know that 6.0 grams of NO release 22.6 kJ of thermal energy.

$$6.0\ g\ NO \longrightarrow 22.6\ kJ$$
$$2\ mol\ NO \longrightarrow ?\ kJ$$

$$2\ mol\ NO \left(\frac{30.0\ g\ NO}{1\ mol\ NO} \right) \left(\frac{22.6\ kJ}{6.0\ g\ NO} \right) = 220\ kJ$$

Thus, 2 moles of NO would **release** 220 kJ of thermal energy.

Because $\Delta H°_{rxn}$ is the amount of thermal energy **absorbed**, its value here is negative.

$$\Delta H°_{rxn}\ =\ -220\ kJ$$

Exercise Three

A. When excess water is added to 0.50 g of Mg_3N_2(s), 3.4 kJ are released as the reaction below occurs at standard state conditions. Find $\Delta H°_{rxn}$ for the equation

$$^1/_2\ Mg_3N_2(s)\ +\ 3\ H_2O(l) \longrightarrow {}^1/_2\ Mg(OH)_2(s)\ +\ NH_3(g)$$

B. When 1.00 g of $H_2S(g)$ reacts with excess $Cl_2(g)$ to form $HCl(g)$ and $S(s)$, 4.85 kJ are evolved at standard state conditions. Find $\Delta H°_{rxn}$ for the reaction represented by the equation below.

$$2 HCl(g) + S(s) \longrightarrow H_2S(g) + Cl_2(g)$$

C. When sodium bromide is formed from sodium and bromine under standard state conditions, the enthalpy of reaction is –360 kJ.

$$Na(s) + \tfrac{1}{2} Br_2(l) \longrightarrow NaBr(s) \qquad \Delta H°_{rxn} = -360 \text{ kJ}$$

How many joules are released by the reaction when 0.23 g of Na react with excess Br_2 at 25°C and 1 atm?

AN EXPERIMENTAL DETERMINATION OF ΔH

Example 4

Two solutions are mixed, 50.0 mL of 0.50 M Fe^{3+} and 50.0 mL of 0.50 M Cu^+, both at a temperature of 21.4°C.

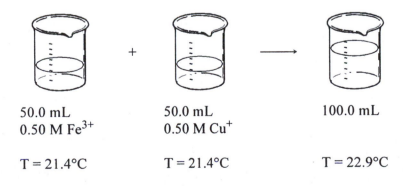

50.0 mL   50.0 mL   100.0 mL
0.50 M Fe^{3+} 0.50 M Cu^+

T = 21.4°C   T = 21.4°C   T = 22.9°C

A reaction occurs in the mixture, and the temperature of the mixture rises to 22.9°C.

The reaction is represented by the equation below. At the conditions of the experiment, what is the value of ΔH_{rxn} for this equation?

$$Fe^{3+}(aq) + Cu^+(aq) \longrightarrow Fe^{2+}(aq) + Cu^{2+}(aq)$$

To find ΔH_{rxn} for this equation, we must find both the amount of thermal energy produced by the reaction and the amount of reactants that produced it.

To use the results of this experiment to find the amount of thermal energy produced by the reaction, we must make several assumptions.
  (1) All thermal energy produced by the reaction is absorbed by the mixture.
  (2) The heat capacity of the mixture is the same as that of water, namely 4.184 J/gK.
  (3) The solutions have the same density as water, and so have a combined mass of 100.0 grams.

Based on these assumptions, we can obtain values for the quantities in the thermal energy equation.

The thermal energy absorbed by the mixture is given by the equation

$$q = c \cdot m \cdot \Delta T$$

and
$$c = 4.184 \text{ J/g K}$$
$$m = 100.0 \text{ g}$$
$$\Delta T = (22.9° + 273.15) - (21.4° + 273.15)$$
$$= (296.05 \text{ K} - 294.55 \text{ K}) = 1.5 \text{ K}$$

Then,
$$q = (4.184 \text{ J/g K})(100.0 \text{ g})(22.9° - 21.4°)$$
$$= (4.184 \text{ J/g K})(100.0 \text{ g})(1.5 \text{ K})$$
$$= 630 \text{ J}$$

This is the thermal energy *absorbed* by the mixture. It is, therefore, the thermal energy *released* by the reaction. Because thermal energy is released by the reaction, the reaction is exothermic, and the value of $\Delta H < 0$.

The chemical equation for the reaction, as given above, involves 1 mole of each reactant, and its ΔH represents the reaction of 1 mole of each. Therefore, we need to find how many moles of Fe^{3+} and Cu^+ reacted in this experiment. We must determine the number of moles of both in the mixture, to see if either is in excess.

$$\text{moles Fe}^{3+} = 50.0 \text{ mL} \left(\frac{1 \text{ L}}{1000 \text{ mL}} \right) \left(\frac{0.50 \text{ mol Fe}^{3+}}{1 \text{ L}} \right) = 0.025 \text{ mol Fe}^{3+}$$

$$\text{moles Cu}^+ = 50.0 \text{ mL} \left(\frac{1 \text{ L}}{1000 \text{ mL}} \right) \left(\frac{0.50 \text{ mol Cu}^+}{1 \text{ L}} \right) = 0.025 \text{ mol Cu}^+$$

The mixture contains the same amounts of Fe^{3+} and Cu^+, and the reaction occurs in a 1-to-1 molar ratio. Therefore, neither Fe^{3+} nor Cu^+ is in excess.

When 0.025 mole of Fe^{3+} reacted, 630 J of thermal energy were released. The chemical equation represents the reaction of 1 mole of Fe^{3+}.

$$1 \text{ mol Fe}^{3+} \left(\frac{630 \text{ J}}{0.025 \text{ mol Fe}^{3+}} \right) = 25000 \text{ J}$$

Because energy is released, ΔH is negative.

$$\Delta H^\circ_{rxn} = -25000 \text{ J} = -25 \text{ kJ}$$

Exercise Four

A 150-mL sample of 0.20 M NaOH solution is mixed with 150 mL of 0.20 M HF solution, both of which have a temperature of 24.5°C. The reaction below occurs, and the temperature of the solution rises to 25.8°C. What is the ΔH of this reaction?

$$NaOH(aq) + HF(aq) \longrightarrow H_2O(l) + NaF(aq)$$

STANDARD ENTHALPY OF FORMATION

The standard enthalpy change of a reaction in which one mole of a compound is formed from its elements in their standard states is called the **standard enthalpy of formation** for that compound.

The standard enthalpy of formation is represented by ΔH°_f

This is the equation for the standard enthalpy of formation of sodium hydroxide.

$$Na(s) + \tfrac{1}{2} O_2(g) + \tfrac{1}{2} H_2(g) \longrightarrow NaOH(s)$$

One mole of NaOH is formed from its elements in their standard states.

The following equations do not represent standard enthalpies of formation:

1. $CaO(s) + H_2O(l) \longrightarrow Ca(OH)_2(s)$
    The reactants are not elements.

2. $Mg(s) + O(g) \longrightarrow MgO(s)$
    The element oxygen is not in its standard state, namely as diatomic molecules.

3. $N_2(g) + O_2(g) \longrightarrow 2\,NO(g)$
    The product contains more than one mole.

Example 5

Write the chemical equation that corresponds to the standard enthalpy of formation of potassium chlorate, $KClO_3(s)$.

This can be done simply by following a sequence of steps. First, write the formula for one mole of $KClO_3(s)$ as a reaction product.

1. $\longrightarrow KClO_3(s)$

Second, write as reactants the formulas for the elements contained in the compound. Be sure that each element is in its standard state form.

2. $K(s) + Cl_2(g) + O_2(g)$ $\longrightarrow KClO_3(s)$

Third, balance the chemical equation. You may use coefficients before even the product, if you like.

3. $2\,K(s) + Cl_2(g) + 3\,O_2(g)$ $\longrightarrow 2\,KClO_3(s)$

If the coefficient of the product is not 1, then divide all coefficients by the product's coefficient. This assures that the product coefficient is equal to 1.

4. $K(s) + \tfrac{1}{2} Cl_2(g) + \tfrac{3}{2} O_2(g) \longrightarrow KClO_3(s)$

This is the chemical equation that corresponds to the standard enthalpy of formation of potassium chlorate.

Example 6

When 0.55 g of Na(s) react with excess $F_2(g)$ to form NaF(s), 13.8 kJ of thermal energy are evolved at standard state conditions. What is the standard enthalpy of formation (ΔH_f°) of NaF(s)?

1. Write the equation for the formation of one mole of NaF(s).
$$Na(s) + \tfrac{1}{2} F_2(g) \longrightarrow NaF(s)$$

2. Find $\Delta H°$ for this chemical equation.
$$0.55 \text{ g Na} \longrightarrow 13.8 \text{ kJ}$$
$$1 \text{ mole Na} \longrightarrow ? \quad \text{kJ}$$

$$1 \text{ mol Na} \left(\frac{22.99 \text{ g Na}}{1 \text{ mol Na}} \right) \left(\frac{13.8 \text{ kJ}}{0.55 \text{ g Na}} \right) = 580 \text{ kJ}$$

$$\Delta H°_{rxn} = -580 \text{ kJ}$$

Because this is the ΔH for a reaction that corresponds to the standard formation of NaF,

$$\Delta H°_f = -580 \text{ kJ/mole NaF}$$

The standard enthalpy of formation of all elements (in their most stable form) is zero joules.

Exercise Five

A. Write the equation for the standard formation of one mole of each of the following compounds.

   1. $NaNO_3(s)$

   2. $C_6H_6(l)$

B. When 0.19 g of $F_2(g)$ react with excess $H_2(g)$ to form HF(g), 2.68 kJ of thermal energy are evolved at standard state conditions. What is $\Delta H°_f$ for HF(g)?

C. For $Li_2O(s)$, $\Delta H_f^\circ = -595.8$ kJ/mole. How many kilojoules are evolved when 10.0 g of lithium react with excess oxygen to form Li_2O at standard state conditions?

HESS'S LAW

Hess's Law: When a chemical equation is the sum of two or more other equations, then the value of ΔH for the first equation is the sum of the ΔH values of the other equations.

See how equations (1), (2), and (3) can be added to give equation (4):

(1) $Ca(s) + 2 H_2O(l) \longrightarrow Ca(OH)_2(s) + H_2(g)$ $\Delta H^\circ_{rxn(1)} = -414.3$ kJ

(2) $CaO(s) + CO_2(g) \longrightarrow CaCO_3(s)$ $\Delta H^\circ_{rxn(2)} = -177.8$ kJ

(3) $Ca(OH)_2(s) \longrightarrow CaO(s) + H_2O(l)$ $\Delta H^\circ_{rxn(3)} = +65.3$ kJ

A. Combine all reactants from (1), (2), and (3) as reactants in a new equation. Similarly combine the products.

reactants	products
$Ca(s) + 2 H_2O(l)$	$Ca(OH)_2(s) + H_2(g)$
$+ CaO(s) + CO_2(g)$	$+ CaCO_3(s)$
$+ Ca(OH)_2(s)$	$+ CaO(s) + H_2O(l)$

B. If some substance appears as both a reactant and a product, remove the same number of moles of it from both sides of the equation, so it no longer appears as both.

reactants	products
$Ca(s) + 2 H_2O(l)$	~~$Ca(OH)_2(s)$~~ $+ H_2(g)$
$+$ ~~$CaO(s)$~~ $+ CO_2(g)$	$+ CaCO_3(s)$
$+$ ~~$Ca(OH)_2(s)$~~	$+$ ~~$CaO(s)$~~ $+$ ~~$H_2O(l)$~~

C. Rewrite the equation using only the remaining reactants and products.

(4) $Ca(s) + H_2O(l) + CO_2(g) \longrightarrow H_2(g) + CaCO_3(s)$

Equation (4) is the sum of equations (1), (2), and (3). Therefore, according to Hess's

Law, the ΔH for equation (4) is the sum of the ΔH values for equations (1), (2), and (3).

$$\Delta H^\circ_{rxn(4)} = \Delta H^\circ_{rxn(1)} + \Delta H^\circ_{rxn(2)} + \Delta H^\circ_{rxn(3)}$$
$$= (-414.3 \text{ kJ}) + (-177.8 \text{ kJ}) + (65.3 \text{ kJ})$$
$$= -526.8 \text{ kJ}$$

Example 7

Find ΔH°_{rxn} for

(1) $C_2H_2(g) + 2 H_2(g) \longrightarrow C_2H_6(g)$

using the standard enthalpies of reaction below.

(2) $2 H_2(g) + O_2(g) \longrightarrow 2 H_2O(l)$ $\Delta H^\circ_{rxn} = -572 \text{ kJ}$

(3) $2 C_2H_2(g) + 5 O_2(g) \longrightarrow 4 CO_2(g) + 2 H_2O(l)$ $\Delta H^\circ_{rxn} = -2598 \text{ kJ}$

(4) $2 C_2H_6(g) + 7 O_2(g) \longrightarrow 4 CO_2(g) + 6 H_2O(l)$ $\Delta H^\circ_{rxn} = -3122 \text{ kJ}$

We need to find a combination of equations (2), (3), and (4) that add to give equation (1).

Equation (1) has 1 mole of $C_2H_2(g)$ as reactant. Only equation (3) also contains $C_2H_2(g)$, but it involves 2 moles as reactant. Therefore, we must multiply equation (3) by ½ to get 1 mole of C_2H_2 as reactant.

Because ΔH is proportional to the amount of reactant, we must also multiply ΔH for equation (3) by ½.

½ × [$2 C_2H_2(g) + 5 O_2(g) \longrightarrow 4 CO_2(g) + 2 H_2O(l)$] $\Delta H^\circ_{rxn} = \frac{1}{2}(-2598 \text{ kJ})$

This gives equation (5), taking care of the reactant C_2H_2 in equation (1).

(5) $C_2H_2(g) + 2½ O_2(g) \longrightarrow 2 CO_2(g) + H_2O(l)$ $\Delta H^\circ_{rxn} = -1299 \text{ kJ})$

Equation (1) also has 2 moles of $H_2(g)$ as reactant. Equation (2) has two moles of $H_2(g)$ as reactant, so we can use it as it is. This becomes equation (6).

(6) $2 H_2(g) + O_2(g) \longrightarrow 2 H_2O(l)$ $\Delta H^\circ_{rxn} = -572 \text{ kJ}$

Equation (1) also has 1 mole of $C_2H_6(g)$ as product. Equation (4) has two moles of $C_2H_6(g)$ as reactant, so we must multiply it by ½ and reverse its direction. When we do this, we must also multiply its ΔH by ½ and reverse its sign.

½ × [$4 CO_2(g) + 6 H_2O(l) \longrightarrow 2 C_2H_6(g) + 7 O_2(g)$] $\Delta H^\circ_{rxn} = -\frac{1}{2}(-3122 \text{ kJ})$

This gives equation (7).

(7) $2 CO_2(g) + 3 H_2O(l) \longrightarrow C_2H_6(g) + 3½ O_2(g)$ $\Delta H^\circ_{rxn} = +1561 \text{ kJ})$

Now we add equations (5), (6), and (7), and their corresponding ΔH values.

reactants	products	$\Delta H°_{rxn}$
$C_2H_2(g)$ + 2½ $O_2(g)$	2 $CO_2(g)$ + $H_2O(l)$	-1299 kJ
2 $H_2(g)$ + $O_2(g)$	2 $H_2O(l)$	-572 kJ
2 $CO_2(g)$ + 3 $H_2O(l)$	$C_2H_6(g)$ + 3½ $O_2(g)$	$+1561$ kJ

$$C_2H_2(g) + 2\,H_2(g) \longrightarrow C_2H_6(g) \qquad \Delta H°_{rxn} = -310 \text{ kJ}$$

Exercise Six

A. Using these standard enthalpies of reaction

$$4\,NH_3(g) + 7\,O_2(g) \longrightarrow 4\,NO_2(g) + 6\,H_2O(l) \qquad \Delta H°_{rxn} = -1395 \text{ kJ}$$

$$N_2(g) + 2\,O_2(g) \longrightarrow 2\,NO_2(g) \qquad \Delta H°_{rxn} = +67.4 \text{ kJ}$$

find the standard enthalpy of reaction for the equation

$$2\,N_2(g) + 6\,H_2O(l) \longrightarrow 4\,NH_3(g) + 3\,O_2(g)$$

B. Using information from part A, determine the standard enthalpy of reaction for the equation

$$N_2(g) + 3\,H_2O(l) \longrightarrow 2\,NH_3(g) + {}^3/_2\,O_2(g) \qquad \Delta H°_{rxn} = ?$$

C. From the standard enthalpies of reaction below, find the standard enthalpy of formation of $N_2H_4(l)$.

$$N_2(g) + 2\,O_2(g) \longrightarrow 2\,NO_2(g) \qquad\qquad \Delta H°_{rxn} = +67.4\ kJ$$

$$N_2H_4(l) + 3\,O_2(g) \longrightarrow 2\,NO_2(g) + 2\,H_2O(l) \qquad \Delta H°_{rxn} = -554.4\ kJ$$

$$2\,H_2(g) + O_2(g) \longrightarrow 2\,H_2O(l) \qquad\qquad \Delta H°_{rxn} = -571.5\ kJ$$

FURTHER APPLICATIONS OF HESS'S LAW

From Hess's Law it follows that the $\Delta H°_{rxn}$ for any chemical reaction is the sum of the $\Delta H°_f$ for all products minus the sum of the $\Delta H°_f$ for all reactants.

To see how this follows from Hess's Law, look at chemical equations (1), (2), and (3) below. All three of these are derived from equations for the $\Delta H°_f$ for some compound. Equation (1), for example, is the reverse of the equation for the $\Delta H°_f$ of P_4O_{10}. Therefore, its $\Delta H°_{rxn}$ is the negative of $\Delta H°_f$ for P_4O_{10}.

(1) $\quad P_4O_{10}(s) \longrightarrow 4\,P(s) + 5\,O_2(g) \qquad\qquad \Delta H°_{rxn(1)} = -\Delta H°_f\,(P_4O_{10})$

(2) $\quad 6\,H_2O(l) \longrightarrow 6\,H_2(g) + 3\,O_2(g) \qquad\qquad \Delta H°_{rxn(2)} = -6\,\Delta H°_f\,(H_2O(l))$

(3) $\quad H_2(g) + 4\,P(s) + 8\,O_2(g) \longrightarrow 4\,H_3PO_4(s) \qquad \Delta H°_{rxn(3)} = 4\,\Delta H°_f\,(H_3PO_4)$

(4) $\quad P_4O_{10}(s) + 6\,H_2O(l) \longrightarrow 4\,H_3PO_4(s)$

Equation (4) is the sum of equations (1), (2), and (3). Therefore, its $\Delta H°_{rxn}$ is the sum of the $\Delta H°_{rxn}$ values for equations (1), (2), and (3).

$$\Delta H°_{rxn(4)} = \Delta H°_{rxn(1)} + \Delta H°_{rxn(2)} + \Delta H°_{rxn(3)}$$

$$= -\Delta H°_f(P_4O_{10}) + [\,-6\,\Delta H°_f(H_2O(l))\,] + 4\,\Delta H°_f(H_3PO_4)$$

$$= 4\,\Delta H°_f(H_3PO_4) - [\,6\,\Delta H°_f(H_2O(l)) + \Delta H°_f(P_4O_{10})\,]$$

$$= \sum \Delta H°_f(products) - \sum \Delta H°_f(reactants)$$

The last line shows that the $\Delta H°_{rxn}$ for this equation is the sum of the $\Delta H_f^°$ for all products minus the sum of the $\Delta H_f^°$ for all reactants.

Note that the enthalpy of formation of each reactant and product is multiplied by its coefficient in the balanced equation.

For any reaction:

$$\Delta H°_{rxn} = \sum \Delta H_f^°(products) - \sum \Delta H_f^°(reactants)$$

Example 8

Calculate the standard enthalpy of reaction for

$$C_2H_4(g) + 3\ O_2(g) \longrightarrow 2\ CO_2(g) + 2\ H_2O(l)$$

using these standard enthalpies of formation

$$\Delta H_f^°(C_2H_4(g)) = +52.3\ kJ/mol$$

$$\Delta H_f^°(CO_2(g)) = -393.7\ kJ/mol$$

$$\Delta H_f^°(H_2O(l)) = -285.8\ kJ/kol$$

The $\Delta H°_{rxn}$ is the sum of the $\Delta H_f^°$ of the products minus the sum of the $\Delta H_f^°$ of the reactants.

$$\Delta H°_{rxn} = \sum \Delta H°_f(products) - \sum \Delta H°_f(reactants)$$

Remember to multiply each $\Delta H°_f$ by the corresponding coefficient in the balanced chemical equation.

$$\Delta H°_{rxn} = [\ 2\ \Delta H_f^°(CO_2) + 2\ \Delta H_f^°(H_2O(l))\ ] - [\ \Delta H_f^°(C_2H_4) + 3\ \Delta H_f^°(O_2)\ ]$$

Substitute the $\Delta H_f^°$ values, remembering that $\Delta H_f^°$ for an element is zero.

$$\Delta H°_{rxn} = [\ (2\ mol\ CO_2)\ (-393.7\ kJ/mol\ CO_2)$$
$$+ (2\ mol\ H_2O(l))\ (-285.8\ kJ/mol\ H2O(l))\ ]$$
$$- [\ (1\ mol\ C_2H_4))\ (+52.3\ kJ/mol\ C_2H_4)$$
$$+ (3\ mol\ O_2)\ (0\ kJmol\ O_2)\ ]$$

$$= [\ (-787.4\ kJ) + (-571.6\ kJ)\ ] - [\ (+52.3\ kJ) + (0\ kJ)\ ]$$

$$= [\ -1359.0\ kJ\ ] - [\ 52.3\ kJ\ ] = -1411.3\ kJ$$

Exercise Seven

Use the standard enthalpies of formation in this table to answer the questions below.

Compound	ΔH_f° (kJ/mol)	Compound	ΔH_f° (kJ/mol)
$CO(g)$	−110.4	$H_2O(l)$	−285.8
$CO_2(g)$	−393.7	$Fe_2O_3(s)$	−822.2

A. Calculate the standard enthalpy of reaction for the equation

$$Fe_2O_3(s) + 3\ CO(g) \longrightarrow 2\ Fe(s) + 3\ CO_2(g)$$

B. The standard enthalpy of reaction for the equation below is −3535 kJ. Calculate the standard enthalpy of formation of $C_5H_{12}(l)$.

$$C_5H_{12}(l) + 8\ O_2(g) \longrightarrow 5\ CO_2(g) + 6\ H_2O(l)$$

PROBLEMS

1. If one mole of ethane gas (C_2H_6) is burned in oxygen to $CO_2(g)$ and $H_2O(l)$, with the simultaneous release of 1560 kJ at standard state conditions, what is the value of ΔH_f° for C_2H_6?

    ΔH_f° for $H_2O(l) = -285.8$ kJ/mol ΔH_f° for $CO_2(g) = -393.7$ kJ/mol

2. The standard enthalpy of formation of $NH_3(g)$ is –46.0 kJ/mol. Calculate the thermal energy evolved when 8.0 g of $H_2(g)$ react with an excess of $N_2(g)$.

3. Find the standard enthalpy of formation of $CH_4(g)$ given the following data.

    $$CH_4(g) + 2\ O_2(g) \longrightarrow CO_2(g) + 2\ H_2O(l) \qquad \Delta H^\circ = -891.2 \text{ kJ}$$
    $$C(s) + O_2(g) \longrightarrow CO_2(g) \qquad \Delta H^\circ = -393.7 \text{ kJ}$$
    $$H_2(g) + {}^1/_2\ O_2(g) \longrightarrow H_2O(l) \qquad \Delta H^\circ = -285.8 \text{ kJ}$$

4. When 50.0 mL of 0.50 M $NH_3(aq)$ is mixed with 50.0 mL of 0.50 M $HCl(aq)$, the temperature of the mixed solutions rises from 21.3°C to 24.4°C. Calculate the ΔH°_{rxn} for
    $$NH_3(aq) + HCl(aq) \longrightarrow NH_4Cl(aq)$$

5. Find ΔH°_{rxn} for the reaction
    $$2\ N_2(g) + 5\ O_2(g) \longrightarrow 2\ N_2O_5(g)$$
    using the standard enthalpies of reaction below.

    $$2\ H_2(g) + O_2(g) \longrightarrow 2\ H_2O(l) \qquad \Delta H^\circ_{rxn} = -571.6 \text{ kJ}$$
    $$N_2O_5(g) + H_2O(l) \longrightarrow 2\ HNO_3(l) \qquad \Delta H^\circ_{rxn} = -76.6 \text{ kJ}$$
    $$N_2(g) + 3\ O_2(g) + H_2(g) \longrightarrow 2\ HNO_3(l) \qquad \Delta H^\circ_{rxn} = -348.2 \text{ kJ}$$

6. Iron metal reacts with oxygen forming iron(III) oxide.
    $$4\ Fe(s) + 3\ O_2(g) \longrightarrow 2\ Fe_2O_3(s) \qquad \Delta H^\circ = -1648.4 \text{ kJ}$$
    How much heat energy is produced when 45.4 g of iron undergo reaction?

7. When 0.800 g of Mg is added to 250.0 mL of 0.40 M $HCl(aq)$, the temperature of the mixture rises from 23.4°C to 37.9°C. Determine the enthalpy change for the reaction represented by the equation below.
    $$Mg(s) + 2\ HCl(aq) \longrightarrow H_2(g) + MgCl_2(aq)$$

8. When 100.0 mL of 1.0 M HCl is mixed with 100.0 mL of 1.0 M NaOH solution, both at a temperature of 20.4°C, they react as described by the chemical equation below. What is the final temperature of the mixture when the reaction is complete?
    $$HCl(aq) + NaOH(aq) \longrightarrow NaCl(aq) + H_2O(l) \qquad \Delta H^\circ = -58.7 \text{ kJ}$$

USING THE IDEAL GAS LAW

This lesson deals with:

1. Using the ideal gas law to find the pressure, temperature, volume, or amount of a gas when three of these quantities are known.
2. Determining the final conditions of a gas when the initial conditions and the changes are known.
3. Calculating the density of a gas at given conditions.
4. Calculating the molar mass of a gas.
5. Finding the pressure or amount of a gas in a mixture using Dalton's Law of Partial Pressures.

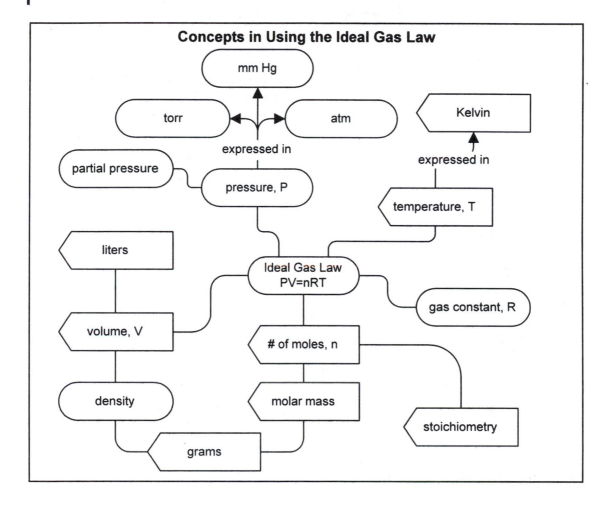

Concepts in Using the Ideal Gas Law

THE PROPERTIES OF GASES

A gas takes both its shape and its volume from its container.
Properties of gases: pressure, volume, temperature.

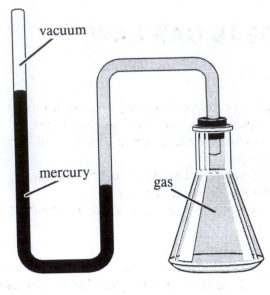

Pressure

Pressure is force per area.

A **manometer** is a pressure-measuring device made of a glass tube containing mercury. Gas exerts pressure on one side of mercury, pushing it up into the other side. The weight of the mercury lifted is equal to the force exerted by the gas.

weight of mercury = force of gas

The weight of mercury is proportional to its height in the tube.
The difference in the levels of the mercury is proportional to the pressure of the gas.

1 mm Hg = 1 torr
1 atmosphere = 760 mm Hg = 760 torr

International System:

1 Newton/meter^2 = 1 Pascal
1 atmosphere = 1.013 × 10^5 Pascal

Volume

Most commonly measured in liters.

Temperature

Temperature must be expressed in absolute units.
The temperature scale must start with zero.
The Kelvin scale will be used.

To convert Celsius to Kelvin, add 273 to the Celsius temperature:

0°C + 273 = 273 K

THE IDEAL GAS EQUATION

$PV = nRT$

P: pressure
V: volume   ⎱ variables
n: moles
T: temperature

R: a constant

The value of R depends on the units used for P, V, and T. When P is in atmospheres, V in liters, and T in Kelvin, then

$$R = 0.0821 \text{ L atm / mol K}$$

Other values of R are listed in the table of constants at the back of the Workbook.

Example 1

A 500.0 mL bottle contains hydrogen gas at 605 torr pressure at 18°C. How many moles of hydrogen gas does the bottle contain?

1. Identify the known values of the variables.

    $P = 605$ torr $T = 18°C$
    $V = 500.0$ mL n: to be found

2. Make sure the units of P, V, and T, agree with those of R.

    $$R = 0.0821 \text{ L atm/mol K}$$

    $$V = 500.0 \text{ mL} \left(\frac{1 \text{ L}}{1000 \text{ mL}} \right) = 0.5000 \text{ L}$$

    $$P = 605 \text{ torr} \left(\frac{1 \text{ atm}}{760 \text{ torr}} \right) = 0.796 \text{ atm}$$

    $$T = 18°C + 273.15 = 291 \text{ K}$$

3. Substitute the known values into the ideal gas law equation, and solve for the unknown variable.

    $$PV = nRT$$

    $$n = \frac{PV}{RT}$$

    $$= \frac{(0.796 \text{ atm})(0.5000 \text{ L})}{(0.0821 \text{ L atm/mol K})(291 \text{ K})}$$

    $$= 1.67 \times 10^{-2} \text{ mol}$$

Exercise One

A 50.0-liter tank designed to hold 250.0 moles of helium gas can withstand a maximum pressure of 200.0 atmospheres. At what temperature would the tank reach its maximum pressure?

PROBLEMS INVOLVING CHANGES IN CONDITIONS

$$PV = nRT$$

Solve the equation for the constant R.

$$R = \frac{PV}{nT}$$

Because R is constant, the ratio $\frac{PV}{nT}$ is also a constant.

Therefore, when a gas undergoes a change from state 1 to state 2, the ratio remains constant.

$$\frac{P_1 V_1}{n_1 T_1} = \frac{P_2 V_2}{n_2 T_2}$$

Example 2

What is the volume of a gas at standard 0°C and 1 atm, if its volume at 142°C and 476 torr is 38.0 mL?

1. Identify the known values of the variables.

state 1	state 2
$P_1 = 476$ torr	$P_2 = 1$ atm
$V_1 = 38.0$ mL	$V_2 = ?$
$T_1 = 142°C = 415$ K	$T_2 = 0°C = 273$ K

Because no gas is added or removed, the number of moles does not change, so $n_1 = n_2$.

2. Convert each variable to the same units.

$$P_1 = 476 \text{ torr} \qquad P_2 = 760 \text{ torr}$$

3. Substitute the values into the equation:

$$\frac{P_1 V_1}{n_1 T_1} = \frac{P_2 V_2}{n_2 T_2}$$

because $n_1 = n_2$,
$$\frac{P_1 V_1}{T_1} = \frac{P_2 V_2}{T_2}$$

$$\frac{(476 \text{ torr})(38.0 \text{ mL})}{415 \text{ K}} = \frac{(760 \text{ torr}) V_2}{273 \text{ K}}$$

$$V_2 = \frac{(476 \text{ torr})(38.0 \text{ mL})(273 \text{ K})}{(415 \text{ K})(760 \text{ torr})} = 15.6 \text{ mL}$$

Exercise Two

A small canister of gas has a pressure of 25 atm at 22°C. What would be the pressure in the canister, if it were taken out of doors on a day when the outside temperature was –26°C?

GAS DENSITY

Density is mass per volume (mass divided by volume).

$$\text{density} = d = \frac{m}{V}$$

The Ideal Gas Law equation relates the number of moles of gas to its pressure, temperature, and volume.

$$n = \frac{PV}{RT}$$

The number of moles of a substance is equal to its mass, m, divided by its molar mass, M. That is, $n = m/M$. Therefore,

$$\frac{m}{M} = \frac{PV}{RT}$$

Then, the density of a gas is

$$d = \frac{m}{V} = \frac{MP}{RT}$$

Example 3

What is the density of CO_2 at 22°C and 1.0 atm?

$$d = \frac{MP}{RT}$$

$P = 1.0$ atm
$T = 22°C = 295$ K
$M = 44$ g/mol
Use $R = 0.0821$ L atm/mol K

$$d = \frac{(44 \text{ g/mol})(1.0 \text{ atm})}{(0.0821 \text{ L atm/mol K})(295 \text{ K})}$$

$$= 1.8 \text{ g/L}$$

Exercise Three

Which is denser, helium gas at 1.0 atm and 0°C, or nitrogen gas at 55 torr and 200°C?

MOLAR MASS OF A GAS

By replacing n in the Ideal Gas Law equation with the ratio of mass to molar mass (m/M), we get an equation that involves the molar mass of a gas.

$$PV = \frac{m}{M}RT$$

This can be solved for an equation that relates the molar mass of a gas to its mass, pressure, temperature, and volume of a gas.

$$M = \frac{mRT}{PV}$$

This allows the determination of the molar mass of a gas from measurements of its mass, temperature, pressure, and volume.

Example 4

Vapor fills a flask with a volume of 109.0 mL at a temperature of 97°C and a pressure of 741.5 torr. The mass of the vapor is 0.1673 grams. What is the molar mass of the vapor?

$$M = \frac{mRT}{PV}$$

Identify the four needed values: mass, temperature, pressure, and volume.

$$m = 0.1673 \text{ g}$$

$$T = 97°C + 273.15 = 370 \text{ K}$$

$$P = 741.5 \text{ torr} \left(\frac{1 \text{ atm}}{760 \text{ torr}} \right) = 0.9756 \text{ atm}$$

$$V = 109.0 \text{ mL} \left(\frac{1 \text{ L}}{1000 \text{ mL}} \right) = 0.1090 \text{ L}$$

Choose a gas constant value with the same units and solve for M.

$$R = 0.0821 \text{ L atm/mol K}$$

$$M = \frac{(0.1673 \text{ g})(0.0821 \text{ L atm/mol K})(370 \text{ K})}{(0.9756 \text{ atm})(0.1090 \text{ L})}$$

$$= 47.8 \text{ g/mol}$$

Exercise Four

When 0.527 gram of a substance is vaporized, it occupies a volume of 135 mL at 98°C and 756.3 torr. What is the molar mass of this substance?

GASEOUS MIXTURES

For a mixture of nitrogen, oxygen, and carbon dioxide, the total number of moles is the sum of the moles of each gas in the mixture.

$$n_{total} = n_{N_2} + n_{O_2} + n_{CO_2}$$

From the Ideal Gas Law Equation for this mixture,

$$P_{total} = n_{total} RT / V$$

$$= \left(n_{N_2} + n_{O_2} + n_{CO_2} \right)(RT/V)$$

$$= n_{N_2}(RT/V) + n_{O_2}(RT/V) + n_{CO_2}(RT/V)$$

$$= P_{N_2} + P_{O_2} + P_{CO_2}$$

Dalton's Law of Partial Pressures:
The total pressure of a mixture of gases is the sum of the pressures of the individual gases.

Example 5

Some solid $CaCO_3$ was sealed into a 250-mL flask along with argon gas at a pressure of 610 torr at 22°C. The flask was heated and some of the $CaCO_3$ decomposed, giving off CO_2 gas.

$$CaCO_3(s) \longrightarrow CaO(s) + CO_2(g)$$

After the flask had cooled to 22°C, the pressure was 840 torr. How many moles of CO_2 were produced?

The moles of CO_2 are related to the pressure of CO_2 in the flask.

$$n_{CO_2} = \frac{P_{CO_2} V}{RT}$$

The increase in pressure is due to the CO_2. The pressure of the CO_2 is

$$P_{CO_2} = 840 \text{ torr} - 610 \text{ torr} = 230 \text{ torr}$$

$$P_{CO_2} = 230 \text{ torr} \left(\frac{1 \text{ atm}}{760 \text{ torr}} \right) = 0.30 \text{ atm}$$

$$V = 250 \text{ mL} \left(\frac{1 \text{ L}}{1000 \text{ mL}} \right) = 0.25 \text{ L}$$

$$T = 22°C + 273.15 = 295 \text{ K}$$

$$n_{CO_2} = \frac{(0.30 \text{ atm})(0.25 \text{ L})}{(0.0821 \text{ L atm/mol K})(295 \text{ K})}$$

$$= 3.1 \times 10^{-3} \text{ moles}$$

Exercise Five

A 40.0-liter tank contains a gas at 1820 torr and 23°C. What would be the pressure in the tank if 0.500 mole of gas were added to the cylinder without changing the temperature?

PROBLEMS

1. A gas occupies 85 mL at 0°C and 710 torr. What temperature would be required to increase the pressure to 760 torr with the volume remaining constant?

2. The volume of a gas mixed with water vapor at 32.0°C at 742 torr is 1350 mL. What would be the volume of the gas at 0.0°C and 760 torr if all the water vapor were removed? The pressure of water vapor at 32°C is 36 torr.

3. A 2.00-g sample of a gas occupies 1.09 liters at 28°C and 702 torr. What is the molar mass of the gas?

4. What volume is occupied by 10.0 g of nitrogen gas at 25°C and 835 torr?

5. A cylinder containing 23.5 moles of nitrogen gas has a volume of 9.81 liters. What is the pressure of the gas at 23°C?

6. A flask contains 28.6 g of SO_2 gas. At 40°C, the pressure of the gas is 850 torr. What is the volume of the flask?

7. When the pressure of argon gas is 6.43 atm, at what Celsius temperature will its density be 10.3 g/L?

8. The density of an unknown gas is 1.31 g/L at 749 torr and 20.0°C. What is the molar mass of the gas?

9. A cruise ship is propelled by steam-driven turbines. Superheated steam (steam at a temperature above the boiling point of water) enters the turbine at 371°C and 51 atm. When it exits the turbine, a liter of this superheated steam will have expanded to 153 liters at 131 torr. What is the Celsius temperature of the steam as it exits the turbine?

ELECTRONIC STRUCTURE OF ATOMS

This lesson deals with:

1. Representing electron distribution in *s*, *p*, and *d* orbitals.
2. Determining the number of electrons in a subshell.
3. Writing electron configurations for ground-state atoms.
4. Writing electron configurations for ground state monatomic ions.
5. Determining a set of quantum numbers for an electron in a particular orbital.

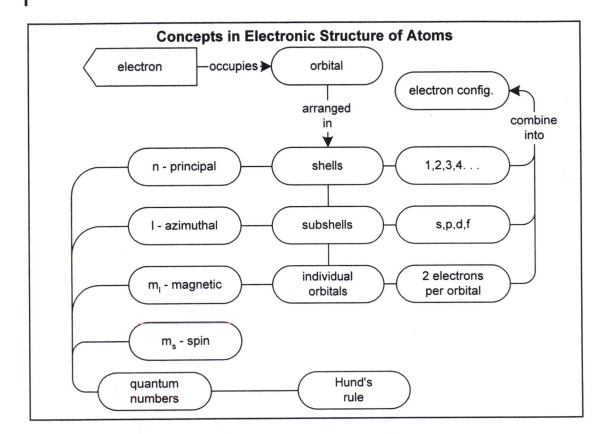

ATOMIC ORBITALS

An atom is composed of three kinds of particles: electrons, protons, and neutrons. The protons and neutrons are gathered in a tiny nucleus at the center of the atom. The electrons surround the nucleus and occupy most of the space inside the atom.

> Components of an atom:
> protons ⎫
> neutrons ⎬ in nucleus
> electrons

The Schrödinger equation relates the energies of electrons to their distribution around the nucleus.

Orbitals are solutions of the Schrodinger equation.

An **orbital** is a region of space around the nucleus in which an electron is likely to be found.

TYPES OF ORBITALS AND THEIR REPRESENTATIONS

There are several types of orbitals, distinguished by their shapes.

An *s* orbital is spherical.

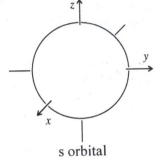

s orbital

A *p* orbital is dumbbell shaped, and are oriented along the x, y, and z axes. Therefore, there are three orientations of *p* orbital.

A set of *p* orbitals consists of three orbitals, and the electrons in these orbitals all have the same energy.

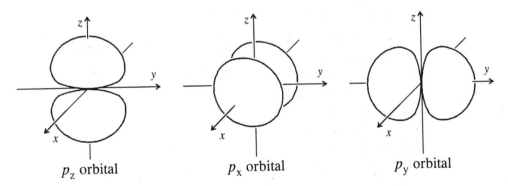

p_z orbital p_x orbital p_y orbital

The shapes of the *d* orbitals are somewhat more complex, and there are five different kinds. A set of *d* orbitals consists of five orbitals, and the electrons in these orbitals all have the same energy.

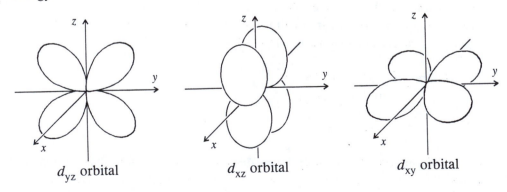

d_{yz} orbital d_{xz} orbital d_{xy} orbital

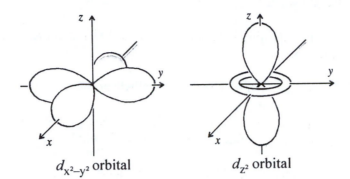

$d_{x^2-y^2}$ orbital d_{z^2} orbital

Even more complex are the shapes of the f orbitals (7 different kinds) and g orbitals (9 different kinds). However, no low-energy atoms have electrons in these kinds of orbitals.

Each orbital is a region of space occupied by no more than two electrons.

Exercise One

A. What is the maximum number of electrons that a p_z orbital can hold?
    A d_{xy} orbital?

B. What is the maximum number of electrons that a set of p orbitals can hold?

    A set of d orbitals?

C. Sketch the representation of the electron distribution for the following (label each axis):

    1. p_x 2. s

    3. p_z 4. d_{xy}

ENERGIES OF ORBITALS

The orbitals in an atom are grouped into levels, called shells, that are numbered 1, 2, 3, etc., in order of increasing energy.

These levels (shells) are divided into sublevels, called subshells.

A subshell is a group of orbitals that have the same energy. A subshell can be one s orbital, or a set of three p orbitals, or a set of five d orbitals, etc.

All orbitals in the same subshell have the same energy.

The division of orbitals among energy shells and subshells is summarized in the chart below.

main energy shell	subshells in main shell
1	$1s$
2	$2s$ $2p$
3	$3s$ $3p$ $3d$
4	$4s$ $4p$ $4d$ $4f$
5	$5s$ $5p$ $5d$ $5f$ $5g$

The number of electrons in an atom is equal to the atomic number of the atom.

The electrons in an atom in its lowest-energy state (the **ground state**) are distributed among the orbitals so that the lower-energy orbitals are filled before higher-energy orbitals. As the atomic number increases from one element to the next, the electrons are arranged as though they add to the next-highest energy orbital with a vacancy. The order of filling of the orbitals is from left to right in the following list of subshells.

$$1s\ 2s\ 2p\ 3s\ 3p\ 4s\ 3d\ 4p\ 5s\ 4d\ 5p\ 6s\ 4f\ 5d \ldots$$

Each orbital holds at most two electrons. An **electron configuration** indicates the arrangement of the electrons in the orbitals.

Atom	no. of electrons	electron configuration
H	1	$1s^1$
He	2	$1s^2$
Li	3	$1s^2\,2s^1$
Be	4	$1s^2\,2s^2$
B	5	$1s^2\,2s^2\,2p_x^{\,1}$

When there are several orbitals of the same energy (such as a set of p orbitals), the electrons spread out, one to an orbital, before they begin to pair up.

C	6	$1s^2\,2s^2\,2p_x^{\,1}\,2p_y^{\,1}$
N	7	$1s^2\,2s^2\,2p_x^{\,1}\,2p_y^{\,1}\,2p_z^{\,1}$
O	8	$1s^2\,2s^2\,2p_x^{\,2}\,2p_y^{\,1}\,2p_z^{\,1}$

Hund's rule: In filling the orbitals of a subshell, one electron must be placed in each orbital prior to placing two electrons in the same orbital.

There is a correlation between an element's electron configuration and its position in the Periodic Table. The correlation is described in Lesson 11, Periodic Properties of the Elements.

Exercise Two

Write the electron configuration of each of the following atoms.

Be:

N:

Na:

S:

FORMATS OF ELECTRON CONFIGURATIONS

Electron configurations are represented in several formats:
    expanded
    condensed
    noble-gas notation

For vanadium (atomic number = 23):

$$V: \ 1s^2 \ 2s^2 \ 2p_x^2 \ 2p_y^2 \ 2p_z^2 \ 3s^2 \ 3p_x^2 \ 3p_y^2 \ 3p_z^2 \ 4s^2 \ 3d_{xy}^1 \ 3d_{yz}^1 \ 3d_{xz}^1$$

This form of the electron configuration is called an **expanded electron configuration**. It shows all occupied orbitals individually.

Another form of the electron configuration shows only subshells and not separate orbitals. This form is called the **condensed electron configuration**.

This is the condensed electron configuration of vanadium:

$$V: \ 1s^2 \ 2s^2 \ 2p^6 \ 3s^2 \ 3p^6 \ 4s^2 \ 3d^3$$

A shorthand notation often used in electron configurations is the **noble-gas notation**. The inner electrons of an atom have the same configuration as the electrons in the previous noble gas. For example, counting down in atomic number from vanadium (atomic number 23), the previous noble gas is argon (atomic number 18). Therefore, the inner electrons of V are represented with the notation [Ar], indicating the electron configuration of argon. The expanded electron configuration of vanadium, using the noble-gas notation is

$$V: \quad [Ar]\ 4s^2\ 3d_{xy}^{\ 1}\ 3d_{yz}^{\ 1}\ 3d_{xz}^{\ 1}$$

The condensed electron configuration of vanadium using noble-gas notation is

$$V: \quad [Ar]\ 4s^2\ 3d^3$$

Electrons in the orbitals beyond the noble-gas core are called **valence electrons**. The valence electrons in vanadium are the $4s^2\ 3d^3$ electrons. These are the electrons that participate in the chemical activity of vanadium atoms.

Because the orbitals in a subshell all have the same energy, the expanded configurations for vanadium given above, in which the valence electrons are represented as

$$4s^2\ 3d_{xy}^{\ 1}\ 3d_{yz}^{\ 1}\ 3d_{xz}^{\ 1}$$

is equivalent to

$$4s^2\ 3d_{z^2}^{\ 1}\ 3d_{x^2-y^2}^{\ 1}\ 3d_{xy}^{\ 1}$$

and to many other distributions among the five d orbitals. The significant aspect of the valence electron configuration of V is that it contains three electrons in three *different 3d* orbitals.

An exception to the systematic filling of orbitals

chromium (atomic number 24), we expect its electron configuration to be

$$1s^2\ 2s^2\ 2p^6\ 3s^2\ 3p^6\ 4s^2\ 3d^4$$

but it actually is $1s^2\ 2s^2\ 2p^6\ 3s^2\ 3p^6\ 4s^1\ 3d^5$

It appears that two half-filled subshells ($4s^1\ 3d^5$) are lower in energy than one filled and one not half-filled ($4s^2\ 3d^4$).

A similar exception occurs in the case of copper, where half-filled and filled subshells are lower in energy than filled and not half-filled.

Exercise Three

Write an electron configuration for each of the following atoms in the formats indicated

A. Ti

   (1) expanded:

   (2) condensed:

   (3) noble-gas notation:

B. As

(1) expanded:

(2) condensed:

(3) noble-gas notation:

C. Cu

(1) expanded:

(2) condensed:

(3) noble-gas notation:

ELECTRON CONFIGURATIONS OF MONATOMIC IONS

A monatomic ion is the ion formed when an atom gains or loses one or more electrons.

For main-group elements, the electron configuration of the ion is the same as that of a neutral atom with the same number of electrons.

O^{2-} ion is formed when two electrons are added to a neutral O atom, which is a main-group element..
> The O^{2-} ion contains 10 electrons, which is the same number as a Ne atom.
>> The electron configuration of O^{2-} is the same as that of Ne.

$$O^{2-}:\ 1s^2\ 2s^2\ 2p_x^{\ 2}\ 2p_y^{\ 2}\ 2p_z^{\ 2}$$

Atoms and ions that have the same electron configuration are called **isoelectronic**.
> O^{2-} and Ne are isoelectronic.

When an electron is removed from an atom, the electron is always removed from the highest shell. If there are occupied subshells in the highest shell of the atom, then the electron is removed from the highest subshell.

A Na^+ ion is formed from a Na atom by removing one electron.
> The electron configuration of Na is $1s^2\ 2s^2\ 2p^6\ 3s^1$
>> The highest shell is the 3 shell. In forming the Na^+ ion, the electron is removed from the $3s$ orbital.
>>> The electron configuration of Na^+ is $1s^2\ 2s^2\ 2p^6$

A S^+ ion is formed from a S atom by removing one electron.
> The electron configuration of S is $1s^2\ 2s^2\ 2p^6\ 3s^2\ 3p^4$
>> The highest shell is the 3 shell, and the highest subshell within that shell is the p subshell. In forming the S^+ ion, the electron is removed from a $3p$ orbital.
>>> The electron configuration of S^+ is $1s^2\ 2s^2\ 2p^6\ 3s^2\ 3p^3$

An Fe^{3+} ion is formed by removing 3 electrons from an Fe atom.
The electron configuration of Fe is $1s^2\, 2s^2\, 2p^6\, 3s^2\, 3p^6\, 4s^2\, 3d^6$
The highest shell is the 4 shell, so electrons are removed from the 4s orbital first.
The electron configuration of Fe^{2+} is $1s^2\, 2s^2\, 2p^6\, 3s^2\, 3p^6\, 3d^6$
To form an Fe^{3+} ion, one more electron must be removed. Now the highest shell is the 3 shell, and the highest subshell within that shell is the d subshell. In forming the Fe^{3+} ion, the electron is removed from a 3d orbital.
The electron configuration of Fe^{3+} is $1s^2\, 2s^2\, 2p^6\; 3s^2\, 3p^6\, 3d^5$

Exercise Four

A. Write a condensed electron configuration for each of the following ions.

   1. S^{2-}

   2. K^+

   3. Al^{3+}

   4. F^-

B. Which of the ions in A are isoelectronic?

C. Write a condensed electron configuration for each of the following ions, using noble gas notation.

   1. Cr^{2+}

   2. I^-

   3. Pb^{2+}

QUANTUM NUMBERS

Each electron in an atom has a unique set of four quantum numbers. Their symbols and names are:

n	principal quantum number
l	azimuthal quantum number
m_l	magnetic quantum number
m_s	spin quantum number

Pauli exclusion principle:
   no two electrons in an atom can have the same set of four quantum numbers.

principal quantum number = number of the shell in which an electron is located

An electron in a $2s$ orbital has $n = 2$.
An electron in a $3d$ orbital has $n = 3$.

azimuth quantum number indicates the subshell in which an electron is located.

for an s subshell	$l = 0$
for a p subshell	$l = 1$
for a d subshell	$l = 2$
etc.	

In any main energy shell with a principal quantum number of n, l has values that range from 0 to $n-1$.

Thus, there are n subshells in each shell with principal quantum number n.

In the $n = 3$ shell, l can be 0, 1, or 2. So the $n = 3$ shell contains only s, p, and d subshells.

magnetic quantum number ranges from $+l$ to $-l$
For an s subshell, $l = 0$, so $m_l = 0$.
There is only one s orbital in a subshell.

For a d subshell, $l = 2$, $m_l = 2$, 1, 0, −1, or −2.
There are five d orbitals in a subshell.

Spin quantum number has only two possible values

$$m_s = +1/2 \text{ or } m_s = -1/2$$

To summarize:

n indicates the main energy shell
l indicates the subshell in the main shell
m_l indicates the orbital in the subshell
m_s indicates which of the two electrons in the orbital

Table 1: Possible Sets of Quantum Numbers for the First Three Values of n

n	1	2				3								
l	0	0	1			0	1			2				
m_l	0	0	1	0	−1	0	1	0	−1	2	1	0	−1	−2
m_s	$\pm^1/_2$	$\pm^1/_2$	$\pm^1/_2$	$\pm^1/_2$	$\pm^1/_2$	$\pm^1/_2$	$\pm^1/_2$	$\pm^1/_2$	$\pm^1/_2$	$\pm^1/_2$	$\pm^1/_2$	$\pm^1/_2$	$\pm^1/_2$	$\pm^1/_2$
possible orbitals	$1s$	$2s$	$2p_x$	$2p_y$	$2p_z$	$3s$	$3p_x$	$3p_y$	$3p_z$	$3d_{xy}$	$3d_{yz}$	$3d_{xz}$	$3d_{x2-y2}$	$3d_{z2}$

What are the quantum numbers of the electrons in the underlined orbital?

O: $1s^2\, 2s^2\, \underline{2p_x^2}\, 2p_y^1\, 2p_z^1$

the second shell, so $n = 2$
subshell with p orbitals, so $l = 1$
one of the three p orbitals, so m_l is either +1, 0, or −1

both electrons have the above three quantum-number values
one electron has $m_s = +^1/_2$
the other has $m_s = -^1/_2$

The unpaired electrons in a partially filled subshell have the same m_s.

O: $1s^2\ 2s^2\ 2p_x^{\ 2}\ 2p_y^{\ 1}\ 2p_z^{\ 1}$

The $2p_y$ and $2p_z$ electrons have the same m_s value.

Exercise Five

A. Give a set of quantum numbers for the $3p_y$ and $3p_z$ electrons in a phosphorus atom.

$3p_y$: $n =$ _____ $l =$ _____ $m_l =$ _____ $m_s =$ _____

$3p_z$: $n =$ _____ $l =$ _____ $m_l =$ _____ $m_s =$ _____

B. Which of the following atoms could have electrons with both of the following sets of quantum numbers?

$n = 3$ $l = 2$ $m_l = -1$ $m_s = +^1/_2$
$n = 3$ $l = 2$ $m_l = 2$ $m_s = -^1/_2$

1. S   2. C   3. Ti   4. Mn   5. Br

PROBLEMS

1. Write an expanded electron configuration for each of the following:
   (a) F   (b) Se   (c) Ar   (d) Cr

2. Write a condensed electron configuration for each of the following:
   (a) Ga^{3+} (b) Zn^{2+} (c) Se^{2-} (d) Ag^+

3. Write a condensed electron configuration using noble gas notation for each of the following:
   (a) Ba   (b) Sn^{2+} (c) Sb   (d) Au^{3+}

4. (a) Give an example of an atom that is isoelectronic with I^-.
   (b) Give an example of a positive ion that is isoelectronic with I^-.
   (c) Give an example of a negative ion that is isoelectronic with I^-.

PERIODIC PROPERTIES OF THE ELEMENTS

This lesson deals with:

1. Classifying elements by their position in the periodic table.
2. Predicting the formula of a compound of an element based on the formula of an analogous compound of another element in the same group.
3. Writing the equation for a reaction based on an analogous equation for a reaction involving other group members.
4. Writing outer electron configurations based on the period table.
5. Predicting relative values of ionization potentials, atomic sizes and electronegativities based on the periodic table.
6. Estimating the properties of elements and compounds based on the values of these properties for other group members.

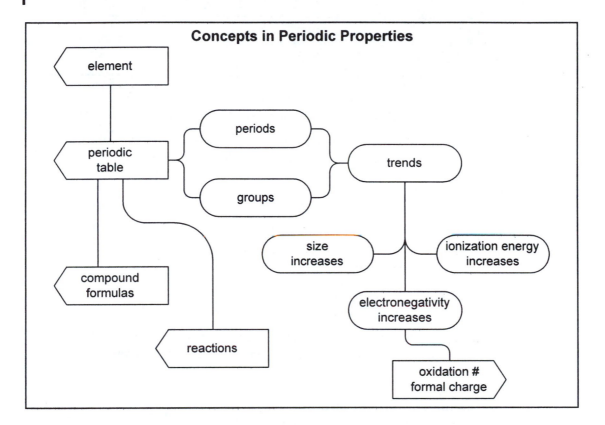

THE PERIODIC TABLE

Columns of elements in the periodic table are called **groups**.
Rows of elements in the periodic table are called **periods**.

Group IA elements are called **alkali metals**.
Group IIA elements are called **alkaline earth metals**.
Group VIIA elements are called **halogens**.
Group VIIIA elements are called **noble gases**.

PERIODIC TABLE OF THE ELEMENTS

IA	IIA												IIIA	IVA	VA	VIA	VIIA	VIIIA
1 **H** 1.008																		2 **He** 4.003
3 **Li** 6.941	4 **Be** 9.012												5 **B** 10.81	6 **C** 12.01	7 **N** 14.01	8 **O** 16.00	9 **F** 19.00	10 **Ne** 20.13
11 **Na** 22.99	12 **Mg** 24.30	IIIB	IVB	VB	VIB	VIIB		VIIIB		IB	IIB		13 **Al** 26.98	14 **Si** 28.09	15 **P** 30.97	16 **S** 32.06	17 **Cl** 35.45	18 **Ar** 39.95
19 **K** 39.10	20 **Ca** 40.08	21 **Sc** 44.96	22 **Ti** 47.87	23 **V** 50.94	24 **Cr** 52.00	25 **Mn** 54.94	26 **Fe** 55.85	27 **Co** 58.93	28 **Ni** 58.69	29 **Cu** 63.55	30 **Zn** 65.39		31 **Ga** 69.72	32 **Ge** 72.61	33 **As** 74.92	34 **Se** 78.96	35 **Br** 79.90	36 **Kr** 83.80
37 **Rb** 85.47	38 **Sr** 87.62	39 **Y** 88.90	40 **Zr** 91.22	41 **Nb** 92.91	42 **Mo** 95.94	43 **Tc** (98)	44 **Ru** 101.1	45 **Rh** 102.9	46 **Pd** 106.4	47 **Ag** 107.9	48 **Cd** 112.4		49 **In** 114.8	50 **Sn** 118.7	51 **Sb** 121.8	52 **Te** 127.6	53 **I** 126.9	54 **Xe** 131.3
55 **Cs** 132.9	56 **Ba** 137.3	57 **La** 138.9	72 **Hf** 178.5	73 **Ta** 180.9	74 **W** 183.8	75 **Re** 186.2	76 **Os** 190.2	77 **Ir** 192.2	78 **Pt** 195.1	79 **Au** 197.0	80 **Hg** 200.6		81 **Tl** 204.4	82 **Pb** 207.2	83 **Bi** 209.0	84 **Po** (209)	85 **At** (210)	86 **Rn** (222)
87 **Fr** (223)	88 **Ra** (226)	89 **Ac** (227)	104 **Rf** (261)	105 **Db** (262)	106 **Sg** (263)	107 **Bh** (262)	108 **Hs** (265)	109 **Mt** (266)	110 **Ds** (271)	111 **Uuu** (272)								

Inner Transition Elements:

Lanthanides:	58 **Ce** 140.1	59 **Pr** 140.9	60 **Nd** 144.2	61 **Pm** (145)	62 **Sm** 150.4	63 **Eu** 152.0	64 **Gd** 157.2	65 **Tb** 158.9	66 **Dy** 162.5	67 **Ho** 164.9	68 **Er** 167.3	69 **Tm** 168.9	70 **Yb** 173.0	71 **Lu** 175.0
Actinides:	90 **Th** 232.0	91 **Pa** 231.0	92 **U** 238.0	93 **Np** (237)	94 **Pu** (244)	95 **Am** (243)	96 **Cm** (247)	97 **Bk** (247)	98 **Cf** (251)	99 **Es** (252)	100 **Fm** (257)	101 **Md** (258)	102 **No** (259)	103 **Lw** (262)

The elements may be divided into three classes:

metals   non-metals   metalloids.

Metals are: shiny,
good conductors of electricity, and
malleable (capable of being hammered into sheets).

Non-metals are: not shiny (dull)
generally poor conductors of electricity, and brittle.

Metalloids have: some properties characteristic of metals (e.g. being shiny),
and some characteristic of non-metals (e.g., being brittle).

Metals:
1. Elements in groups IA and IIA.
2. Heavier elements in groups IIIA (Al, Ga, In, Tl),
IVA (Sn, Pb), and VA(Bi).
3. Transition (B groups) and inner transition elements (lanthanides, elements 58–71, and actinides, elements 90–103).

Non-metals:
Elements in the region at the upper right of the periodic table.

Metalloids:
Elements adjacent to a diagonal line from boron to tellurium on the periodic table.

Exercise One

A. Give the symbols for two examples of each of the following:

transition metals _____   halogens _____

alkaline earth metals _____   lanthanides _____

inner transition elements _____   alkali metals _____

B. Identify each of the following:

the alkaline earth metal in the 3rd period _____

the heaviest lanthanide element _____

the 6th period element in group VIB _____

the lightest noble gas _____

PREDICTING FORMULAS OF COMPOUNDS

Elements in the same group have similar properties.
    Their compounds have similar formulas.

Example 1

The formula of calcium phosphate is $Ca_3(PO_4)_2$. Predict the formulas of calcium arsenate, magnesium phosphate, and barium arsenate.

calcium arsenate
            contains arsenic, As.
            As is in group VA with P.
            Substitute As for P:
                $Ca_3(AsO_4)_2$

magnesium phosphate
            contains magnesium, Mg.
            Mg is in group IIA with Ca.
            Substitute Mg for Ca:
                $Mg_3(PO_4)_2$

barium arsenate
            Substitute Ba for Ca
            and As for P:
                $Ba_3(AsO_4)_2$

Caution: Substituting the symbol of an element for that of one of its group members does not always yield the correct formula of a compound.

calcium nitrate

contains nitrogen, N.
N is in group VA with P.
Substitute N for P:
$$Ca_3(NO_4)_2$$

THIS IS NOT THE CORRECT FORMULA OF CALCIUM NITRATE.
The correct formula is $Ca(NO_3)_2$.

An incorrect formula often results when the substitution involves the symbol of a non-metallic element in the second period. (N is in the second period.)

Exercise Two

A. The formula of sodium sulfate is Na_2SO_4. Write the formula of:

1. potassium sulfate _____

2. sodium tellurate _____

3. lithium selenate _____

B. Given the following names and formulas
   potassium permanganate: $KMnO_4$ phosphine: PH_3
   calcium chromate: $CaCrO_4$ carbon tetrachloride: CCl_4
   write the formulas of:

1. magnesium tungstate _____

2. silicon tetrafluoride _____

3. sodium perrhenate _____

4. arsine _____

C. The formula of potassium chlorate is $KClO_3$. Which of the following formulas corresponds to a substance which is likely to exist: KFO_3 or $KBrO_3$? Explain.

PREDICTING EQUATIONS FOR REACTIONS

Example 2

Magnesium metal reacts with hydrogen chloride gas to form solid magnesium chloride ($MgCl_2$) and hydrogen gas. Write the equation for the reaction of calcium metal with hydrogen bromide gas.

$$Mg(s) + 2\,HCl(g) \longrightarrow MgCl_2(s) + H_2(g)$$

Substitute Ca for Mg and Br for Cl:

$$Ca(s) + 2\,HBr(g) \longrightarrow CaBr_2(s) + H_2(g)$$

Exercise Three

A. Titanium tetrachloride reacts with water according to the equation:

$$TiCl_4(l) + 2\,H_2O(l) \longrightarrow TiO_2(s) + 4\,HCl(aq)$$

Write the equation for the reaction of solid zirconium tetrabromide with water.

B. When solid potassium chlorate ($KClO_3$) is heated, it decomposes to solid potassium chloride (KCl) and oxygen gas. Write the equation for the decomposition of rubidium bromate.

THE PERIODIC TABLE AND ELECTRONIC STRUCTURE OF ATOMS

Elements in the same main group have the same outer electron configuration.

Table 1. Electron Configurations of Group IIIA Elements

Element	Electron Configuration	Valence Electrons
B	$1s^2\,2s^2\,2p^1$	$2s^2\,2p^1$
Al	$1s^2\,2s^2\,2p^6\,3s^2\,3p^1$	$3s^2\,3p^1$
Ga	$1s^2\,2s^2\,2p^6\,3s^2\,3p^6\,3d^{10}\,4s^2\,4p^1$	$4s^2\,4p^1$
In	$1s^2\,2s^2\,2p^6\,3s^2\,3p^6\,3d^{10}\,4s^2\,4p^6\,4d^{10}\,5s^2\,5p^1$	$5s^2\,5p^1$
Tl	$1s^2\,2s^2\,2p^6\,3s^2\,3p^6\,3d^{10}\,4s^2\,4p^6\,4d^{10}\,4f^{14}\,5s^2\,5p^6\,5d^{10}\,6s^2\,6p^1$	$6s^2\,6p^1$

Each IIIA element has an outer electron configuration of ns^2np^1 (n is the number of the period in which the element is located.)

Table 2. Outer Electron Configurations of Main Group Elements

IA	IIA	IIIA	IVA	VA	VIA	VIIA	VIIIA
ns^1	ns^2	$ns^2\,np^1$	$ns^2\,np^2$	$ns^2\,np^3$	$ns^2\,np^4$	$ns^2\,np^5$	$ns^2\,np^6$

Example 3

Write the outer electron configuration of bromine and lead.

Br is in group VIIA: $ns^2\,np^5$
Br is in the 4th period: $4s^2\,4p^5$

Pb is in group IVA: $ns^2\,np^2$
Pb is in the 6th period: $6s^2\,6p^2$

Exercise Four

A. Give the outer electron configuration of

Ge: _____   Te: _____

Ba: _____   Xe: _____

B. Identify each of the following:

the fifth period element with a p^2 configuration _____

the fourth period element having one d electron _____

the element that has one electron in the $3s$ orbital _____

PERIODIC TRENDS IN PROPERTIES OF THE ELEMENTS

An **atomic radius** is one half the distance between the centers of the closest atoms in the solid element.

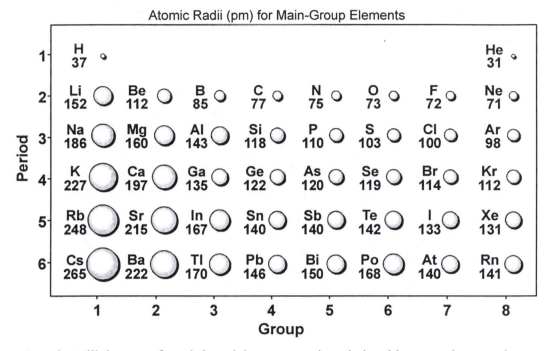

Atomic Radii (pm) for Main-Group Elements

Atomic radii decrease from left to right across each period and increase down each group. (The radii in the chart are given in picometers, where 1 pm = 1×10^{-12} m.)

The **first ionization energy** of an element is the amount of energy required to remove one electron from a neutral atom of the element.

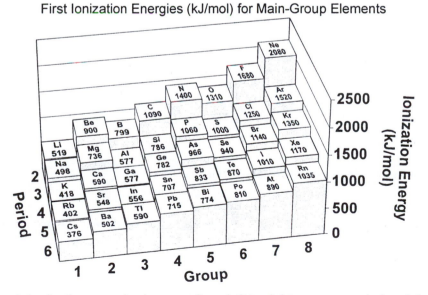

First Ionization Energies (kJ/mol) for Main-Group Elements

In general, ionization energies increase from left to right across a period and decrease down a group.

The **electronegativity** of an element is an indication of the attraction an atom of that element has for the pair of electrons in a covalent bond. (The greater the electronegativity, the greater the attraction for electrons.)

Chart of Electronegativities (Pauling Scale) for Main-Group Elements

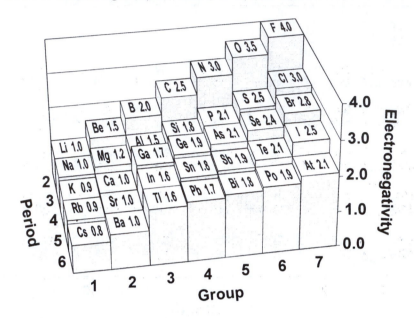

Electronegativities tend to increase from left to right across each period and decrease down each group. Therefore, the most electronegative element is fluorine.

Example 4

Arrange the elements below in order of decreasing electronegativity, referring only to a periodic table.

Al   C   N   Si

Locate the elements in periodic table:

Group: IIIA   IVA   VA

		C	N
Al	Si		

Electronegativities increase from left to right in each period.

N > C   and   Si > Al

Electronegativities decrease down each group.

C > Si

Therefore:

greatest electronegativity = N > C > Si > Al = lowest electronegativity

Exercise Five

To answer these questions, refer only to the periodic table.

A. Compare the atoms O, P, and Sn in reference to

	largest	smallest
atomic size	_____	_____
ionization energy	_____	_____
electronegativity	_____	_____

B. In each set of elements below, circle the one having the largest value of the indicated property:

ionization energy:	B	F	N
electronegativity:	K	Mg	Na
atomic size:	I	P	Sn

ESTIMATING VALUES OF PROPERTIES OF ELEMENTS AND COMPOUNDS

Example 5

A. From the data below, estimate the melting point, boiling point, and density of strontium.

	melting point	boiling point	density
Ca	845°C	1484°C	1.54 g/cm³
Ba	725°C	1140°C	3.51 g/cm³

Locate Ca, Ba, and Sr in periodic table:

IIA
Ca
Sr   Strontium is between calcium and barium in group IIA.
Ba

Estimate the properties of Sr to be the average of those of Ca and Ba:

$$\text{estimated melting point of Sr} = \frac{845°C + 725°C}{2} = 785°C$$

$$\text{estimated boiling point of Sr} = \frac{1484°C + 1140°C}{2} = 1312°C$$

$$\text{estimated density of Sr} = \frac{1.54 \text{ g/cm}^3 + 3.51 \text{ g/cm}^3}{2} = 2.53 \text{ g/cm}^3$$

To judge the effectiveness of this method for estimating properties of elements, compare these estimates to the measured values for Sr.

	melting point	boiling point	density
measured:	769°C	1384°C	2.60 g/cm^3
estimated:	785°C	1312°C	2.53 g/cm^3

The estimate are very good, differing only 2% to 5% from the measured values.

B. Estimate the melting point, boiling point, and density of $SrCl_2$ from the data below:

	melting point	boiling point	density
$CaCl_2$	772°C	1650°C	2.15 g/cm^3
$BaCl_2$	963°C	1560°C	3.92 g/cm^3

Estimate the properties of $SrCl_2$ to be the average of those of $CaCl_2$ and $BaCl_2$.

$$\text{estimated melting point of } SrCl_2 = \frac{772°C + 963°C}{2} = 868°C$$

$$\text{estimated boiling point of } SrCl_2 = \frac{1650°C + 1560°C}{2} = 1605°C$$

$$\text{estimated density of } SrCl_2 = \frac{2.15 \text{ g/cm}^3 + 3.92 \text{ g/cm}^3}{2} = 3.04 \text{ g/cm}^3$$

Compare these estimates to the measured values for $SrCl_2$:

	melting point	boiling point	density
measured:	873°C	1250°C	3.05 g/cm^3
estimated:	868°C	1605°C	3.03 g/cm^3

Here the melting point and density estimate are very good, but the boiling point estimate is about 30% too high.

C. From the data below, estimate the melting point, boiling point, and density of rubidium.

	melting point	boiling point	density
Na	97.8°C	892°C	0.97 g/cm^3
K	63.7°C	760°C	0.86 g/cm^3

Locate Na, K, and Rb in periodic table:

IA
Na
K
Rb   Rubidium is in group IA immediately below K.

To estimate the properties of rubidium, we can suppose that the differences between sodium and potassium will be similar to the differences between potassium and rubidium.

boiling point difference between Na and K: 892°C – 760°C = 132°C

We suppose the difference between potassium and rubidium to be the same.
760°C – bp(Rb) = 132°C
estimated boiling point of Rb = 760°C – 132°C = 628°C

Estimate the melting point and density of Rb in the same way.

melting point difference between Na and K: 97.8°C – 63.7°C = 34.1°C
estimated melting point of Rb = 63.7°C – 34.1°C = 29.6°C

density difference between Na and K:
$0.97 \text{ g/cm}^3 – 0.86 \text{ g/cm}^3 = 0.11 \text{ g/cm}^3$
estimated density of Rb:
$0.86 \text{ g/cm}^3 – 0.11 \text{ g/cm}^3 = 0.75 \text{ g/cm}^3$

Compare these estimates to the measured values for Rb:

	melting point	boiling point	density
measured:	38.9°C	688°C	1.53 g/cm^3
estimated:	29.6°C	628°C	0.75 g/cm^3

In this case, the estimated melting and boiling points are quite accurate, but the estimated density is 50% of the actual value.

Exercise Six

A. Estimate the melting point and boiling point of lithium chloride from the data below.

	melting point	boiling point
LiF	814°C	1676°C
LiBr	547°C	1265°C

B. Estimate the boiling point of phosphine from these data:

for SbH_3, b.p. = $-17°C$

for AsH_3, b.p. = $-55°C$

C. The heat of vaporization of lead is 180 kJ/mole and that of tin is 230 kJ/mole. Estimate the heat of vaporization of germanium.

PROBLEMS

1. Few compounds containing element 85, astatine, have ever been prepared. Which element does astatine most resemble in its chemical behavior?

2. Zinc plus aqueous HCl react according to the equation

$$Zn(s) + 2\ HCl(aq) \longrightarrow ZnCl_2(aq) + H_2(g)$$

Write the chemical equation for cadmium's reaction with aqueous HBr.

3. Calcium telluride's chemical formula is CaTe. Write the chemical formula for each of the following:
    a. barium selenide
    b. calcium oxide
    c. strontium telluride
    d. magnesium selenide
    e. beryllium sulfide

4. Use the information in the following table to estimate the melting point of $GeCl_4$ and the boiling point of $SnCl_4$.

Material	Melting Point (°C)	Boiling Point (°C)
$SiCl_4$	−70.0	57.6
$GeCl_4$		84.0
$SnCl_4$	−33.0	

5. Sodium metal reacts with water to form aqueous sodium hydroxide and hydrogen gas. Write the equation for the reaction of potassium metal with water.

6. In each of the following sets, arrange the atoms or ions in order of increasing (smallest to largest) value of the indicated property. Consult only the Periodic Table; do not use the values given in the various tables in this lesson.

    a. size of atom: O   Si   S   Ge

    b. ionization energy: F   Ne   Na   Mg

    c. size of atom: Al   B   C   K

    d. electronegativity: Li   K   C   N

7. Name the element that corresponds to each of the following descriptions.

    a. The element whose atoms have the electron configuration $1s^2 2s^2 2p^6 3s^2 3p^4$.

    b. The element in the alkaline earth group that has the largest atomic radius.

    c. The element in Group VA whose atoms have the largest ionization energy.

    d. The element whose atoms have the electron configuration $[Ar]3d^{10}4s^1$.

    e. The element whose 2+ ions have the electron configuration $[Kr]4d^6$.

LEWIS STRUCTURES AND THE OCTET RULE

This lesson deals with:

1. Determining the total number of valence electrons of the atoms in a molecule.
2. Determining the number of bonds in a molecule.
3. Drawing the Lewis structure of a neutral molecule.
4. Drawing the Lewis Structure of a polyatomic ion.
5. Drawing the Lewis structure of an ionic compound.
6. Drawing Lewis structures for expanded-octet and odd-electron molecules and ions.

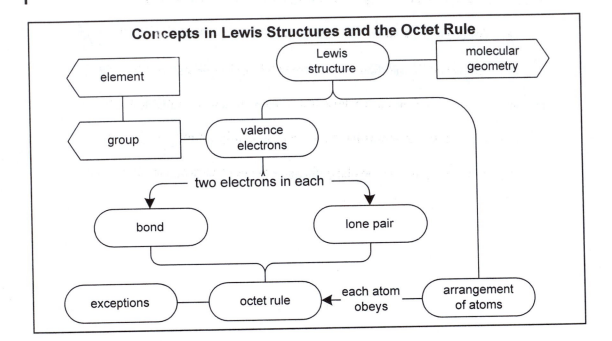

LEWIS STRUCTURES

Lewis structures represent the distribution of valence electrons in molecules. The valence electrons are those in the outermost shell of electrons.

Counting Electrons

How many valence electrons does an atom have? There are several ways to determine this.

1. Write the electron configuration.

$$P:\ 1s^2\ 2s^2\ 2p_x^2\ 2p_y^2\ 2p_z^2\ \underbrace{3s^2\ 3p_x^1\ 3p_y^1\ 3p_z^1}_{\text{valence shell}}$$

The outermost shell of a P atom contains 5 valence electrons.

2. Find the group in the periodic table that contains the atom.
    P is in group VA. P has 5 valence electrons.

How many valence electrons does a molecule have?

Add up the number of valence electrons in all the atoms in the molecule.

PH_3 (P) + 3(H) = (5) + 3(1) = 8 valence electrons

SO_2 (S) + 2(O) = (6) + 2(6) = 18 valence electrons

Exercise One

Give the number of valence electrons in each of the following molecules:

A. NH_3 B. C_2H_4

C. CO_2 D. PCl_3

DRAWING LEWIS STRUCTURES

The Octet Rule: Atoms in molecules usually have eight valence electrons around them.
(Exception: hydrogen atoms have only 2 electrons around them.)

The octet rule is a consequence of the tendency of atoms to fill their outer-shell orbitals.

$$s \quad p_x \quad p_y \quad p_z$$

These four orbitals hold a total of eight electrons. Hydrogen's outer shell is the 1s
orbital, which holds only two electrons.

In a Lewis structure, a straight line represents a bonding pair of electrons, and two dots
represent non-bonding pairs of electrons.

H— F̈:◄——— This pair of dots represents a non-bonding pair of electrons
                    (also called a lone pair of electrons).

——————— This line represents a bonding pair of electrons
                    (also called a shared pair of electrons).

A pair of dots can also be used to represent a pair of bonding electrons. Alternatively,
both bonding pairs and lone pairs of electrons may be represented with lines. The
structures below are also Lewis structures for HF.

H:F̈:   H—F̄ˌ

Example 1

Draw the Lewis structure of NI$_3$.

1. Count the valence electrons.

    (N)+ 3(I) = (5) + 3(7) = 26 valence electrons

2. Connect the atoms together with bonds. In molecules with a single atom of one element and several atoms of another element, the single atom is generally in the center with the other atoms attached to it.

    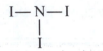

3. Add lone pairs to give each atom 8 electrons.
    a. Give the central atom (nitrogen) eight electrons.

    b. Give each iodine atom eight electrons.

    $$:\ddot{I}-\ddot{N}-\ddot{I}:$$
    $$:\ddot{I}:$$

4. Count the electrons used in the Lewis structure.

    3 bonds × 2 electrons/bond + 10 lone pairs × 2 electrons/pair = 26 electrons

    This number must be the same as the number of valence electrons counted in step 1. It is.

Exercise Two

A. Draw the Lewis structure for CCl$_4$
    1. How many valence electrons does a molecule of CCl$_4$ have?

    2. Connect the atoms together with bonds (put the unique atom in the center).

    3. In the structure above, give each atom 8 electrons.

4. How many electrons did you use in the Lewis structure?

Is this the same as your answer to (1)?

B. Draw the Lewis structure of ammonia, NH_3.

DETERMINING THE CENTRAL ATOM

When a molecule contains more than two elements, or there is no element with a single atom, then consider the number of bonds typically formed by the elements. In most of their compounds, the atoms of many elements typically form the same number of bonds.

In neutral molecules:

C atoms form 4 bonds
N atoms form 3 bonds
O atoms form 2 bonds
H atoms forms 1 bond
F atoms form 1 bond

Other halogen atoms (Cl, Br, and I) frequently, but not always, form 1 bond.

In a molecule, the atom that typically forms the greatest number of bonds is in the center, with other atoms attached to it.

Example 2

A. Draw the Lewis structure for CF_2Cl_2.

1. Count the valence electrons.

$(C) + 2(F) + 2(Cl) = (4) + 2(7) + 2(7) = 32$ valence electrons.

2. Connect the atoms together with bonds.
The C atom is the single atom of its element; place it in the center with F atoms and Cl atoms bonded to it.

$$\begin{array}{c} Cl \\ | \\ F-C-Cl \\ | \\ F \end{array}$$

3. Add lone pairs to give each atom 8 valence electrons. The C atom already has 8 valence electrons. Each F and Cl atom requires 6 more electrons.

4. Count the electrons used in the Lewis structure.
    4 bond × 2 electrons/bond + 12 lone pairs × 2 electrons/pair = 32 electrons
    This is the correct number of valence electrons.

B. Draw the Lewis structure for CNH$_5$

1. Count the valence electrons.
    (C) + (N) + 5(H) = 4 + 5 + 5(1) = 14 electrons

2. Connect the atoms together with bonds.
    H forms 1 bond
    C forms 4 bonds
    N forms 3 bonds

Carbon forms the greatest number of bonds. Put C in center.

Does this agree with the typical number of bonds for these atoms? No. Here C has six bonds and N has one, but C typically forms 4 bonds and N forms 3 bonds. So rearrange the atoms. Move H from C to N until each atom has the proper number of bonds.

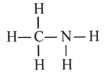

3. Add lone pairs to give each atom 8 electrons (except hydrogen, which needs only 2 valence electrons).

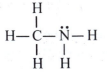

4. Count the number of electrons used in the Lewis structure.
    6 bonds × 2 electrons/bond + 1 lone pair × 2 electrons/pair = 14 electrons
    This is the correct number of electrons.

Exercise Three

A. Draw the Lewis structure for NH_2F.

B. Draw the Lewis structure for COH_4.

WHEN THERE ARE FEWER VALENCE ELECTRONS

Sometimes it happens that the number of electrons used in step 3 is more than the number of valence electrons counted in step 1. Here's what to do in that case.

Example 3

Draw the Lewis structure of H_2CO.

1. Count valence electrons.

$$2(H) + (C) + (O) = 2(1) + (4) + (6) = 12 \text{ valence electrons.}$$

2. Connect the atoms together with bonds.
   H forms 1 bond
   C forms 4 bonds
   O forms 2 bonds
   Carbon forms the greatest number of bonds.

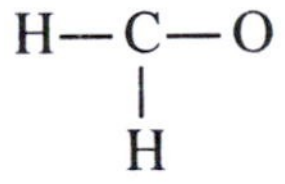

$$H—C—O$$
$$\mid$$
$$H$$

3. Add lone pairs to give each atom 8 electrons (except hydrogen, which gets only 2 electrons).

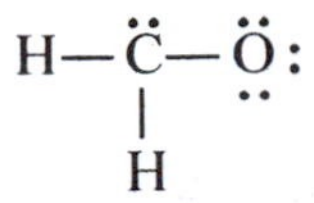

$$H—\ddot{C}—\ddot{O}:$$
$$\mid \quad \ddot{}$$
$$H$$

4. Count the number of electrons used in step 3. Used 14 electrons. However, this is more than the number of valence electrons counted in step 1.

    Used 2 electrons too many.

    2 electrons corresponds to 1 bond. So add another bond in the molecule.

    The bond must go between C and O, because it cannot go between C and H (H forms only one bond).

$$H—C=O$$
$$|$$
$$H$$

    Now, both C and O have their typical numbers of bonds, C has 4 bonds and O has 2.

Add lone pairs to give each atom 8 electrons.

$$H—C=\ddot{\underset{..}{O}}$$
$$|$$
$$H$$

Count electrons used in Lewis structure. Used 12 electrons, which is the correct number.

When you find that too many electrons were used in step 3, divide the number of excess electrons by 2 and add that many bonds in step 2.

2 electrons too many — add 1 bond
4 electrons too many — add 2 bonds
6 electrons too many — add 3 bonds
etc.

Exercise Four

A. Draw the Lewis structure for NOF.

B. Draw the Lewis structure for CO.

RESONANCE STRUCTURES

When drawing some Lewis structures, we find we need a double bond, but it can be placed in more than one position. Such a situation leads to resonance structures.

Example 4

Draw the Lewis structure for SO_2.

1. Count the valence electrons.
$$(S) + 2(O) = (6) + 2(6) = 18 \text{ valence electrons}$$

2. Connect the atoms with bonds.
$$O—S—O$$

3. Add lone pairs to give each atom 8 electrons.
$$\ddot{:O}—\ddot{S}—\ddot{O:}$$

4. Count the electrons used in step 3.
Step 3 used 20 electrons. This is 2 more than the number of valence electrons. Add one more bond.

There are two possibilities:

$$O{=}S—O \quad \text{and} \quad O—S{=}O$$

Neither of these structures by itself is an accurate representation of the bonding in SO_2. One of these structures would indicate that the two bonds in SO_2 are different, but in fact the bonds are the same. The actual arrangement of electrons is an average of the two possibilities. To indicate this, both structures are drawn with a double-headed arrow between them.

$$O{=}S—O \quad \longleftrightarrow \quad O—S{=}O$$

Add lone pairs to give each atom 8 electrons.

$$\ddot{O}{=}\ddot{S}—\ddot{O}: \quad \longleftrightarrow \quad :\ddot{O}—\ddot{S}{=}\ddot{O}$$

Count electrons used. We used 18 electrons this time, which is the proper number.

The two representations are called **resonance structures**. Neither structure, by itself, represents the bonding in SO_2. The bonding is a combination or an average of the resonance structures.

When you find in Step 4 that you need to add double bonds to a Lewis structure, and there is more than one location in which to put the needed bonds, draw resonance structures showing all possibilities. Only the arrangement of the electrons differ in resonance structures; the arrangement of the atoms must remain the same.

Exercise Five

Draw the Lewis structure of SO_3.

LEWIS STRUCTURES FOR POLYATOMIC IONS

Lewis structures can be drawn for both cations and anions.

$$NH_4^+ \qquad\qquad CO_3^{2-}$$

Follow the same rules to draw the Lewis structure.
   Take into account the charge when counting valence electrons.

Example 5

Draw the Lewis structure for CO_3^{2-}

1. Count the valence electrons.
       For the atoms:
             $(C) + 3(O) = 4 + 3(6) = 22$ valence electrons
       For the charge:
             charge indicates there are 2 more electrons.
       total valence electrons $= 22 + 2 = 24$

2. Connect atoms together with bonds.
$$O-C-O$$
$$\mid$$
$$O$$

3. Add lone pairs to give each atom 8 electrons.
$$:\overset{..}{O}-\overset{..}{C}-\overset{..}{O}:$$
$$\mid$$
$$:\overset{..}{O}:$$

4. Count electrons used in step 3. Used 26 electrons.
       Used $26 - 24 = 2$ electrons too many. Need one more bond.

Add one more bond.

$$O\!=\!C\!-\!O \quad \longleftrightarrow \quad O\!-\!C\!-\!O \quad \longleftrightarrow \quad O\!-\!C\!=\!O$$

Add lone pairs to give each atom 8 electrons.

Count electrons used. Used 24 electrons, which is the correct number.

Special notation is used to indicate that this Lewis structure represents an ion. Brackets are drawn around each structure, and the charge is indicated.

Exercise Six

A. Draw the Lewis structure of the sulfate ion, SO_4^{2-}.

B. Draw the Lewis structure of the nitrite ion, NO_2^-.

LEWIS STRUCTURES OF IONIC COMPOUNDS

The Lewis structure of an ionic compound is just the collection of the Lewis structures of the ions that compose it.

Example 6

Draw the Lewis structure for NaCl.

We need the Lewis structures for Na^+ and Cl^-.

Na^+ has lost the one electron in the valence shell of sodium. Therefore, it has an empty valence shell.

Lewis structure for Na^+: $[Na]^+$

Cl^- has eight electrons in its valence shell.

$$\left[:\overset{..}{\underset{..}{Cl}}:\right]^-$$

Lewis structure of NaCl:

$$[Na]^+ \quad \left[:\overset{..}{\underset{..}{Cl}}:\right]^-$$

Example 7

Draw the Lewis structure for Na_2SO_4.

Na_2SO_4 is an ionic compound containing Na^+ ions and SO_4^{2-} ions.

We need the Lewis structures for Na^+ and SO_4^{2-}.

The Lewis structure for Na^+ is in Example 6.

The Lewis structure for SO_4^{2-} was drawn in Exercise Six.

$$\left[\begin{array}{c} :\overset{..}{O}: \\ | \\ :\overset{..}{\underset{..}{O}}-S-\overset{..}{\underset{..}{O}}: \\ | \\ :\overset{..}{\underset{..}{O}}: \end{array}\right]^{2-}$$

Because there are two sodium ions for each sulfate ion, the Lewis structure for Na_2SO_4 contains two sodium ions and one sulfate ion:

$$[Na]^+ \quad \left[\begin{array}{c} :\overset{..}{O}: \\ | \\ :\overset{..}{\underset{..}{O}}-S-\overset{..}{\underset{..}{O}}: \\ | \\ :\overset{..}{\underset{..}{O}}: \end{array}\right]^{2-} \quad [Na]^+$$

Exercise Seven

Draw Lewis structures for the following

A. Na_3PO_4

B. $Ba(OH)_2$

EXPANDED VALENCE SHELLS

This is the Lewis structure of PCl_5:

In this compound, phosphorus has more than 8 electrons in its valence shell.

Phosphorus can accommodate more than 8 electrons in its valence shell, because it has more than 4 orbitals in its valence shell (the 3rd shell).

$$P:\ 1s^2 2s^2 2p^6 3s^2 3p^3 3d^0$$

This is the case with elements whose valence shell is the 3rd period or later. These elements can have an **expanded valence shell**, one containing more than 8 electrons.

Example 8

Draw the Lewis structure for I_3^-.

1. Count valence electrons.

$$3(I) = 3(7) = 21 \text{ electrons}$$

The −1 charge indicates 1 more electron.

total valence electrons = 21 + 1 = 22

2. Join atoms with bonds.

$$I—I—I$$

3. Add lone pairs to give each atom 8 electrons.

$$:\ddot{I}—\ddot{I}—\ddot{I}:$$

4. Count electrons used in step 3. Used 20 electrons.
    Used 2 electrons too few.
    This indicates that the octet rule is not followed by all atoms in this ion.
    Add one pair of electrons to the central atom in the structure of step 3.

$$:\ddot{I}—\overset{..}{\underset{..}{I}}—\ddot{I}:$$

Add brackets and charge of ion.

$$\left[:\ddot{I}—\overset{..}{\underset{..}{I}}—\ddot{I}:\right]^{-}$$

Exercise Eight

Draw the Lewis structure for each of the following.

A. SF_6 B. SF_4

ODD-ELECTRON MOLECULES

Some molecules contain an odd number of electrons, so it is not possible for all electrons to be paired.

Example 9

Draw the Lewis structure for nitric oxide, NO.

1. Count valence electrons.

$$(N) + (O) = (5) + (6) = 11 \text{ electrons}$$

This molecule has an odd number of valence electrons.
Assume temporarily that it has 12 electrons.

2. Join the atoms together with bonds.

    $$N \!-\! O$$

3. Give each atom 8 electrons.

    $$:\!\ddot{N}\!-\!\ddot{O}\!:$$

4. Count the number of electrons used in step 3. Used 14 electrons.
    Used $14 - 12 = 2$ electrons too many.
    Add one bond.

    $$N \!=\! O$$

    Give each atom 8 electrons.

    $$\ddot{N} \!=\! \ddot{O}$$

Used 12 electrons, which is the number we temporarily assumed.

REMOVE ONE ELECTRON.
    The electron may be removed from either the N atom or the O atom. This produces two resonance structures.

$$\dot{N} \!=\! \ddot{O} \quad \longleftrightarrow \quad \ddot{N} \!=\! \dot{O}$$

Another Failure of the Octet Rule: O_2

Experiments show that molecular oxygen has two unpaired electrons. The Lewis structure does not explain this.

Exercise Nine

Draw the Lewis structure for each of the following.

A. BrO

B. ClO_2

PROBLEMS

Draw the Lewis structure for each of the following materials.

A. HCN

B. XeF_2

C. SF_4^{2-}

D. C_2H_4

E. NO_2

F. NH_4Br

OXIDATION NUMBERS AND FORMAL CHARGE

This lesson deals with:

1. Determining the oxidation numbers of the elements in a compound.
2. Determining the possible oxidation numbers of an element.
3. Deciding whether a formula for a compound is consistent with the possible oxidation numbers of its elements.
4. Assigning formal charges to the atoms in a molecule.

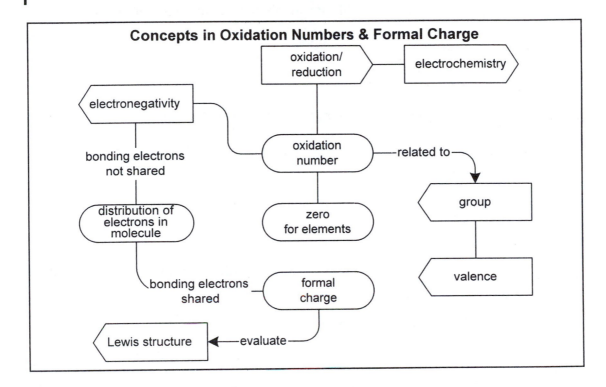

OXIDATION NUMBERS

The oxidation number of an element designates its positive or negative character. Oxidation numbers are a "book-keeping" device for valence electrons.

A *positive* oxidation number indicates that an atom has *lost* electrons, or that the electrons in a bond to the atom are shifted *away* from the atom.

A *negative* oxidation number indicates that an atom has *gained* electrons, or that the electrons in a bond to the atom are shifted *toward* the atom.

The process for assigning oxidation numbers follows a series of rules that are applied in order until oxidation numbers have been determined for all of the elements in a material.

Rules for Assigning Oxidation Numbers

I. **The sum of the oxidation numbers of all atoms in a molecule or ion is equal to the charge of the molecule or ion.**

In H_2SO_4: 2(ox.no. hydrogen) + (ox.no. sulfur) + 4(ox.no. oxygen) = 0

In $Cr_2O_7{}^{2-}$: 2(ox.no. chromium) + 7(ox.no. oxygen) = –2

II. **The oxidation number of the atoms in a pure element is zero.**

In P_4: (ox.no. phosphorus) = 0

III. **Some elements have the same oxidation numbers in all of their compounds.**

Group IA elements have ox.no. = +1
Group IIA elements have ox.no. = +2
Fluorine has an ox.no. = –1

IV. **Hydrogen *usually* has an oxidation number of +1 in its compounds.**

However, in compounds where all other elements are metals (e.g., in CaH_2), its oxidation number is –1.

V. **Oxygen *usually* has an oxidation number of –2 in its compounds.**

However, in peroxides its oxidation number is –1.

VI. **The most electronegative element in a compound takes its lowest oxidation number, which is its group number minus eight.**

In PCl_3: ox.no. chlorine = 7 – 8 = –1

Example 1

Find the oxidation numbers of all atoms in each of the following:

A. $KMnO_4$
     Rule I: (ox.no. K) + (ox.no. Mn) + 4(ox.no. O) = 0
     Rule II: doesn't apply
     Rule III: ox.no. K = +1
        (+1) + (ox.no. Mn) + 4(ox.no. O) = 0
     Rule IV: doesn't apply
     Rule V: ox.no. O = –2
        (+1) + (ox.no. Mn) + 4(–2) = 0
     We now have all of the oxidation numbers except one.
     Solve for the last one.
        (+1) + (ox.no. Mn) + (–8) = 0
        ox.no. Mn = +7

B. $Cr_2O_7^{2-}$

    Rule I: $2(ox.no.\ Cr) + 7(ox.no.\ O) = -2$

    Rules II, III, and IV do not apply.

    Rule V: ox.no. $O = -2$

              $2(ox.no.\ Cr) + 7(-2) = -2$

              $2(ox.no.\ Cr) + (-14) = -2$

              $2(ox.no.\ Cr) = +12$

              $(ox.no.\ Cr) = +6$

C. S_8

    Rule I: $8(ox.no.\ S) = 0$

    Rule II: ox.no. $S = 0$

D. Cr^{3+}

    Rule I: $(ox.no.\ Cr) = +3$

E. H_2S

    Rule I: $2(ox.no.\ H) + (ox.no.\ S) = 0$

    Rules II and III don't apply.

    Rule IV: ox.no. $H = +1$

              $2(+1) + (ox.no.\ S) = 0$

              ox.no. $S = -2$

F. $S_4O_6^{2-}$

    Rule I: $4(ox.no.\ S) + 6(ox.no.\ O) = -2$

    Rules II, III, and IV don't apply.

    Rule V: ox.no. $O = -2$

              $4(ox.no.\ S) + 6(-2) = -2$

              $4(ox.no.\ S) = +10$

              ox.no. $S = +2\frac{1}{2}$

When an element has an oxidation number that is not a whole number, the element has atoms with at least two different oxidation numbers, and the one calculated using the rules is the average of these. We will see how this happens later in the lesson.

Exercise One

Find the oxidation numbers of all atoms in each of the following.

A. $HClO_3$
    ox.no. $H =$ ox.no. $Cl =$ ox.no. $O =$

B. SO_4^{2-}
    ox.no. $S =$ ox.no. $O =$

C. $Na_2S_2O_3$
    ox.no. $Na =$ ox.no. $S =$ ox.no. $O =$

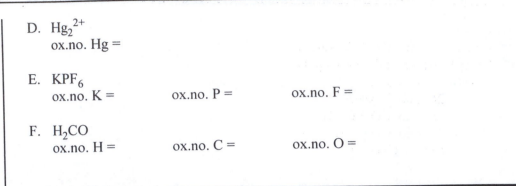

D. Hg_2^{2+}
    ox.no. Hg =

E. KPF_6
    ox.no. K =   ox.no. P =   ox.no. F =

F. H_2CO
    ox.no. H =   ox.no. C =   ox.no. O =

USING RULE SIX

Example 2

Find the oxidation numbers of all atoms in each of the following:

A. H_2PtCl_6

> Rule I: 2(ox.no. H) + (ox.no. Pt) + 6(ox.no. Cl) = 0
> Rules II and III don't apply.
> Rule IV: ox.no. hydrogen = +1
>   2(+1) + (ox.no. Pt) = 6(ox.no. Cl) = 0
> Rule V doesn't apply.
> Rule VI.   Chlorine is the most electronegative element in this
>   compound. It has its lowest oxidation number.
>
>   ox.no. Cl = (group no. of Cl) –8
>   = 7 – 8 = –1
>   2(+1) + (ox.no. Pt) + 6(–1) = 0
>   ox.no. Pt = +4

B. As_2S_3

> Rule I: 2(ox.no. As) + 3(ox.no. S) = 0
> Rules II, III, IV, and V do not apply.
> Rule VI: sulfur is the more electronegative element.
>
>   ox.no. S = 6 – 8 = –2
>   2(ox.no. As) + 3(–2) = 0
>   ox.no. As = +3

Exercise Two

Find the oxidation numbers of all atoms in each of the following.

A. $AlBr_3$
    ox.no. Al =   ox.no. Br =

B. As_4N_4

    ox.no. As =   ox.no. N =

C. CCl_4

    ox.no. C =   ox.no. Cl =

D. NaCN

    ox.no. Na =   ox.no. C =   ox.no. N =

E. K_2SnS_3

    ox.no. K =   ox.no. Sn =   ox.no. S =

EXCEPTIONS TO RULES IV AND V

Example 3

Find the oxidation numbers of all atoms in each of the following.

A. H_2O_2

        Rule I: 2(ox.no. H) + 2(ox.no. O) = 0
        Rules II and III do not apply.
        Rule IV: ox.no. hydrogen = +1
                    2(+1) + 2(ox.no. O) = 0
                    ox.no. oxygen = −1

        An exception to Rule V, a peroxide.

B. $LiAlH_4$

        Rule I: (ox.no. Li) + (ox.no. Al) + 4(ox.no. H) = 0
        Rule II: doesn't apply
        Rule III: ox.no. lithium = +1
        Rule IV: ox.no. hydrogen = −1
                    (+1) + (ox.no. Al) + 4(−1) = 0
                    ox.no. aluminum = +3

Exercise Three

Find the oxidation numbers of all elements in each of the following.

A. OF_2

    ox.no. O =   ox.no. F =

B. $MgCl_2$

    ox.no. Mg =   ox.no. Cl =

C. $Mg(ClO_3)_2$

ox.no. Mg =   ox.no. Cl =   ox.no. O =

D. MgH_2

ox.no. Mg =   ox.no. H =

E. KO_2

ox.no. K =   ox.no. O =

OXIDATION NUMBERS OF THE ELEMENTS

The periodic table on the next page lists the common oxidation numbers of the elements in their compounds.

Metals exhibit positive oxidation numbers.
Negative oxidation numbers distinguish nonmetals.
The *highest* possible oxidation number is equal to the group number.

Group	Highest Oxidation Number	Example
VIIA	+7	NaClO$_4$
VA	+5	HNO$_3$
IVB	+4	TiO$_2$
IA	+1	NaCl

The *lowest* oxidation number is equal to the group number minus eight.

Group	Lowest Oxidation Number	Example
VA	5 – 8 = –3	NH$_3$
VIIA	7 – 8 = –1	HCl

The positive oxidation numbers of metals reflect their tendency to lose electrons.
   Group IA metals lose one valence electron.

The negative oxidation numbers of non-metals reflect their tendency to gain electrons.

   Halogens may have an oxidation number of –1, which indicates the *gain* of one electron.

Non-metals can have positive oxidation numbers when they are combined with a more electronegative non-metal.

   The oxidation number of bromine in BrO_3^- is +5.
   The oxidation number of nitrogen in NO_2 is +4.
   The oxidation number of sulfur in SCl_2 is +2.

IA	IIA	IIIB	IVB	VB	VIB	VIIB		VIIIB		IB	IIB	IIIA	IVA	VA	VIA	VIIA	VIIIA
																	He 0
Li +1	Be +2											B +3	C +4 +2 −4	N +5 +4 +3 +2 −3	O −1 −2	F −1	Ne 0
Na +1	Mg +2											Al +3 −4	Si +4 −4	P +5 +3 −3	S +6 +4 +2 −1 −2	Cl +7 +5 +3 +1 −1	Ar 0
K +1	Ca +2	Sc +3	Ti +4 +3 +2	V +5 +4 +3 +2	Cr +6 +3 +2	Mn +7 +4 +3 +2	Fe +3 +2	Co +3 +2	Ni +2	Cu +2 +1	Zn +2	Ga +3	Ge +4 −4	As +5 +3 −3	Se +6 +4 −2	Br +5 +1 −1	Kr +4 +2
Rb +1	Sr +2	Y +3	Zr +4	Nb +5 +3	Mo +6 +4 +3	Tc +7 +6 +4	Ru +8 +6 +4 +3 +2	Rh +4 +3 +2	Pd +4 +2	Ag +1	Cd +2	In +3	Sn +4 +2	Sb +5 +3 −3	Te +6 +4 −2	I +7 +5 +1 −1	Xe +6 +4 +2
Cs +1	Ba +2	La +3	Hf +4	Ta +5	W +6 +4	Re +7 +6 +4	Os +8 +4	Ir +4 +3	Pt +4 +2	Au +3 +1	Hg +2 +1	Tl +3 +1	Pb +4 +2	Bi +5 +3	Po +4 +2	At −1	Rn 0

H +1 −1

Exercise Four

A. What are the highest and lowest oxidation numbers of phosphorus?

   highest: lowest:

B. Which of the following formulas do not correspond to existing compounds of phosphorus?

$$P_4O_{10} \qquad H_2PO_4 \qquad KPF_6 \qquad Na_2HPO_3 \qquad Ca_2P$$

C. Indicate which of the following are formulas for possible oxides of vanadium.

$$VO \qquad VO_2 \qquad VO_3 \qquad V_2O_5$$

D. What are the oxidation numbers of the elements in sodium peroxydisulfate, $Na_2S_2O_8$?

FORMAL CHARGE

The formal charge is another way of keeping track of electrons. Formal charge is most useful for understanding the bonding in a molecule, while oxidation numbers are more useful for keeping track of electrons in a chemical reaction.

An oxidation number is the charge an atom would have if the electrons in a bond were given to the more electronegative atom in the bond.

Thus, in PF_3, the electrons that bond the fluorine atoms to the phosphorus atom are assigned to fluorine, because fluorine is more electronegative than phosphorus.

This leaves each F with eight valence electrons and P with only 2. Because F atoms start with 7 and they now have 1 more, their oxidation number is −1. Because the P atom starts with 5 electrons and now it has only 2, its oxidation number is +3.

In PH_3, the electrons in the bonds are assigned to P, because it is more electronegative than H.

Now, P has eight electrons. Because it starts with five, it has 3 more, so its oxidation number is −3. Because H starts with 1 and now has none, its oxidation number is +1.

A formal charge is the charge an atom would have if the electrons in a bond were shared equally between the atoms in the bond.

$$:\!\overset{..}{F}\,\underset{\frown}{S}\,P\,\underset{\frown}{2}\,\overset{..}{F}\!:$$
$$:\!\overset{..}{F}\!:$$

In PF_3, the electrons in each bond are divided, one to F and one to P. Because the F atoms start with 7 electrons and still have 7 electrons, their formal charges are zero. Because the P atom starts with 5 electrons and still has 5 electrons, its formal charge is also zero.

As with oxidation numbers, the formal charges of all the atoms in a molecule or ion add up to the charge of the molecule or ion.

Example 4

What are the formal charges of the atoms in the structure of the sulfate ion?

$$\left[\begin{array}{c} :\ddot{O}: \\ :\ddot{O}:S:\ddot{O}: \\ :\ddot{O}: \end{array}\right]^{2-}$$

Divide the electrons in the bonds between the atoms.

$$\begin{array}{c} :\ddot{O}: \\ :\ddot{O}\,S\,\ddot{O}: \\ :\ddot{O}: \end{array}$$

Compare the number of electrons assigned in the structure to the number of valence electrons in the free atoms.

In SO_4^{2-}, the S atom has 4 electrons, and each O atom has 7.
A free S atom has 6 valence electrons, and a free O also has 6.

Assign the formal charges.

The S atom has lost 2 electrons, so it has a formal charge of +2.
Each oxygen atom has gained 1 electron, so each has a formal charge of –1.

Check that the formal charges add to the charge of the ion.

$$(S) + 4\,(O) = +2 + 4\,(-1) = -2$$

Exercise Five

A. Assign formal charges to the atoms in each of the following.

1. H_2O

2. $H-C\equiv N:$

3. $\ddot{O}=\ddot{S}-\ddot{O}:$

B. Use the Lewis structure of the dithionate ion, which follows, to assign the oxidation numbers to the sulfur atoms. We saw earlier, in Example 1 F, that the average oxidation number of sulfur in this ion is 2½.

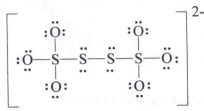

ASSESSING LEWIS STRUCTURES BY FORMAL CHARGE

Formal charges are useful for judging which of several possible Lewis structures is most appropriate for a particular molecule or ion.

The most appropriate Lewis structure will have the formal charges closest to zero, meaning that the electrons are most uniformly distributed in the structure.

Example 5

Which of the following Lewis structures better represents the distribution of electrons in CO_2?

$$:\ddot{O}-C\equiv O: \qquad \ddot{O}=C=\ddot{O}$$

To assess the structures by formal charge, assign shared electrons equally between bonded atoms.

$$:\ddot{O}\,\S\,C\,\ddot{\xi}\,O: \qquad \ddot{O}\,)\!(\,C\,)\!(\,\ddot{O}$$

In the structure on the left,
C has four electrons. It started with 4, so its formal charge is zero. The O on the left has 7 electrons, it started with 6, so its formal charge is −1. The O on the right has 5 electrons, it started with 6, so its formal charge is +1. The formal charges add to zero, which is the charge of the molecule.

In the structure on the right,
C has four electrons. It started with 4, so its formal charge is zero. The O on the left has 6 electrons, it started with 6, so its formal charge is zero. The O on the right has 6 electrons, it started with 6, so its formal charge is also zero. The formal charges add to zero, which is the charge of the molecule.

To summarize, the formal charges are:

$$\begin{array}{ccc} -1 & 0 & +1 \\ :\ddot{O}-C\equiv O: & & \end{array} \qquad \begin{array}{ccc} 0 & 0 & 0 \\ \ddot{O}=C=\ddot{O} & & \end{array}$$

The structure with the formal charges closest to zero best represents the bonding in CO_2, namely the structure on the right.

Exercise Six

Determine using formal charges which Lewis structure of each set below is the better description of bonding.

A. $\ddot{O}=\ddot{N}-\ddot{F}:$ or $:\ddot{O}-\ddot{N}=\ddot{F}$

B. $\left[:\ddot{O}-\overset{\displaystyle :\ddot{O}:}{\underset{\displaystyle :\ddot{O}:}{S}}-\ddot{O}:\right]^{2-}$ or $\left[\ddot{O}=\overset{\displaystyle :\ddot{O}:}{\underset{\displaystyle :\ddot{O}:}{S}}=\ddot{O}\right]^{2-}$

PROBLEMS

1. Find the oxidation number of each element in the following substances.

   (a) N_2O (e) NO (i) HCO_2H

   (b) NO_2 (f) N_2O_5 (j) NH_4I

   (c) $HClO_2$ (g) N_2H_4 (k) $Li_4P_2O_8$

   (d) Cl_2O (h) CH_3OH (l) $Mg(ClO)_2$

2. A molecule of hydrocyanic acid contains one atom of hydrogen, one atom of nitrogen, and one atom of carbon. Use formal charge to predict whether a molecule of hydrocyanic acid is H–C–N or H–N–C.

3. Nitric acid, HNO_3, has three resonance structures. However, two of them contribute more to the bonding than the third one. Draw the three resonance structures and determine which one contributes least.

GEOMETRY OF SIMPLE MOLECULES

This lesson deals with:

1. Describing the structure of molecules in terms of their geometry and bond angles.
2. Applying VSEPR to predict the geometry and bond angles of molecules and ions.

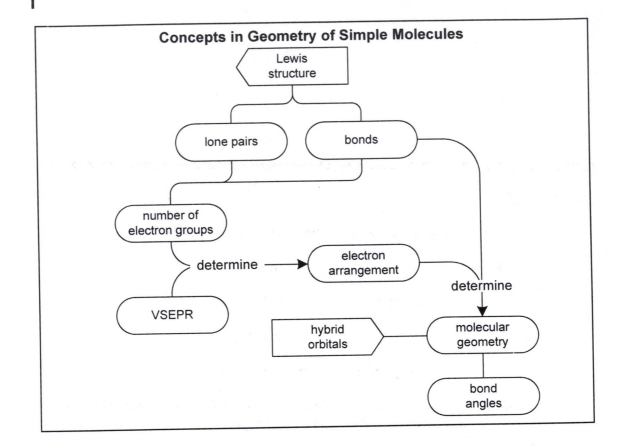

DESCRIBING THE GEOMETRY OF A SIMPLE MOLECULE

The structure of a molecule is described by
    (1) its bond angles and
    (2) its geometry.

A *bond angle* is the angle between two bonds to the same atom.

       angle Y—A—Z (or Z—A—Y)

Bond angles are always less than or equal to 180°.

The geometry of a molecule is described with the geometrical figure that has one of the atoms of the molecule at each of its corners. A molecule may also have an atom at the center of the figure.

triangle

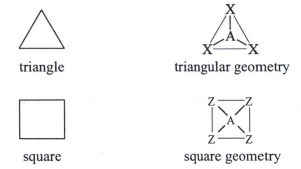

triangular geometry

square

square geometry

Example 1

Describe the geometry of CH_4, NH_3, and H_2O.

CH_4 is methane. All H—C—H bond angles are found by experiments to be 109.5°. The geometry of a methane molecule is tetrahedral.

tetrahedron

tetrahedral geometry

The tetrahedron is a 3-dimensional figure. To represent the 3-dimensional geometry of the methane molecule, a heavy, triangular bond line is used to connect the C to the H in front of it. A dashed bond line connects the C to the H behind it. Regular bond lines connect the C to the H atoms in the same plane.

representation of structure of methane

NH_3 is ammonia. The H—N—H bond angles are 107°.

triangular pyramid

triangular pyramidal
geometry of ammonia

structure
of ammonia

H_2O is water. The H—O—H bond angle is 104.5°.

A water molecule contains three atoms.

There are only two possible geometries for molecules containing three atoms:

A—B—C   linear   A—B—C bond angle = 180°

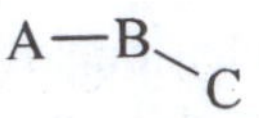

A—B⟍C   non-linear   A—B—C bond angle < 180°

H_2O is non-linear (sometimes called "bent")

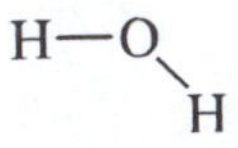

H—O⟍H   non-linear   A—B—C bond angle < 180°

bond angle = 104.5°

Exercise One

A. Silicon tetrachloride ($SiCl_4$) is a molecule with tetrahedral geometry. What is the angle of a Cl—Si—Cl bond?

Draw a representation of the tetrahedral geometry of silicon tetrachloride.

B. In carbon dioxide, the O—C—O bond angle is 180°. What is the geometry of CO_2?

Draw a representation of the geometry of carbon dioxide.

VALENCE SHELL ELECTRON PAIR REPULSION (VSEPR)

The pairs of valence electrons around an atom repel each other and tend to be as far apart as possible.

Example 2

Apply VSEPR to CH_4, NH_3, and H_2O.

1. Draw the Lewis structures:

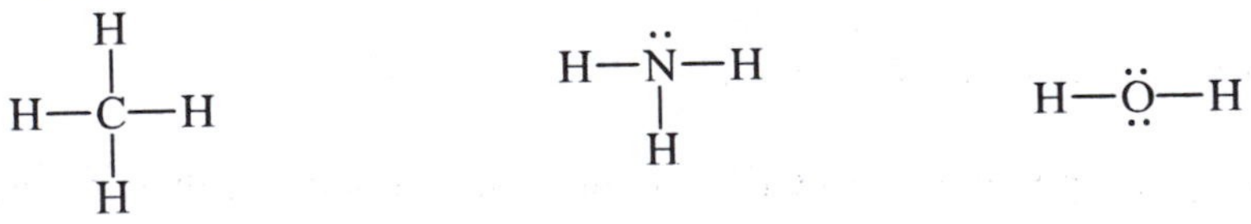

$$H-\underset{\underset{H}{|}}{\overset{\overset{H}{|}}{C}}-H \qquad H-\overset{\cdot\cdot}{\underset{\underset{H}{|}}{N}}-H \qquad H-\overset{\cdot\cdot}{\underset{\cdot\cdot}{O}}-H$$

2. The central atom in each has four electron pairs around it. These four pairs will be as far

apart as possible when they are located at the corners of a tetrahedron around the central atom.

This arrangement of the electron pairs (bonding pairs and lone pairs) is the same in all three molecules, CH_4, NH_3, and H_2O. This arrangement determines the arrangement of the atoms, and therefore, the geometry of the molecules.

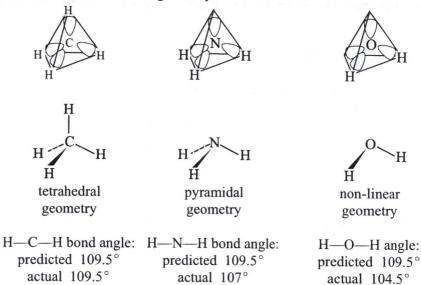

tetrahedral geometry	pyramidal geometry	non-linear geometry
H—C—H bond angle: predicted 109.5° actual 109.5°	H—N—H bond angle: predicted 109.5° actual 107°	H—O—H angle: predicted 109.5° actual 104.5°

Lone pairs take up more space at the central atom than do bonding pairs, because a lone pair is attracted only to the nucleus of the central atom. Because lone pairs take up more space than bonding pairs, the bond angles in NH_3 and H_2O are reduced from the tetrahedral angle of 109.5°. This reduction occurs only with the small central atoms from the second row of the periodic table.

Exercise Two

Use VSEPR to predict the geometry and bond angles of the following.

A. PI_3

B. SCl_2

C. $CHCl_3$

FURTHER CONSEQUENCES OF VSEPR

In applying VSEPR to predict the geometry of a molecule,
1. draw its Lewis structure.
2. count the number of electron groups associated with the central atom
    (An electron group is a lone pair or a bond. A multiple bond counts as only one bond.)
3. Consult Table 1.

Table 1. Geometries and Bond Angles Predicted by VSEPR

Total Number of Electron Groups	Arrangement of Bonds & Lone Pairs	Bond Angles	Number of bonds	Number of Lone Pairs	Molecular Geometry
2	linear	180°	2	0	linear
3	triangular	120°	3	0	triangular
			2	1	non-linear
4	tetrahedral	109.5°	4	0	tetrahedral
			3	1	pyramidal
			2	2	non-linear
5	triangular bipyramidal	120° and 90°	5	0	triangular bipyramidal
		120° and 90°	4	1	see-saw
		90°	3	2	T-shaped
		180°	2	3	linear
6	octahedral	90°	6	0	octahedral
			5	1	square pyramidal
			4	2	square

Example 3

Apply VSEPR to predict the geometry of formaldehyde, H_2CO.

1. Draw the Lewis structure.

$$H-\overset{|}{\underset{|}{C}}=\overset{..}{\underset{..}{O}}$$
$$\overset{|}{H}$$

Shapes of Simple Molecules

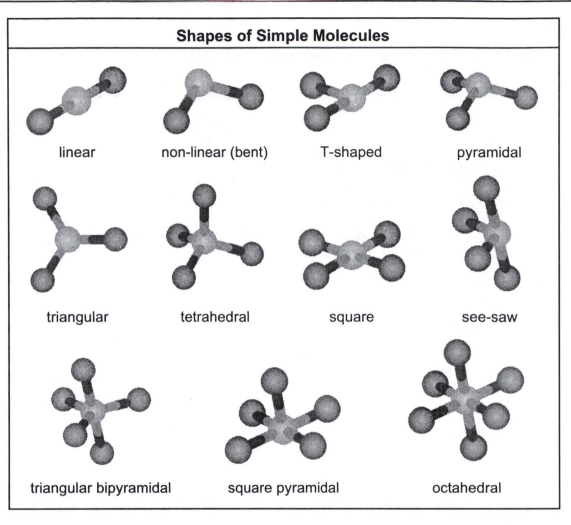

linear   non-linear (bent)   T-shaped   pyramidal

triangular   tetrahedral   square   see-saw

triangular bipyramidal   square pyramidal   octahedral

2. Count the number of electron groups

> There are *three* bonds to C (Remember, a double bond counts as only one bond).
> There are *no* lone pairs.
> The number of electron groups is $3 + 0 = 3$.

3. Consult Table I.

The electron groups (bonds) are arranged in a triangle around the carbon atom. The geometry of the molecule is triangular. The predicted H—C—H and H—C—O bond angles are 120°. (The H—C—O bond angles are actually 126°, and the H—C—H angle is 108°, because the double bond between C and O takes more space than the single bonds between C and H.)

Example 4

Apply VSEPR to predict the geometry of sulfur dioxide, SO_2.

1. Draw the Lewis structure.

$$:\ddot{O}—\ddot{S}=\ddot{O} \longleftrightarrow \ddot{O}=\ddot{S}—\ddot{O}:$$

2. Count the number of electron groups.
    When there are resonance forms, only one form needs to be examined. Each resonance form (if correctly drawn) will give the same result.

    There are *two* bonds to S (one single and one double) and one lone pair.
    There are 2 + 1 or 3 electron groups.

3. Consult Table I.

    The electron groups (bonds and lone pairs) are arranged in a triangle around the S atom. The lone pair occupies one corner of the triange. The bonds occupy the other two corners. Therefore, SO_2 has a non-linear geometry, and its predicted O—S—O bond angle is 120°. (The actual bond angle in SO_2 is 119°, because the lone pair occupies more space than the bonds.

Exercise Three

Use VSEPR to predict the geometry and bond angles of the following molecules.

A. ONF

B. HCN

VSEPR FOR POLYATOMIC IONS

Example 5

Apply VSEPR to the nitrate ion, NO_3^-.

1. Draw the Lewis structure.

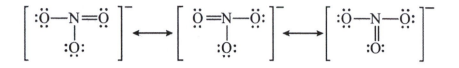

2. There are three electron groups around nitrogen.

3. Consult Table I.

triangular geometry
predict O–N–O bond angle is 120°
(actual bond angle is 120°)

Exercise Four

Apply VSEPR to the following ions to predict their geometry and bond angles.

A. SO_4^{2-} B. SO_3^{2-}

C. NO_2^- D. NH_4^+

VSEPR FOR NON-OCTET RULE MOLECULES

Example 6

Apply VSEPR to determine the geometry and bond angles of PCl_5.

1. Draw the Lewis structure.

2. There are *five* bonds to P and no lone pairs. So, there are electron groups.
3. Consult Table I.

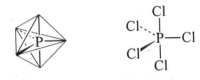

geometry is triangular bipyramidal
predicted bond angles 120° and 90°
(actual bond angles 120° and 90°)

Example 7

Apply VSEPR to determine the geometry and bond angles of the SF_4^{2-} ion.

1. Draw its Lewis structure.

2. There are *four* bonds to S and there are *two* lone pairs on S, for a total of six electron groups.
3. Consult Table I.

The arrangement of the electron groups (bonds and lone pairs) around the S atom is octahedral. The lone pairs, because they are larger than bonding pairs, move as far apart as possible on opposite sides of the S atom. Therefore, the bonding pairs and the fluorine atoms are arranged in a square around the S atom. The bond angles are 90°.

Exercise Five

Apply VSEPR to the following molecules and ions to predict their geometries and bond angles.

A. SF_6 B. BrF_3

C. XeF_4 D. I_3^-

PROBLEMS

1. Predict the shape of each of the following molecules.

   (a) HClO   (b) CS_2 (c) O_3 (d) NO_2

2. Predict the shape of each of the following ions.

   (a) CO_3^{2-} (b) ClO_3^- (c) PF_4^+ (d) H_3O^+

3. Estimate the bond angles in each of the following.

   (a) O_3 (b) $SnCl_4$ (c) H_2CO (d) C_2H_2

VALENCE BONDS AND HYBRID ORBITALS

This lesson deals with:

1. Determining which atomic orbitals overlap in forming a given bond.
2. Determining the proper hybridization of an atom from the bond angles around that atom.
3. Identifying sigma and pi bonds in a molecule.
4. Postulating the hybridization of an atom in a molecule from the Lewis structure of the molecule.

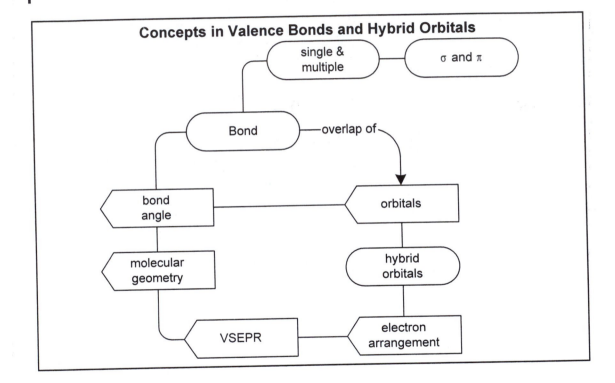

COVALENT BONDS

According to valence bond theory, a **covalent bond** is a shared pair of electrons. A more complete description of a bond indicates which electrons are involved.

Electron configurations show how electrons are arranged in atoms.
In atoms, electrons are arranged in orbitals.

H: $1s^1$
F: $1s^2\, 2s^2\, 2p_x{}^2\, 2p_y{}^2\, 2p_z{}^1$

How are electrons arranged in molecules?
Valence bond theory says that valence electrons are in either
      1. bonds formed from overlapping atomic orbitals
  or
      2. atomic orbitals.

All electrons that are not in the valence shell orbitals of an atom remain in atomic orbitals when the atom is in a molecule. Only electrons in the valence shell form bonds.

Valence shell orbitals are those in the outermost shell of an atom.

H: $\boxed{1s^1}$

F: $1s^2$ $\boxed{2s^2\,2p_x^{\,2}\,2p_y^{\,2}\,2p_z^{\,1}}$

Valence shell orbitals are those that overlap when a covalent bond is formed between two atoms.

H—H

This bond is formed by overlap of the $1s$ orbital of one H atom with the $1s$ orbital of the other H atom. This can be represented as
$$1s\,(H) + 1s\,(H)$$

Example 1

Which orbitals overlap in forming the bond in the HCl molecule?

H—Cl

The bond is formed from two overlapping atomic orbitals.

1. One orbital from each atom.

    H: $1s^1$
    Cl: $1s^2\,2s^2\,2p^6\,3s^2\,3p_x^{\,2}\,3p_y^{\,2}\,3p_z^{\,1}$

2. The two atomic orbitals used in forming the bond must together contain 2 electrons.

    The bond forms from $1s$ from H and $3p_z$ from Cl:

    $$1s\,(H) \;+\; 3p_z\,(Cl)$$

Exercise One

Determine which atomic orbitals are used in forming the bonds in

A. HBr

B. ICl

HYBRID ATOMIC ORBITALS

The standard s, p, and d atomic orbitals are insufficient to explain the bonding in many molecules.

Methane, CH_4, is such a case.

Which atomic orbitals are used in forming the bonds between the hydrogen atoms and the carbon atom in methane, CH_4?

H: $1s^1$

C: $1s^2\ \underbrace{2s^2 2p_x^1 2p_y^1 2p_z^0}$
     valence shell

Expect 2 bonds: $1s$ from H + $2p_x$ from C
                    $1s$ from H + $2p_y$ from C

This sort of bonding would lead to a molecule with formula CH_2. However, there is no such molecule. The compound containing one carbon atom and hydrogen is CH_4

The electron configuration for C given above is appropriate only for an isolated carbon atom. In methane, the carbon atom is not isolated; it is very close to hydrogen atoms.

When an atom is near other atoms, its orbitals may rearrange. This rearrangement is call *hybridization*.

In methane, the four valence orbitals ($2s$, $2p_x$, $2p_y$, and $2p_z$) rearrange (hybridize) into four different orbitals.

$$2s + 2p_x + 2p_y + 2p_z \longrightarrow \boxed{\text{HYBRIDIZE}} \longrightarrow sp^3 + sp^3 + sp^3 + sp^3$$

These new orbitals are called sp^3 hybrid orbitals.

The shape of an sp^3 orbital is like a p orbital, but having one large lobe and one small one.

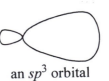

an sp^3 orbital

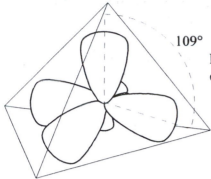

109°

In the set of four sp^3 hybrid orbitals,
each one points toward the corner of a tetrahedron.

The angle between the sp^3 hybrid orbitals is 109 degrees.

When surrounded by 4 other atoms, carbon has the electron configuration:

$$1s^2 \underbrace{(sp^3)^1 (sp^3)^1 (sp^3)^1 (sp^3)^1}_{\text{valence shell}}$$

Exercise Two

Determine which atomic orbitals are used in forming each of the four bonds in CH_4.
(Remember carbon's new electronic configuration!)

Example 2

The bond angles in ammonia are 107 degrees. What atomic orbitals are involved in the
bonding in NH_3?

$$H:\ 1s^1$$
$$N:\ 1s^2\, 2s^2\, 2p_x^{\,1}\, 2p_y^{\,1}\, 2p_z^{\,1}$$

If the p orbitals of nitrogen were involved, the bond angles would be 90 degrees, because this
is the angle between the p orbitals.

Nitrogen atoms, like carbon atoms, also form hybrid orbitals when near other atoms. When N is near 3 other atoms, its valence shell orbitals hybridize.

$$2s + 2p_x + 2p_y + 2p_z \longrightarrow \boxed{\text{HYBRIDIZE}} \longrightarrow sp^3 + sp^3 + sp^3 + sp^3$$

The new electron configuration is:

$$\text{N: } 1s^2\,(sp^3)^2\,(sp^3)^1\,(sp^3)^1\,(sp^3)^1$$

This hybrid orbital contains nitrogen's lone pair of electrons.

These three hybrid orbitals form bonds.

$$\text{H}-\overset{\cdot\cdot}{\text{N}}-\text{H}$$
$$|$$
$$\text{H}$$

The 3 bonds in NH_3 are all $1s$ (H) + sp^3 (N).
The lone pair of electrons is in the remaining sp^3 orbital.
The bond angles are close to the angles between the sp^3 hybrid orbitals.

Does the fluorine atom in HF hybridize?

Because there are only two atoms in HF, there is no bond angle. Without a bond angle, there is no information to indicate whether F in HF uses hybrid orbitals.

However, the bond angle in the $\text{H}-\text{F}-\text{H}^+$ ion indicates that the orbitals of F are hybridized.

The elements of the second period generally use hybrid orbitals in forming covalent bonds. Heavier elements, such as S in H_2S, frequently do not use hybrid orbitals.

Exercise Three

A. The bond angle in H_2O is 105°. What atomic orbitals are used in forming the bonds?

B. The bond angle in H_2S is close to 90°. What atomic orbitals are used in forming the bonds?

OTHER TYPES OF HYBRID ORBITALS

Example 3

What orbitals are involved in the bonding in ethylene, in which the bond angles are 120 degrees?

ethylene, C_2H_4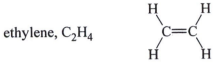

When a carbon atom is bonded to 3 other atoms, 3 of its valence shell orbitals are hybridized. One hybrid orbital is used for each bonded atom.

$$2s + 2p_x + 2p_y \longrightarrow \boxed{\text{HYBRIDIZE}} \longrightarrow sp^2 + sp^2 + sp^2$$

The $2p_z$ orbital is not hybridized.

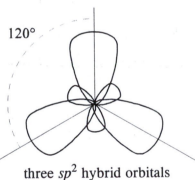

120°

The three sp^2 hybrid orbitals are arranged in a triangle around the atom. The angle between them is 120°.

three sp^2 hybrid orbitals

New valence shell electron configuration:

$$(sp^2)^1 (sp^2)^1 (sp^2)^1 2p_z{}^1$$

A single bond, such as the H—C bond, involves one pair of overlapped orbitals.

H—C bond: $1s$ (H) $+$ sp^2 (C)

A double bond, such as the C=C bond, involves two pairs of overlapped orbitals.

C=C bond: sp^2 (C) $+$ sp^2 (C)

$2p_z$ (C) $+$ $2p_z$ (C)

Hybrid orbitals overlap end-to-end, with the overlap directly between the bonded atoms. This kind of overlap, directly between the atoms, is called sigma (σ) overlap.

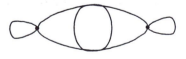

sigma overlap
of two hybrid orbitals

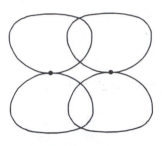

pi overlap
of two *p* orbitals

The overlap of *p* orbitals can occur side-to-side, with the overlap above and below the line between the atoms. This kind of overlap is called pi (π) overlap.

A double bond contains two components: a sigma bond and a pi bond. The sigma bond forms from hybrid atomic orbitals, the pi bond forms from side-to-side overlap of *p* orbitals.

Whenever any two atoms bond together:
1. the first bond is a sigma (σ) bond;
2. all additional bonds are pi (π) bonds.

The triple bond in N_2 has three components: one σ bond and two π bonds.

Exercise Four

Describe each of the indicated bonds by indicating whether it is sigma or pi, and by indicating the atomic orbitals from which it is formed.

$$\begin{array}{ccccc} & \overset{\displaystyle Br}{\displaystyle |} & & & \\ H- & C- & C= & C- & H \\ & | & | & | & \\ & H & H & H & \end{array}$$

A. The Br—C bond.

B. The H—C bond on the left of the molecule.

C. The H—C bond on the right of the molecule.

D. The C—C single bond.

E. The C=C double bond. (This has two components.)

Example 4

What orbitals form the bonds in carbon dioxide, which is a linear molecule (180° bond angle)?

$$\ddot{O} = C = \ddot{O}$$

The carbon has two double bonds. Each double bond involves one hybrid orbital and one p orbital. Therefore, the carbon atom must have two hybrid orbitals and two unhybridized p orbitals.

When bonded to two other atoms, carbon atoms form two hybrid orbitals.

$$2s + 2p_x \quad \longrightarrow \boxed{\text{HYBRIDIZE}} \longrightarrow \quad sp + sp$$

The $2p_y$ and $2p_z$ orbitals do not hybridize.

The two sp hybrid orbitals are oriented in opposite directions to each other, 180° apart.

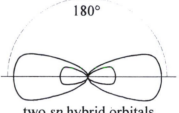

two sp hybrid orbitals

New valence shell electronic configuration for C: $(sp)^1(sp)^1 2p_y{}^1 2p_z{}^1$

Each oxygen atom has a double bond and two lone pairs. The double bond involves one hybrid orbital and one p orbital. Each lone pair occupies one hybrid orbital. Therefore, oxygen has three hybrid orbitals and one unhybridized p orbital.

$$2s + 2p_x + 2p_y \quad \longrightarrow \boxed{\text{HYBRIDIZE}} \longrightarrow \quad sp^2 + sp^2 + sp^2$$

The $2p_z$ orbital does not hybridize.

Valence shell electron configuration for O: $(sp^2)^2(sp^2)^2(sp^2)^1 2p_z{}^1$

Then, the bonds in carbon dioxide are formed as follows:

$$O = C = O$$

sigma (σ): sp^2 (O) + sp (C)
pi (π): $2p_z$ (O) + $2p_z$ (C)

sigma (σ): sp (C) + sp^2 (O)
pi (π): $2p_y$ (C) + $2p_y$ (O)

Exercise Five

Answer the following questions about acetylene.

$$H—C≡C—H$$

A. To how many atoms is each carbon atom bonded?

B. What is the hybridization of the orbitals in each carbon?

C. Indicate the atomic orbitals used in forming each of the following bonds and the type(s) of each bond.

1. The H—C bond on the left.

2. The C ≡ C triple bond (3 components).

3. The C—H bond on the right.

HYBRID ORBITALS OF ATOMS
WITH MORE THAN EIGHT VALENCE ELECTRONS

Example 5

Describe the bonding in PCl_5 and SF_6.

PCl_5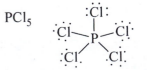

In atoms with more than eight valence electrons, the *d* orbitals are involved in bonding.

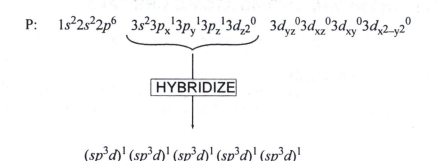

P: $1s^2 2s^2 2p^6$ $\underbrace{3s^2 3p_x^{\ 1} 3p_y^{\ 1} 3p_z^{\ 1} 3d_{z^2}^{\ 0}}$ $3d_{yz}^{\ 0} 3d_{xz}^{\ 0} 3d_{xy}^{\ 0} 3d_{x^2-y^2}^{\ 0}$

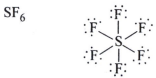

HYBRIDIZE

$(sp^3d)^1 (sp^3d)^1 (sp^3d)^1 (sp^3d)^1 (sp^3d)^1$

Each P—Cl bond is formed by overlap of an *sp3d* orbital of P with a $3p_z$ orbital of Cl.

SF_6

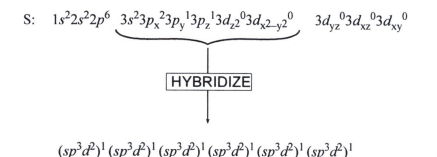

S: $1s^2 2s^2 2p^6$ $\underbrace{3s^2 3p_x^{\ 2} 3p_y^{\ 1} 3p_z^{\ 1} 3d_{z^2}^{\ 0} 3d_{x^2-y^2}^{\ 0}}$ $3d_{yz}^{\ 0} 3d_{xz}^{\ 0} 3d_{xy}^{\ 0}$

HYBRIDIZE

$(sp^3d^2)^1 (sp^3d^2)^1 (sp^3d^2)^1 (sp^3d^2)^1 (sp^3d^2)^1 (sp^3d^2)^1$

Each S—F bond is formed by overlap of an sp^3d^2 orbital of S with an sp^3 orbital of F.

Exercise Six

Indicate which orbitals are used in forming the bonds in each molecule below.

A. SeF_6 B. SbI_5

MOLECULAR GEOMETRY AND HYBRID ATOMIC ORBITALS

To determine the orbital hybridization of an atom in a molecule or ion:

1. draw the Lewis structure of the molecule or ion.
2. count the total number of sigma bonds and lone pairs around the atom in question.
3. consult the table to predict the geometry of the atom and the corresponding orbital hybridization.

Total Number of σ Bonds & Lone Pairs	Hybridization	Geometry of Orbitals	Number of Lone Pairs	Geometry of Bonded Atoms	Predicted Bond Angle
2	sp	Linear	0	Linear	180°
3	sp^2	Triangular	0	Triangular	120°
			1	Non-linear	120°
4	sp^3	Tetrahedral	0	Tetrahedral	109.5°
			1	Pyramidal	109.5°
			2	Non-linear	109.5°
5	sp^3d	Triangular bipyramidal	0	Triangular bipyramidal	120° and 90°
			1	See-saw shape	120° and 90°
			2	T-shaped	90°
			3	Linear	180°
6	sp^3d^2	Octahedral	0	Octahedral	90°
			1	Square pyramidal	90°
			2	Square	90°

Example 6

Determine the hybridization of nitrogen in NOBr (nitrosyl bromide).

1. Draw the Lewis structure

$$:\ddot{Br}\!-\!\ddot{N}\!=\!\ddot{O}$$

2. Count the number of sigma bonds and lone pairs around nitrogen.

N in NOBr = 2 sigma bonds + 1 lone pair
= 3

3. Consult the table to find hybridization of nitrogen.

sp^2 hybrid orbitals — predict Br—N—O bond angle of 120°

(measured Br—N—O bond angle is 117°)

$$:\ddot{Br} - N = \ddot{O}$$

sigma (σ): $4p_z$ (Br) + sp^2 (N)

sigma (σ): sp^2 (N) + sp^2 (O)
pi (π): $2p_z$ (N) + $2p_z$ (O)

lone pair in sp^2 of N

Caution: for elements beyond the second period, hybrid orbitals are not always used.

Use information about bond angles to determine whether hybrid orbitals are used. For example, consider PI_3 and PH_3:

$$:\ddot{I} - \ddot{P} - \ddot{I}:$$
$$:\ddot{I}:$$

$$H - \ddot{P} - H$$
$$H$$

3 σ bonds + 1 lone pair
predict 109.5° bond angles
actual angles: 102°
Therefore, phosphorus has
sp^3 hybrid orbitals.

3 σ bonds + 1 lone pair
predict 109.5° bond angles
actual angles: 94°
Therefore, phosphorus does not
use hybrid orbitals.

Hybrid orbitals are more likely to be used if very large atoms are crowded around the central atom, such as the iodine atoms around phosphorus in PI_3.

Exercise Seven

Give the hybridization of the indicated atom in each molecule or ion below.

A. H_2O_2 (oxygen hybridization)

B. NO_2^- (nitrogen hybridization)

C. XeF_4 (xenon hybridization)

D. ICl_2^- (iodine hybridization)

E. CO_3^{2-} (carbon hybridization)

F. NH_4^+ (nitrogen hybridization)

PROBLEMS

1. Answer the following questions for
   - A. the hydrogen cyanide molecule, HCN
   - B. the formaldehyde molecule, H_2CO
   - C. the acetylene molecule, C_2H_2

   (a) What is the hybridization of the carbon atom?
   (b) What is the shape of the molecule?
   (c) Which atomic orbitals are used to form the H—C bond?
   (d) Which atomic orbitals are used in the multiple bond? Indicate whether each bond is a sigma or a pi bond.

2. What atomic orbitals overlap in forming the bond in HBr? Classify the bond as σ or π.

3. Indicate the hybrid orbitals used by the underlined atom in each of the following molecules. Remember that each lone pair of electrons occupies a hybrid orbital.

   (a) $\underline{S}F_4$　　(b) $\underline{N}H_3$　　(c) $\underline{B}F_3$　　(d) $\underline{Sn}Cl_4$

COLLIGATIVE PROPERTIES OF SOLUTIONS

This lesson deals with:

1. Calculating the vapor pressure of a solvent.
2. Calculating the freezing point depression and boiling point elevation of a solvent.
3. Calculating the molar mass of a solute from freezing point depression data.
4. Calculating the value of the mole number (i) of a solute from freezing point depression data.
5. Calculating the osmotic pressure of a solvent.
6. Calculating the molar mass of a solute from osmotic pressure data.

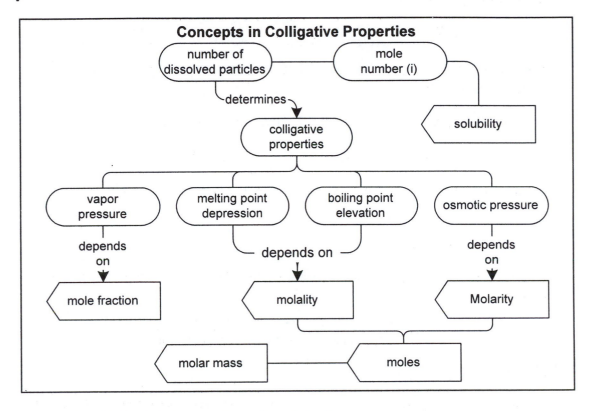

COLLIGATIVE PROPERTIES OF SOLUTIONS

Colligative properties of a solution depend primarily upon the *number* of solute particles in solution, and not on the identity of the particles.

These are the colligative properties of a solution:
1. vapor pressure lowering
2. boiling point elevation
3. freezing point depression
4. osmotic pressure

Vapor Pressure Lowering

The vapor pressure of the solvent in a solution is less than the vapor pressure of the pure solvent.

Raoult's law relates the vapor pressure of the solvent to the concentration of the solution.

$$\text{Raoult's Law:} \qquad P_{solvent} = X_{solvent}\, P^{\circ}_{solvent}$$

$P_{solvent}$:   vapor pressure of solvent in a solution
$X_{solvent}$:   mole fraction of *solvent* in a solution
$P^{\circ}_{solvent}$: vapor pressure of *pure* solvent

The mole fraction takes into account the number of solute particles.

$$X_{solvent} = \frac{\text{moles of solvent}}{\text{moles of solvent + moles of solute particles}}$$

If the solute is an electrolyte, then the number of moles of solute particles is greater than the number of moles of solute.

For example, $Na_2SO_4(aq)$ is $2\, Na^+(aq)$ and $SO_4^{2-}(aq)$

Moles of particles $= 3 \times (\text{moles of } Na_2SO_4)$.

The mole number (*i*) of the solute indicates the number of moles of particles per mole of solute.

For $Na_2SO_4(aq)$, $i = 3$.

In Raoult's law for electrolytes, it is necessary to use the mole number in calculating the mole fraction of solvent.

$$X_{solvent} = \frac{(\text{moles solvent})}{(\text{moles solvent}) + i\,(\text{moles solute})}$$

Example 1

Find the vapor pressure of water at 25°C for a solution made by dissolving 100.0 grams of Na_2SO_4 in 500.0 grams of water. (The vapor pressure of pure water at 25°C is 23.8 mm Hg.)

$$P_{solvent} = X_{solvent}\, P^{\circ}_{solvent}$$

$$P^{\circ}_{solvent} = 23.8 \text{ mm Hg}$$

$$X_{H_2O} = \frac{(\text{moles } H_2O)}{(\text{moles } H_2O) + i\,(\text{moles } Na_2SO_4)}$$

$$\text{moles } Na_2SO_4 = 100.0 \text{ g } Na_2SO_4 \left(\frac{1 \text{ mol } Na_2SO_4}{142.04 \text{ g } Na_2SO_4} \right) = 0.7040 \text{ mol } Na_2SO_4$$

$$\text{moles } H_2O = 500.0 \text{ g } H_2O \left(\frac{1 \text{ mol } H_2O}{18.02 \text{ g } H_2O} \right) = 27.75 \text{ mol } H_2O$$

For Na_2SO_4, $i = 3$.

$$X_{H_2O} = \frac{(27.75)}{(27.75) + 3 \, (0.7040)} = 0.9293$$

$$P_{solvent} = (0.9293)(23.8 \text{ mm Hg}) = 22.1 \text{ mm Hg}$$

Exercise One

A. What is the vapor pressure of water from an aqueous solution containing 50.0 grams of NaCl and 1.00 kg of water at 22°C? (The vapor pressure of pure water at 22°C is 19.8 mm Hg.)

B. A solution is made by mixing 150.0 grams of benzene, C_6H_6, and 150.0 grams of pentane, C_5H_{12}.

1. What is the vapor pressure of benzene, at 21°C, from this solution? (The vapor pressure of pure benzene is 77.4 mm Hg at 21°C.)

2. What is the vapor pressure of pentane, at 21°C, from this solution?   (The vapor pressure of pure pentane at 21°C is 442 mm Hg.)

3. What is the total vapor pressure of the two components of this mixture?

FREEZING POINT DEPRESSION AND BOILING POINT ELEVATION OF SOLUTIONS

The freezing point of a solution is lower than the freezing point of the pure solvent. The boiling point of a solution is higher than the boiling point of the pure solvent.

Freezing Point Depression

ΔT_f = (freezing point of pure solvent) – (freezing point of solvent in solution)

$$\Delta T_f = k_f\, m\, i$$

ΔT_f	=	freezing point depression
k_f	=	freezing point depression constant (depends on solvent)
m	=	molality of the solution
i	=	mole number of solute

Boiling Point Elevation

ΔT_b = (boiling point of solvent in solution) – (boiling point of pure solvent)

$$\Delta T_b = k_b\, m\, i$$

ΔT_b	=	boiling point elevation
k_b	=	boiling point elevation constant (depends on solvent)
m	=	molality of solution
i	=	mole number of solute

Table 1. Freezing Point, Boiling Point, k_f and k_b for Various Solvents

Solvent	fp (°C)	bp (°C)	k_f (K/m)	k_b (K/m)
Benzene	5.51	80.1	5.12	2.53
Carbon tetrachloride	−22.8	76.8	29.8	5.02
Ethanol	−114.6	78.4	1.99	1.22
Water	0.00	100.0	1.86	0.512

Example 2

Find the boiling point and freezing point of the solvent ethanol in a solution made by dissolving 18.0 grams of the non-electrolyte fructose ($C_6H_{12}O_6$) in 200.0 grams of ethanol (C_2H_5OH).

1. Boiling point elevation:

$$\Delta T_b = k_b \, m \, i$$

$$k_b = 1.22 \text{ K/m}$$

$$m = \frac{\text{mol fructose}}{\text{kg ethanol}}$$

$$\text{mol fructose} = 18.0 \text{ g} \left(\frac{1 \text{ mol}}{180.2 \text{ g}} \right) = 0.100 \text{ mol}$$

$$\text{kg ethanol} = 200.0 \text{ g} \left(\frac{1 \text{ kg}}{1000 \text{ g}} \right) = 0.2000 \text{ kg}$$

$$m = \frac{0.100 \text{ mol fructose}}{0.2000 \text{ kg ethanol}} = 0.500 \text{ m}$$

$i = 1$ (fructose is a non-electrolyte)

$$\Delta T_b = (1.22 \text{ K/m})(0.500 \text{ m})(1) = 0.610 \text{ K}$$

Celsius degrees are the same size as Kelvin temperature units. Therefore, ΔT in Celsius is the same magnitude as ΔT in Kelvin.

$$\Delta T_b = 0.610°C$$

$$\begin{aligned} \text{boiling point} &= \text{(boiling point of pure ethanol)} + \Delta T_b \\ &= 78.4°C + 0.610°C = 79.0°C \end{aligned}$$

2. Freezing point depression:

$$\Delta T_f = k_f \, m \, i$$

$$\Delta T_f = (1.99 \text{ K/m})(0.500 \text{ m})(1) = 0.995 \text{ K} = 0.995°C$$

$$\begin{aligned} \text{freezing point} &= \text{(freezing point of pure solvent)} - \Delta T_f \\ &= -114.6°C - 0.995°C = -115.6°C \end{aligned}$$

Exercise Two

Calculate the freezing point and boiling point of the solvent carbon tetrachloride in a solution containing 40.0 grams of the non-electrolyte resorcinol ($C_6H_6O_2$) dissolved in 150.0 grams of carbon tetrachloride.

DETERMINATION OF MOLAR MASS BY FREEZING POINT DEPRESSION

Example 3

Find the molar mass of a non-electrolyte that causes the freezing point of the solvent dibromoethane ($C_2H_4Br_2$) to decrease from 10.06°C to 7.81°C when 0.466 gram of solute is dissolved in 20.0 grams of solvent. (For dibromoethane, $k_f = 11.80$ K/m.)

The molar mass of a substance is equal to the mass of a sample divided by the number of moles in that sample.

$$\text{molar mass} = \frac{\text{grams of sample}}{\text{moles in sample}} = \frac{0.466 \text{ g}}{x \text{ mol}}$$

The number of moles of the solute can be determined from the molality of the solution, and the molality can be determined from the freezing-point depression.

1. Calculate molality.

$$m = \frac{\Delta T_f}{k_f \, i}$$

$$\Delta T_f = 10.06°C - 7.81°C = 2.25°C$$
$$k_f = 11.80 \text{ K/m}$$
$$i = 1$$

$$m = \frac{2.25 \text{ K}}{(11.80 \text{ K/m})(1)} = 0.191 \text{ m}$$

2. Calculate moles of solute.

$$\text{moles solute} = \left(0.191 \; \frac{\text{mol solute}}{\text{kg solvent}} \right) (0.0200 \; \text{kg solvent})$$

$$= 3.82 \times 10^{-3} \; \text{mol}$$

3. Calculate molar mass.

$$\text{molar mass} = \frac{0.466 \; \text{g}}{3.82 \times 10^{-3} \; \text{mol}} = 122 \; \text{g/mol}$$

Exercise Three

A solution is prepared by dissolving 2.14 grams of an unknown compound (non-electrolyte) in 20.0 grams of benzene. The freezing point of benzene in this solution is –0.37°C. What is the molar mass of the unknown compound? (See Table 1 for additional data.)

DETERMINING THE VALUE OF THE MOLE NUMBER OF A SOLUTE

Example 4

When 5.63 grams of cadmium bromide, $CdBr_2$, are dissolved in 100.0 grams of water, the observed freezing point depression is 0.54°C. Is $CdBr_2$ a strong, weak, or non-electrolyte?

$$i = \frac{\Delta T_f}{k_f \; m}$$

$$\Delta T_f = 0.54°C$$

$$k_f = 1.86 \; \text{K/m}$$

$$m = \frac{\text{mol } CdBr_2}{\text{kg } H_2O}$$

$$\text{mol } CdBr_2 = (5.63 \text{ g})\left(\frac{1 \text{ mol}}{272.2 \text{ g}}\right) = 0.0207 \text{ mol}$$

$$m = \frac{0.0207 \text{ mol } CdBr_2}{0.100 \text{ kg } H_2O} = 0.207 \text{ m}$$

$$i = \frac{0.54 \text{ K}}{(1.86 \text{ K/m})(0.207 \text{ m})} = 1.40$$

If $CdBr_2$ dissociated completely, its mole number would be 3.

$$CdBr_2(aq) \longrightarrow Cd^{2+}(aq) + 2 \, Br^-(aq)$$

If $CdBr_2$ were a non-electrolyte, its mole number would be 1.

Because its observed mole number is less than 3 but greater than 1, $CdBr_2$ is a weak electrolyte in aqueous solution.

Exercise Four

A 0.10 m aqueous solution of $Hg_2(NO_3)_2$ has a freezing point of –0.56°C. Which equation below better describes the dissociation of $Hg_2(NO_3)_2$ in aqueous solution? (See Table 1 for additional data.)

(1) $Hg_2(NO_3)_2(aq) \longrightarrow Hg_2^{2+}(aq) + 2 \, NO_3^-(aq)$

(2) $Hg_2(NO_3)_2(aq) \longrightarrow 2 \, Hg^+(aq) + 2 \, NO_3^-(aq)$

OSMOTIC PRESSURE

Osmosis is the movement of **solvent** from a region of higher **solvent concentration** through a semipermeable membrane to a region of lower solvent concentration. The membrane is called semipermeable because solvent molecules can move through it, but solute molecules (or ions) cannot. In the figure, solvent moves by osmosis from the pure solvent into the bag of solution. The external pressure that is just sufficient to prevent osmosis is called the **osmotic pressure**.

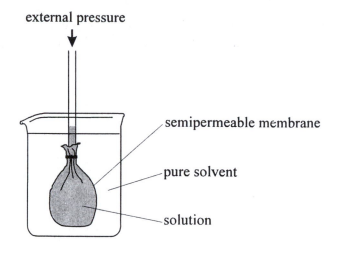

Osmotic pressure is represented by Π (capital Greek pi).

$$\Pi = iMRT$$

i = mole number of solute
M = molarity of solution
R = ideal gas constant
T = Kelvin temperature

Note that molarity (M) is used rather than molality (m).

Example 5

Calculate the osmotic pressure of 0.10 M aqueous NaCl at 25°C.

$$\Pi = iMRT$$

i = 2
M = 0.10 M
R = 0.0821 L-atm/mole-K
T = 25°C + 273 = 298 K

$$\Pi = 2\left(0.10 \ \frac{mol}{L}\right)\left(0.0821 \ \frac{L\text{-}atm}{mol\text{-}K}\right)(298 \ K)$$

$$= 4.9 \ atm$$

Exercise Five

A solution is prepared by dissolving 100.0 grams of $(NH_4)_2SO_4$ in a sufficient volume of water to make 250.0 mL of solution. What is the osmotic pressure of this solution at 0°C?

DETERMINATION OF MOLAR MASS BY OSMOTIC PRESSURE

Example 6

Find the molar mass of a protein, a non-electrolyte, if a solution prepared by dissolving 0.240 gram of the protein in 10.00 mL of solution has an osmotic pressure of 6.83 mm Hg at 37°C.

The molar mass of a substance is equal to the mass of a sample divided by the number of moles in that sample.

$$\text{molar mass} = \frac{\text{grams of sample}}{\text{moles in sample}} = \frac{0.240 \text{ g}}{x \text{ mol}}$$

The number of moles of protein can be determined from the molarity and volume of the solution, and the molarity can be determined from the osmotic pressure.

$$M = \frac{\Pi}{iRT}$$

$$R = 0.0821 \text{ L-atm/mole K}$$

$$\Pi = (6.83 \text{ mm Hg})\left(\frac{1 \text{ atm}}{760 \text{ mm Hg}}\right) = 8.99 \times 10^{-3} \text{ atm}$$

$$i = 1$$

$$T = 37 + 273 \text{ K} = 310 \text{ K}$$

$$M = \frac{8.99 \times 10^{-3} \text{ atm}}{(1)(0.0821 \text{ L-atm/mol-K})(310 \text{ K})} = 3.53 \times 10^{-4} \text{ M}$$

$$\text{moles protein} = (3.53 \times 10^{-4} \text{ mole/L})(0.01000 \text{ L}) = 3.53 \times 10^{-6} \text{ mol}$$

$$\text{molar mass of protein} = \frac{0.240 \text{ g}}{3.53 \times 10^{-6} \text{ mol}} = 6.80 \times 10^{4} \text{ g/mol}$$

Exercise Six

A piece of synthetic polymer (a non-electrolyte) weighing 2.463 grams is dissolved in toluene to make 50.00 mL of solution. The osmotic pressure of this solution is 7.43 mm Hg at 22°C. What is the average molar mass of the polymer?

PROBLEMS

1. The freezing point of carbon tetrachloride solvent in a certain solution is $-63.8°C$. What is the boiling point of this solution?

2. Ethylene glycol, $C_2H_6O_2$, can be used as an antifreeze in an automobile radiator. How many grams of ethylene glycol would be needed to prevent 4.00 kg of water from freezing above $-5.0°C$?

3. The vapor pressure of pure benzene, C_6H_6, at 20°C is 75 mm Hg, while the vapor pressure of toluene, C_7H_8, at this temperature is 22 mm Hg. What is the total vapor pressure (i.e., the vapor pressure of benzene plus the vapor pressure of toluene) of a solution that contains 50.0 g of benzene and 100.0 g of toluene?

4. A sample of cholesterol (a non-electrolyte) weighing 86.4 mg is dissolved in 10.00 mL of solution. The osmotic pressure of this solution is 0.534 atm at 19°C. What is the molar mass of cholesterol?

5. A solution is prepared by dissolving 6.00 grams of urea, $CO(NH_2)_2$, in 100.0 grams of chloroform, $CHCl_3$. The freezing point of chloroform in this solution is $-68.2°C$. The freezing point of pure chloroform is $-63.5°C$. What is the value of k_f for $CHCl_3$?

6. Glucose, $C_6H_{12}O_6$, is a nonvolatile nonelectrolyte when dissolved in water. An aqueous solution of glucose is prepared by dissolving 50.0 grams of glucose in 100.0 grams of water. For this solution calculate
   (a) the vapor pressure at 25°C,
   (b) the boiling point,
   (c) the freezing point.

7. How many grams of glycerol, $C_3H_8O_3$, must be dissolved in 8.50 kg of water to make a solution that freezes at $-32°C$?

8. A solution is prepared by dissolving 26.0 g of an unknown nonelectrolyte in 380 g of water. The solution freezes at $-1.18°C$. What is the molar mass of the nonelectrolyte?

9. A solution is prepared by dissolving 3.0 g of KBr in 250 mL of water. What is the freezing point of this solution?

10. What is the osmotic pressure at 25°C of a solution prepared by dissolving 4.00 g of sugar, $C_{12}H_{22}O_{11}$, in 500.0 mL of solution?

CHEMICAL EQUILIBRIUM IN REACTIONS OF GASES

This lesson deals with:

1. Writing the equilibrium constant expression for a given reaction.
2. Calculating the value of the equilibrium constant from the number of moles of reactants and products at equilibrium.
3. Calculating the value of K_P for an equilibrium system given the value of its K_C.
4. Calculating the amounts of substances present at equilibrium, given the initial conditions and the value of the equilibrium constant.
5. Calculating the effect on equilibrium concentrations of adding additional material involved in the reaction.
6. Using Le Chatelier's principle to predict the direction an equilibrium will shift in response to a change in its conditions.

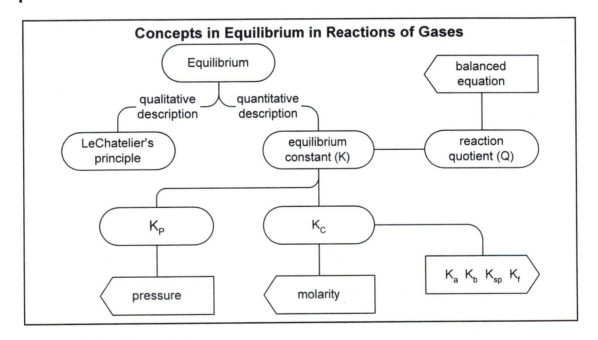

THE REACTION QUOTIENT

For every chemical reaction, there is an expression that indicates the completeness of the reaction. This is called the **reaction quotient**.

For a reaction represented by the generalized chemical equation

$$p\ A(g)\ +\ q\ B(g)\ \longrightarrow\ x\ C(g)\ +\ y\ D(g)$$

the reaction quotient has the form

$$Q = \frac{[C]^x\,[D]^y}{[A]^p\,[B]^q}$$

In this expression, the square brackets represent concentrations in moles per liter (molarity).

The value of the reaction quotient expresses quantitatively the extent of the reaction. For a reaction that has gone almost to completion, there will be very little reactant and much product. In this case, Q will be very large. For a reaction that has produced very little product and in which most of the reactants remain unreacted, Q will be very small. The value of Q increases as a reaction proceeds.

A **chemical equilibrium** exists when a chemical reaction and its reverse reaction occur simultaneously at the same rate, so that the molar concentrations of reactants and products do not change with time.

$$p\ A(g)\ +\ q\ B(g)\ \rightleftharpoons\ x\ C(g)\ +\ y\ D(g)$$

When the concentrations of reactants and products do not change, the value of the reaction quotient does not change. This value of Q, when the reaction is at equilibrium, is called the equilibrium constant for the reaction, and is represented by K. For a reaction at equilibrium,

$$K = \frac{[C]^x [D]^y}{[A]^p [B]^q}$$

where [C], [D], [A], and [B] are the concentrations of the products and reactants when the reaction is at equilibrium.

For reactions that involve liquids and solids, these are not included in the reaction quotient or equilibrium constant expression.

$$p\ E(s)\ +\ q\ F(g)\ \rightleftharpoons\ x\ G(l)\ +\ y\ H(g)$$

$$K = \frac{[H]^y}{[F]^q}$$

The concentrations of solids and liquids are constant and do change during a reaction. To simplify calculations involving equilibrium, these factors that do not change are omitted.

In writing an equilibrium constant expression (or reaction quotient):

1. Balance the chemical equation.
2. Write the concentration of each product in the numerator of K. Use the coefficient of each product in the balanced equation as the exponent of its concentration.
3. Write the concentration of each reactant in the denominator of K. Use the coefficient of each reactant in the balanced equation as the exponent of its concentration.
4. Delete all pure solids and pure liquids from the equilibrium constant expression.

The equilibrium constants for specific types of reactions are given special names, such as acid ionization constant (K_a) or solubility product (K_{sp}). However, no matter what the name, the process for writing the equilibrium constant expression is always the same.

Example 1

Write an equilibrium constant expression for each of the following reactions.

A. $NO_2(g) + SO_2(g) \rightleftharpoons NO(g) + SO_3(g)$

   1. The equation is balanced.

   2. $K = \dfrac{[NO] \, [SO_3]}{}$

   3. $K = \dfrac{[NO] \, [SO_3]}{[NO_2] \, [SO_2]}$

   4. There are no solids or liquids in the reaction.

$$K = \frac{[NO] \, [SO_3]}{[NO_2] \, [SO_2]}$$

B. $C_2H_4(g) + O_2(g) \rightleftharpoons CO_2(g) + H_2O(g)$

   1. Balance the equation: $C_2H_4(g) + 3\,O_2(g) \rightleftharpoons 2\,CO_2(g) + 2\,H_2O(g)$

   2. $K = \dfrac{[CO_2]^2 \, [H_2O]^2}{}$

   3. $K = \dfrac{[CO_2]^2 \, [H_2O]^2}{[C_2H_4] \, [O_2]^3}$

   4. There are no solids or liquids in the reaction.

$$K = \frac{[CO_2]^2 \, [H_2O]^2}{[C_2H_4] \, [O_2]^3}$$

C. $Ca(OH)_2(s) + C_2H_2(g) \rightleftharpoons CaC_2(s) + 2\,H_2O(l)$

   1. The equation is balanced.

   2. $K = \dfrac{[CaC_2] \, [H_2O]^2}{}$

   3. $K = \dfrac{[CaC_2] \, [H_2O]^2}{[Ca(OH)_2] \, [C_2H_2]}$

   4. Delete $Ca(OH)_2$ from denominator, CaC_2 and H_2O from numerator.

$$K = \frac{1}{[C_2H_2]}$$

Exercise One

Write equilibrium constant expressions for the following equilibrium reactions.

A. $CH_4(g) + H_2S(g) \rightleftharpoons H_2(g) + CS_2(g)$

B. $Si(CH_3)_4(g) + 12\ CO_2(g) \rightleftharpoons SiO_2(s) + 16\ CO(g) + 6\ H_2O(g)$

C. $CaCO_3(s) \rightleftharpoons CaO(s) + CO_2(g)$

D. $2\ PbS(s) + 3\ O_2(g) \rightleftharpoons 2\ PbO(s) + 2\ SO_2(g)$

CALCULATING THE VALUE OF THE EQUILIBRIUM CONSTANT

Example 2

A chemist put 4.95 moles of SO_3 in a 500.0 mL container and heated it to 527°C. When equilibrium was established, there were 0.300 mole of SO_2 in the container. What is the value of the equilibrium constant at 527°C for the equation below?

$$2\ SO_2(g) + O_2(g) \rightleftharpoons 2\ SO_3(g)$$

1. Write the equilibrium constant expression.

$$K = \frac{[SO_3]^2}{[SO_2]^2 [O_2]}$$

2. Arrange the data in tabular form under the equation.

	2 SO$_2$(g)	+	O$_2$(g)	$\rightleftharpoons$	2 SO$_3$(g)
initial moles	0		0		4.95
initial conc	0		0		9.90
change in conc	+2x		+x		−2x
equil conc	2x		x		9.90−2x

a. In this table, the "initial moles" line gives the number of moles at the start of the reaction.

b. The "initial conc" line gives the concentrations at the start of the reaction. Because the container has a volume of 0.500 L, the moles above are divided by 0.500 to get the molar concentrations.

c. The "change in conc" line gives the concentration change that occurs as the reaction goes from its initial condition to the equilibrium condition. The coefficients of the balanced equation are used, because for each mole of O$_2$ formed, 2 moles of SO$_2$ are also formed, and 2 moles of SO$_3$ decompose.

d. The "equil conc" line contains expressions for the equilibrium concentrations of each substance in the K expression. These expressions are the sums of the two previous lines.

3. Find the values of the equilibrium concentrations.

0.300 mole of SO$_2$ is present at equilibrium
Therefore, the equilibrium concentration of SO$_2$ is

$$\frac{0.300 \text{ mol}}{0.5000 \text{ L}} = 0.600 \text{ M}$$

According to the table above, the equilibrium concentration of SO$_2$ is 2x.
Therefore, $x = 0.300$ M.

Then, the equilibrium concentrations are

[SO$_2$] = 2x = 0.600 M
[O$_2$] = x = 0.300 M
[SO$_3$] = 9.90 − 2x = 9.90 − 0.600 = 9.30 M

4. Calculate the value of K.

$$K = \frac{[SO_3]^2}{[SO_2]^2 [O_2]} = \frac{(9.30)^2}{(0.600)^2 (0.300)} = 8.01 \times 10^2$$

Exercise Two

A mixture of 2.5 moles of N_2 and 2.5 moles of H_2 are sealed in a 1.0-liter container and heated to 400°C. When equilibrium is established, the container holds 1.0 mole of NH_3. Calculate K for the equilibrium equation below at 400°C.

$$N_2(g) + 3\,H_2(g) \rightleftharpoons 2\,NH_3(g)$$

CALCULATING EQUILIBRIUM CONCENTRATIONS

Example 3

For the reaction $2\,HI(g) \rightleftharpoons H_2(g) + I_2(g)$, $K = 0.016$ at 520°C. A 4.0-mole sample of HI is placed in a 10.0 liter container and heated to 520°C. Calculate the numbers of moles of HI, I_2, and H_2 in the container when equilibrium is reached.

1. Write the equilibrium constant expression.

$$K = \frac{[H_2]\,[I_2]}{[HI]^2} = 0.016$$

2. Arrange the data in tabular form.

	2 HI(g) $\rightleftharpoons$	H$_2$(g) +	I$_2$(g)
initial moles	4.0	0	0
initial conc	0.40	0	0
change in conc	−2x	+x	+x
equil conc	0.40−2x	x	x

3. Substitute the equilibrium concentrations into the K expression, and solve for x.

$$\frac{(x)(x)}{(0.40-2x)^2} = 0.016$$

$$\frac{(x)^2}{(0.40-2x)^2} = 0.016$$

Take the square root of both sides.

$$\frac{(x)}{(0.40-2x)} = 0.126$$

Solve for x.

$$x = 0.126\,(\,0.40 - 2x\,)$$

$$x = 0.0504 - 0.252\,x$$

$$1.252\,x = 0.0504$$

$$x = \frac{0.0504}{1.252} = 0.040$$

4. Determine the values of the equilibrium concentrations and amounts.

[HI] = 0.40 − 2x = 0.40 − 0.080 = 0.32 mol/L.
moles of HI = 0.32 mol/L × 10.0 L = 3.2 moles

$[H_2] = [I_2] = x$ moles/liter = 0.040 mole/liter
moles of H_2 and moles of I_2 = 0.040 mol/L × 10.0 L = 0.40 moles

---

Exercise Three

A mixture of 6.0 moles of H_2 and 6.0 moles of CO_2 is placed in a 10.0-liter container and heated to 750°C. For the reaction below, the value of K is 0.77 at 750°C. Calculate the number of moles of H_2, CO_2, H_2O, and CO in the mixture when equilibrium is established.

$$H_2(g) + CO_2(g) \rightleftharpoons H_2O(g) + CO(g)$$

CALCULATING THE EFFECT OF A CHANGE IN CONCENTRATION

Example 4

An equilibrium mixture at 50°C in a 1.00-liter container consists of 0.20 mole of NO_2 and 2.00 moles of N_2O_4.

$$N_2O_4(g) \rightleftharpoons 2\,NO_2(g)$$

One more mole of N_2O_4 is added to this mixture. Calculate the numbers of moles of NO_2 and N_2O_4 in the mixture once equilibrium is reestablished.

1. Write the equilibrium constant expression.

$$K = \frac{[NO_2]^2}{[N_2O_4]}$$

   Calculate its value from the given information.

$$K = \frac{(0.20)^2}{(2.00)} = 0.020$$

2. Summarize the data in tabular form.

	$N_2O_4(g)$	$\rightleftharpoons$	$2\,NO_2(g)$
initial conc	2.00		0.20
added conc	1.00		
change in conc	$-x$		$+2x$
equil conc	$3.00-x$		$0.20+2x$

3. Substitute the equilibrium concentrations into the K expression and solve for x.

$$\frac{(0.20 + 2x)^2}{(3.00 - x)} = 0.020$$

$$(0.20 + 2x)^2 = 0.020\,(3.00 - x)$$

$$0.040 + 0.80x + 4x^2 = 0.060 - 0.020x$$

$$4x^2 + 0.82x - 0.020 = 0$$

The last equation is a quadratic equation, and can be solved using the quadratic formula, namely, when

$$a\,x^2 + b\,x + c = 0$$

then

$$x = \frac{1}{2a}\left(-b \pm \sqrt{b^2 - 4ac}\right)$$

Therefore, in this case,

$$x = \frac{1}{2(4)}\left(-0.82 \pm \sqrt{(0.82)^2 - 4(4)(-0.020)} \right)$$

$$= \frac{1}{8}\left(-0.82 \pm \sqrt{0.672 + 0.32} \right)$$

$$= \frac{1}{8}\left(-0.82 \pm 0.996 \right)$$

$$= 0.022 \text{ mol/L}$$

4. Determine the values of the equilibrium concentrations and amounts.

$[N_2O_4] = 3.00 - x = 3.00 - 0.022 = 2.98$ mol/L
   moles N_2O_4 = (2.98 mol/L) (1.00 L) = 2.98 mol

$[NO_2] = 0.20 + 2x = 0.20 + 2 (0.022) = 0.24$ mol/L
   moles NO_2 = (0.24 mol/L) (1.00 L) = 0.24 mol

Exercise Four

A mixture of hydrogen, iodine, and hydrogen iodide gases in a 1.00-liter container at 450°C contained 0.71 mol H_2, 0.71 mol I_2, and 5.0 mol HI at equilibrium. How many moles of each gas are in the container after 1.0 mole of HI is added and the reaction returns to equilibrium? The equation for the reaction is

$$H_2(g) + I_2(g) \rightleftharpoons 2 HI(g)$$

LE CHATELIER'S PRINCIPLE

Le Chatelier's Principle: If a system at equilibrium is disturbed by some change, the system will shift to reduce the immediate effect of the change.

Example 5

Use Le Chatelier's Principle to predict the direction in which the equilibrium will shift when the changes A, B, and C are made.

$$N_2(g) + 3\ H_2(g) \rightleftharpoons 2\ NH_3(g) \qquad \Delta H = -94.2\ kJ$$

If N_2 and H_2 react to form more NH_3, the equilibrium is said to shift in the forward direction.

If NH_3 dissociates to produce more N_2 and H_2, the equilibrium is said to shift in the reverse direction.

A. add NH_3

Le Chatelier's Principle says that the equilibrium will react to reduce the immediate effect, an increase in the amount of ammonia. Therefore, the reaction consumes ammonia, and it shifts in the reverse direction.

B. reduce volume of container

$$PV = nRT$$

Reducing the volume will increase the pressure. Le Chatelier's Principle says the equilibrium will react to decrease the pressure. Pressure is proportional to the amount (moles) of gas. Therefore, the equilibrium will shift to reduce the moles of gas. The left side of the equation involves 4 moles of gas, while the right side only 2 moles. By shifting in the forward direction, the equilibrium reduces the number of moles of gas.

C. increase temperature

Increasing the temperature corresponds to adding heat. The equilibrium will respond in order to reduce the amount of heat in the system. The reaction is exothermic (its ΔH is negative). This means that a shift in the forward direction would release heat. Therefore, the equilibrium shifts in the reverse direction.

Exercise Five

Predict the direction (forward, reverse, no change) of the shift in each of the following equilibrium reactions for each indicated operation.

A. $2\,SO_2(g) + O_2(g) \rightleftharpoons 2\,SO_3(g)$

    1. remove SO_3

    2. increase volume

B. $4\,NH_3(g) + 5\,O_2(g) \rightleftharpoons 4\,NO(g) + 6\,H_2O(g)$ $\Delta H = -8000\ kJ$

    1. add O_2

    2. lower temperature

    3. decrease volume

C. $CO(g) + H_2O(g) \rightleftharpoons CO_2(g) + H_2(g)$

    1. remove H_2

    2. increase volume

THE K_p EXPRESSION

The equilibrium constant for a reaction involving gases is sometimes expressed in terms of gas pressures instead of concentrations. For a gas, the concentration is related to its pressure. The relationship can be obtained from the ideal gas equation.

$$PV = nRT$$

Concentration is moles per volume, n/V.

$$\frac{n}{V} = \frac{P}{RT}$$

To see how this relationship can be used to express an equilibrium constant in terms of gas pressures, consider the reaction

$$2\,SO_2(g) + O_2(g) \rightleftharpoons 2\,SO_3(g)$$

For this reaction,

$$K = \frac{[SO_3]^2}{[SO_2]^2\,[O_2]}$$

From the ideal gas equation,

$$[SO_3] = \frac{n_{SO_3}}{V} = \frac{P_{SO_3}}{RT}$$

$$[SO_2] = \frac{n_{SO_2}}{V} = \frac{P_{SO_2}}{RT}$$

$$[O_2] = \frac{n_{O_2}}{V} = \frac{P_{O_2}}{RT}$$

Therefore,

$$K = \frac{(P_{SO_3}/RT)^2}{(P_{SO_2}/RT)^2 (P_{O_2}/RT)} = \frac{P_{SO_3}^2 (1/RT)^2}{P_{SO_2}^2 (1/RT)^2 P_{O_2}(1/RT)} = \frac{(P_{SO_3})^2}{(P_{SO_2})^2 (P_{O_2})(1/RT)}$$

$$= \frac{(P_{SO_3})^2}{(P_{SO_2})^2 (P_{O_2})} (RT)$$

The equilibrium constant expressed with pressures is

$$K_P = \frac{(P_{SO_3})^2}{(P_{SO_2})^2 (P_{O_2})}$$

In general,

$$K_P = K (RT)^x$$

where

$x =$ (sum of the coefficients of the products) −
(sum of the coefficients of the reactants)

Example 6

Calculate the value of K_P for the equilibrium system of Example 2.

$$2\ SO_2(g) + O_2(g) \rightleftharpoons 2\ SO_3(g)$$

$$K_P = K (RT)^x$$

$x = 2 - (2+1) = -1$
$T = 527°C = 800\ K$
$R = 0.0821\ L\ atm/mol\ K$
$K = 8.01 \times 10^2$

$$K_P = (8.01 \times 10^2) [(0.0821) (800)]^{-1}$$

$$K_P = 12.2$$

Exercise Six

Calculate the value of K_P for the equilibrium system of Exercise Two.

$$N_2(g) + 3\,H_2(g) \rightleftharpoons 2\,NH_3(g)$$

PROBLEMS

1. A sample containing 2.00 moles of H_2 is placed with 2.00 moles of I_2 in a 1.00-liter flask at 500°C. What will be the final concentration (in moles/liter) of HI, H_2, and I_2 in the flask, when equilibrium is established?

$$H_2(g) + I_2(g) \rightleftharpoons 2\,HI(g) \qquad K = 47.3 \text{ at } 500°C$$

2. The equations for the reaction between hydrogen and bromine at various temperatures are given below.

$$100°C: \qquad H_2(g) + Br_2(g) \rightleftharpoons 2\,HBr(g)$$

$$0°C: \qquad H_2(g) + Br_2(l) \rightleftharpoons 2\,HBr(g)$$

$$-100°C: \qquad H_2(g) + Br_2(s) \rightleftharpoons 2\,HBr(s)$$

Write the equilibrium constant expression for the reaction at each temperature.

3. A sample of 2.00 grams of ammonia was placed in a 25.0 mL container and heated to 300°C. After equilibrium had been reached, the concentration of hydrogen was 1.58 moles/liter. What is the value of the equilibrium constant at 300°C for the reaction below?

$$2 NH_3(g) \rightleftharpoons N_2(g) + 3 H_2(g)$$

4. At 1000°C, the value of equilibrium constant for the reaction below is 1.62. How many moles of each compound will there be at equilibrium if 1.00 mole of H_2, 2.00 moles of CO_2, and 3.00 moles of CO are placed in a 10.0-liter tank and heated to 1000°C?

$$H_2(g) + CO_2(g) \rightleftharpoons H_2O(g) + CO(g)$$

5. For the reaction

$$PCl_5(g) \rightleftharpoons PCl_3(g) + Cl_2(g)$$

$K = 0.050$ at 150°C. A quantity of PCl_5 is placed in a 5.0-liter container at 150°C. When equilibrium is established, the container holds 0.50 mole of Cl_2. How many moles of PCl_5 were originally placed in the container?

6. A mixture of 30.0 mol of NO(g) and 18.0 mol of O_2(g) is placed in an empty 3.00 liter flask held at a constant temperature. These gases react until the reaction represented below comes to equilibrium. At equilibrium, the vessel contains 26.4 moles of NO_2. What is the value of the equilibrium constant at the constant temperature?

$$2 NO(g) + O_2(g) \rightleftharpoons 2 NO_2(g)$$

7. At a certain temperature, the value of the equilibrium constant for the reaction represented by the equation below is 51.5.

$$H_2(g) + I_2(g) \rightleftharpoons 2 HI(g) \qquad K = 51.5$$

Suppose 3.00 mol of H2 and 3.00 mol of I2 are combined in a 3.00 L flask at that temperature. What are the values of the concentrations, $[H_2]$, $[I_2]$, and $[HI]$, at equilibrium?

8. The reaction represented by the equation below is at equilibrium in a container with $[CO] = 0.30$ mol/L, $[H_2O] = 0.10$ mol/L, $[CO_2] = 0.20$ mol/L, and $[H_2] = 0.60$ mol/L.

$$CO(g) + H_2O(g) \rightleftharpoons CO_2(g) + H_2(g)$$

How many moles per liter of H_2O must be added to increase $[H_2]$ to 0.70 mol/L?

9. After 48 moles of PCl_5 are placed in a 4.00 L reactor, it is heated to a temperature at which the reaction below has an equilibrium constant equal to 0.050. What are the concentrations of PCl_5, PCl_3, and Cl_2 at equilibrium in the reactor?

$$PCl_5(g) \rightleftharpoons PCl_3(g) + Cl_2(g)$$

ACIDS AND BASES

This lesson deals with:

1. Identifying Arrhenius acids and bases.
2. Relating the hydrogen ion and hydroxide ion concentrations to each other using the K_w expression.
3. Calculating hydrogen ion and hydroxide ion concentrations in solutions of acids and bases.
4. Calculating the pH and pOH of solutions of strong acids and bases.
5. Calculating the pH and pOH of a solution resulting from the mixture of an acid and a base.
6. Identifying Brønsted acids and bases.
7. Identifying Lewis acids and bases in a given reaction equation.

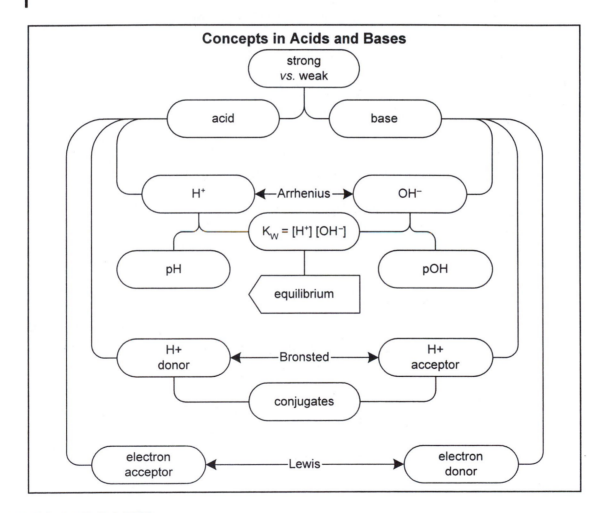

ACIDS AND BASES

Acids are substances that share a common set of properties. Likewise, bases share a different set of common properties.

Classic Properties of Acids and Bases

Acids: 1.   have a sour taste.
        2.   react with certain metals to produce $H_2(g)$.
        3.   react with carbonates to produce $CO_2(g)$.
        4.   change the color of certain vegetable dyes (indicators).
        5.   neutralize a base.

Bases: 1.   have a bitter taste.
        2.   have a soapy feel.
        3.   reverse the indicator color change caused by acids.
        4.   neutralize an acid.

Arrhenius Chemical Definitions of Acids and Bases

Acid: an aqueous solution that contains more H^+ than OH^-.

Base: an aqueous solution that contains more OH^- than H^+.

Ionization of Water

In water, some of the molecules form ions.

$$H_2O(l) \rightleftharpoons H^+(aq) + OH^-(aq)$$

The ionization of water is an equilibrium, with an equilibrium constant called K_w.

$$K_w = [H^+][OH^-] = 1.0 \times 10^{-14} \text{ at } 25°C$$

(The concentration of liquid H_2O is not used in the K_w expression.
Pure liquids do not appear in equilibrium constant expressions.)

Example 1

A. What is $[H^+]$ in pure water at 25°C?

In pure water, the H^+ ions and OH^- ions are only from the ionization of water. When water ionizes, it produces an equal number of H^+ ions and OH^- ions. Therefore,
$$[H^+] = [OH^-]$$

In water at 25°C,
$$[H^+][OH^-] = 1.0 \times 10^{-14}$$

Therefore, in pure water at 25°C, $[H^+][OH^-] = [H^+]^2$, and

$$[H^+]^2 = 1.0 \times 10^{-14}$$

$$[H^+] = \sqrt{1.0 \times 10^{-14}}$$

$$[H^+] = 1.0 \times 10^{-7} \text{ M}$$

B. What is $[H^+]$ in an aqueous solution at 25°C, that contains 0.0025 M OH⁻? Is this solution acidic or basic?

In water at 25°C,

$$[H^+][OH^-] \quad = \quad 1.0 \times 10^{-14}$$

In this solution,

$$[OH^-] = 0.0025 \text{ M}$$

$$[H^+](0.0025 \text{ M}) = 1.0 \times 10^{-14}$$

$$[H^+] = \frac{1.0 \times 10^{-14}}{0.0025} = 4.0 \times 10^{-12} \text{ M}$$

The hydrogen-ion concentration (4.0×10^{-12} M) is much less than the hydroxide-ion concentration (0.0025 M).
Therefore, the solution is basic.

Exercise One

A. A 0.10 M solution of ammonium chloride has a hydrogen ion concentration of 2.36×10^{-5} M. What is $[OH^-]$? Is this solution an acid or a base?

B. At 65°C, the value of K_w is 1.0×10^{-13}. What is $[H^+]$ in water at this temperature?

STRONG AND WEAK ACIDS AND BASES

Strong acid: all molecules of a strong acid are ionized in solution

$$HCl(g) \xrightarrow{\text{dissolve in water}} H^+(aq) + Cl^-(aq)$$

In 0.10 M HCl(aq), $[H^+] = 0.10$ M and $[Cl^-] = 0.10$ M

Weak acid: not all molecules form ions in solution, some (many) remain as molecules

$$HC_2H_3O_2(aq) \rightleftharpoons H^+(aq) + C_2H_3O_2^-(aq)$$

In 0.10 M $HC_2H_3O_2$, $[H^+] \ll 0.10$ M and $[C_2H_3O_2^-] \ll 0.10$ M

Strong acids are distinguished from weak acids by electrical conductivity.
  Strong acids are strong electrolytes, and weak acids are weak electrolytes.
  In solution, strong acids are 100% ionized.
  There are seven common strong acids. These are listed in Table 1.

Table 1. Common Strong Acids

HCl(aq)	hydrochloric acid	H_2SO_4(aq)	sulfuric acid
HBr(aq)	hydrobromic acid	$HClO_3$(aq)	chloric acid
HI(aq)	hydroiodic acid	$HClO_4$(aq)	perchloric acid
HNO_3(aq)	nitric acid		

Weak acids are less than 100% ionized in aqueous solution.

Strong bases completely dissociate into aqueous hydroxide ions and aqueous cations. There are many strong bases. Table 2 lists some of them, the hydroxides of the group IA and IIA metals.

Table 2. Common Strong Bases

LiOH	lithium hydroxide	$Mg(OH)_2$	magnesium hydroxide
NaOH	sodium hydroxide	$Ca(OH)_2$	calcium hydroxide
KOH	potassium hydroxide	$Ba(OH)_2$	barium hydroxide
RbOH	rubidium hydroxide	$Sr(OH)_2$	strontium hydroxide
CsOH	cesium hydroxide		

In 0.10 M NaOH(aq), $[OH^-] = 0.10$ M and $[Na^+] = 0.10$ M

Weak base: aqueous solutions of base that is not completely ionized.

$$NH_3(aq) + H_2O(l) \rightleftharpoons NH_4^+(aq) + OH^-(aq)$$

In 0.10 M NH_3(aq), $[OH^-] \ll 0.10$ M and $[NH_4^+] \ll 0.10$ M

Example 2

A.   What are $[H^+]$ and $[OH^-]$ in 0.045 M NaOH?

NaOH is a strong base, so $[OH^-] = 0.045$ M.

$$K_w = [H^+][OH^-] = 1.0 \times 10^{-14}$$

$$[H^+] = \frac{1.0 \times 10^{-14}}{[OH^-]}$$

$$= \frac{1.0 \times 10^{-14}}{0.045} = 2.2 \times 10^{-13}$$

B. A substance has a molar mass of 125 g/mol. When 0.040 g is dissolved in 100 mL of water, $[H^+] = 5 \times 10^{-9}$ M. Is this a strong or weak acid or base?

$$[OH^-] = \frac{1.0 \times 10^{-14}}{5 \times 10^{-9}} = 2 \times 10^{-6}$$

$[OH^-] > [H^+]$ Therefore, the solution is a base.

Concentration of solution:

$$\frac{(0.040 \text{ g}) \left(\dfrac{1 \text{ mol}}{125 \text{ g}} \right)}{0.100 \text{ L}} = 3.2 \times 10^{-3} \text{ M}$$

If this were a strong base, $[OH^-]$ would be at least 3.2×10^{-3} M. Because $[OH^-] < 3.2 \times 10^{-3}$ M, the solution is a weak base.

Exercise Two

A. What are $[H^+]$ and $[OH^-]$ in an aqueous solution prepared by dissolved 0.50 g of $HClO_4$ in 750 mL of solution?

B.   When 1.0 liter of aqueous solution containing 0.10 g CH_2O_2 is prepared, the hydroxide-ion concentration of the solution is 1.6×10^{-11} M. Is this a strong or weak acid or base?

pH AND pOH

Definition of pH: $pH = -\log [H^+]$

If $[H^+] = 0.025$ M,
        then $pH = -\log [H^+] = -\log (0.025) = 1.6$

At 25°C,
        in an acidic solution, pH < 7,
        in a basic solution, pH > 7,
        in a neutral solution, pH = 7.

Definition of pOH: $pOH = -\log [OH^-]$

If $[OH^-] = 4 \times 10^{-5}$ M,
        then $pOH = -\log [OH^-] = -\log (4 \times 10^{-5}) = 4.4$

A relationship between pH and pOH:

$$[H^+][OH^-] \quad = \quad 1.0 \times 10^{-14} \text{ at } 25°C$$

$$\log([H^+][OH^-]) \quad = \quad \log(1.0 \times 10^{-14})$$

$$\log[H^+] + \log[OH^-] \quad = \quad -14.00$$

$$(-\log[H^+]) + (-\log[OH^-]) \quad = \quad 14.00$$

$$pH \quad + \quad pOH \quad = \quad 14.00 \quad \text{at } 25°C$$

Example 3

A. What is the pH and pOH of a solution containing 0.082 g $Ba(OH)_2$ in 500.0 mL of solution?

> $Ba(OH)_2$ is a strong base.
>
> $Ba(OH)_2$ in solution is $Ba^{2+}(aq) + 2\ OH^-(aq)$
>
> Concentration $Ba(OH)_2$:
>
> $$\frac{(0.082\ g)\left(\dfrac{1\ mol}{171\ g}\right)}{0.500\ L} = 9.6 \times 10^{-4}\ M$$
>
> $[OH^-] = 2\ (9.6 \times 10^{-4}\ M) = 1.92 \times 10^{-3}\ M$
>
> $pOH = -\log[OH^-] = -\log(1.92 \times 10^{-3}) = 2.72$
>
> $pH = 14.00 - 2.72 = 11.28$
>
> (These pH and pOH values have 2 significant figures, namely the 2 digits following the decimal point. The digits to the left of the decimal point are *not* significant; they merely indicate the exponent of 10 in the original number.)

B. What is the concentration of a nitric acid solution that has a pH = 4.30?

> Because nitric acid is a strong acid,
>
> | | | |
> |---|---|---|
> | $[H^+]$ | = | concentration of acid |
> | pH | = | $-\log[H^+]$ |
> | $\log[H^+]$ | = | $-pH$ |
> | $[H^+]$ | = | 10^{-pH} |
> | $[H^+]$ | = | $10^{-4.30} = 5.0 \times 10^{-5}\ M$ |
>
> Therefore, the solution contains $5.0 \times 10^{-5}\ M\ HNO_3$.

Exercise Three

A. What are the pH and pOH of a solution with $[OH^-] = 4.2 \times 10^{-5}\ M$?

B. What are pH and pOH of an aqueous solution prepared by dissolving 0.36 g of KOH in 250 mL of solution?

ACID-BASE NEUTRALIZATION

Arrhenius acid: aqueous solution with $[H^+] > [OH^-]$.

Arrhenius base: aqueous solution with $[OH^-] > [H^+]$.

Strong acid-strong base reaction:

$$H^+(aq) + OH^-(aq) \longrightarrow H_2O(l)$$

Example:

$$HCl(aq) \quad + \quad NaOH(aq) \longrightarrow NaCl(aq) \quad + \quad H_2O(l)$$

$$acid \quad + \quad base \longrightarrow salt \quad + \quad water$$

neutral
(neither acid nor base)

Example 4

What is the pH of a solution produced by mixing 24 mL of 0.020 M HNO_3 with 26 mL of 0.020 M KOH?

$$HNO_3(aq) + KOH(aq) \longrightarrow KNO_3(aq) + H_2O(l)$$

We want to find the pH of the mixture. We can do this either from $[H^+]$ or from $[OH^-]$ and the relationship between pH and pOH.

$$pH = -\log[H^+] \qquad or \qquad pH = 14.00 - pOH = 14.00 - (-\log[OH^-])$$

$$[H^+] = \frac{mol\ H^+}{L\ soln} \qquad or \qquad [OH^-] = \frac{mol\ OH^-}{L\ soln}$$

Therefore, we need to know the volume of the solution.

$$volume\ of\ solution\ =\ 24\ mL\ +\ 26\ mL\ =\ 50\ mL\ =\ 0.050\ L$$

To find moles of H^+ or OH^- in the mixture, we must know the moles of the reactants.

$$\text{moles HNO}_3 = (0.020 \text{ M})(0.024 \text{ L}) = 4.8 \times 10^{-4} \text{ mol}$$

$$\text{moles KOH} = (0.020 \text{ M})(0.026 \text{ L}) = 5.2 \times 10^{-4} \text{ mol}$$

More KOH is used in the mixture than HNO_3. Therefore, KOH is in excess. (The chemical equation shows that 1 mole of HNO_3 reacts with 1 mole of KOH.)

$$\text{excess KOH} = \text{mol KOH} - \text{mol HNO}_3$$

$$= 5.2 \times 10^{-4} \text{ mol} - 4.8 \times 10^{-4} \text{ mol}$$

$$= 4 \times 10^{-5} \text{ mole}$$

Therefore, the mixture contains 4×10^{-5} mole of OH^-.

$$[OH^-] = \frac{4 \times 10^{-5} \text{ mol OH}^-}{0.050 \text{ L}} = 8 \times 10^{-4} \text{ M}$$

$$pOH = -\log(8 \times 10^{-4}) = 3.1$$

$$pH = 14.00 - 3.1 = 10.9$$

Exercise Four

What is the pH of a solution prepared by adding 0.36 g NaOH to 240 mL of 0.055 M HCl?

BRØNSTED ACID-BASE CONCEPT

Brønsted acid: species that donates a hydrogen ion (H^+) in a reaction.

Brønsted base: species that accepts a hydrogen ion in a reaction.

$$NH_3(aq) + H_2O(l) \rightleftharpoons NH_4^+(aq) + OH^-(aq)$$
$$\text{base} \qquad \text{acid}$$

Brønsted acid-base reaction: transfer of a hydrogen ion from one species to another.

This means that if a reaction is a Brønsted acid-base reaction, then the reverse reaction is also a Brønsted acid-base reaction.

$$NH_4^+(aq) + OH^-(aq) \rightleftharpoons NH_3(aq) + H_2O(l)$$
$$\text{acid} \qquad \text{base}$$

Brønsted acid reacts to form a base.

$$NH_4^+ \longrightarrow NH_3 \qquad + \quad H^+$$
$$\text{acid} \qquad\qquad \text{conjugate base}$$

Brønsted base reacts to form an acid.

$$OH^- + H^+ \longrightarrow H_2O$$
$$\text{base} \qquad\qquad \text{conjugate acid}$$

Example 5

A.   Find the conjugate base of HSO_4^-.

   HSO_4^- is the acid. An acid gives up a hydrogen ion.

$$HSO_4^- \longrightarrow H^+ + SO_4^{2-}$$
$$\text{acid} \qquad\qquad \text{conjugate base}$$

B.   Find the conjugate acid of HSO_4^-.

   HSO_4^- is the base. A base gains a hydrogen ion.

$$HSO_4^- + H^+ \longrightarrow H_2SO_4$$
$$\text{base} \qquad\qquad \text{conjugate acid}$$

C.   Identify the acids and bases in the following equation.

$$CN^- + H_2O \rightleftharpoons HCN + OH^-$$

CN^- gains H^+ to form HCN. Therefore, CN^- is the base, and HCN is the conjugate acid.
H_2O loses H^+ to form OH^-. Therefore, H_2O is the acid, and OH^- is the conjugate base.

Exercise Five

A. What is the conjugate base of each of the following?

$$HCl \qquad H_2PO_4^- \qquad CCl_3CO_2H$$

B. What is the conjugate acid of each of the following?

$$H_2PO_4^- \qquad NH_3 \qquad CO_3^{2-}$$

C. Identify the Brønsted acids and bases in the following equations.

$$ClO_2^- + H_2O \rightleftharpoons HClO_2 + OH^-$$

$$HSO_4^- + H_2O \rightleftharpoons H_3O^+ + SO_4^{2-}$$

$$NH_4^+ + NH_2^- \rightleftharpoons NH_3 + NH_3$$

LEWIS ACID-BASE CONCEPT

Lewis acid: accepts a pair of electrons.

Lewis base: donates a pair of electrons.

$$NH_3 + BF_3 \longrightarrow H_3NBF_3$$

Examine the valence electrons:

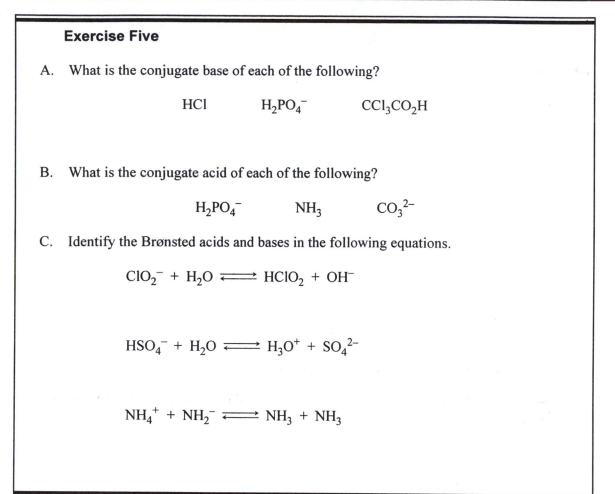

N has 1 lone pair
and 3 bonding pairs

B has 3 electron pairs

N has four bonding pairs
B has four bonding pairs

B has accepted a pair of electrons from N.

A Lewis base must have a lone pair of electrons.
A Lewis acid must have an empty orbital in its valence shell.

Example 6

Identify the Lewis acid and base in the following reaction.

$$PCl_5 + NaCl \longrightarrow Na[PCl_6]$$

To see what is happening to the electron pairs in this reaction, draw the Lewis structures.

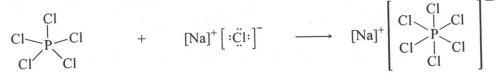

P has 5 bonding pairs   Cl has 4 lone pairs   P has 6 bonding pairs
Cl has 3 lone pairs

P has gained an electron pair. Therefore, PCl_5 is a Lewis acid.

Cl^- has donated an electron pair. Therefore, Cl^- is a Lewis base.

Exercise Six

Identify the Lewis acid and base in each of the following reactions.

A. $Ag^+(aq) + 2 CN^-(aq) \longrightarrow Ag(CN)_2^-(aq)$

B. $BCl_3 + CH_3OCH_3 \longrightarrow (CH_3)_2OBCl_3$

C. $2 H_2CO + SnCl_4 \longrightarrow SnCl_4(OCH_2)_2$

PROBLEMS

1. Calculate the pH of a 0.020 M solution of each of the following.

    (a) HCl
    (b) KOH
    (c) $Ba(OH)_2$

2. Calculate the concentration of H^+ in each of the following.

    (a) a sample of milk with a pH of 6.82
    (b) a sample of lemon juice with a pH of 2.3
    (c) a sample of coffee with a pH of 5.38
    (d) a sample of $Mg(OH)_2$ solution with a pH of 10.52

3. Calculate the pH of the solution that results when 25.0 mL of 0.120 M HCl is mixed
    with
    (a) 40.0 mL of H_2O
    (b) 70.0 mL of 0.0500 M HBr
    (c) 30.0 mL of 0.100 M KOH
    (d) 100.0 mL of 0.0200 M $Ca(OH)_2$

4. How many milliliters of 0.040 M NaOH are needed to exactly neutralize 100.0 mL of an
    HNO_3 solution with a pH of 2.50?

5. (a) Give the conjugate base of: H_2O, NH_3, $HC_2O_4^-$, $Fe(H_2O)_6^{3+}$
    (b) Give the conjugate acid of: H_2O, NH_3, O^{2-}, HS^-, $N(CH_3)_3$

6. Identify each of the following as a Lewis acid or Lewis base.

    (a) NH_3
    (b) H^+
    (c) Al^{3+}
    (d) CN^-

IONIZATION OF WEAK ACIDS AND BASES

This lesson deals with:

1. Writing the K_a and K_b expressions for weak acids and bases.
2. Calculating the pH of a solution of known concentration from the value of K_a or K_b.
3. Calculating the value of K_a and K_b from pH and concentration data.
4. Calculating the percent ionization in a solution of known concentration from the value of K_a or K_b.
5. Calculating the concentration of a solution from the pH and the value of K_a or K_b.
6. Calculating the concentrations of the ions in a solution of a diprotic acid.

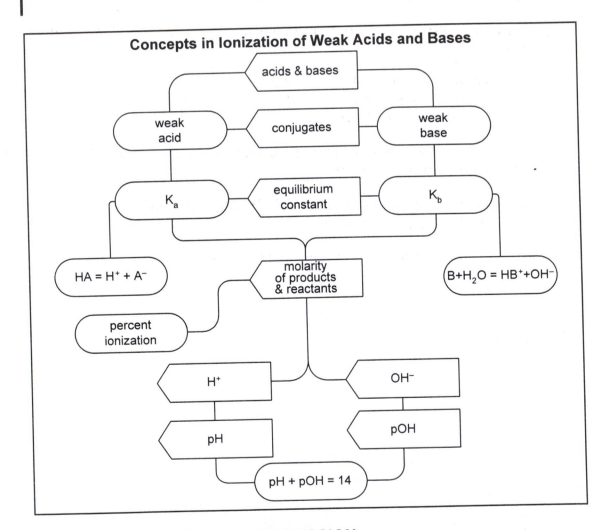

Concepts in Ionization of Weak Acids and Bases

THE EQUILIBRIUM CONSTANT EXPRESSION

A weak acid is a compound that, when dissolved in water, ionizes only partially, forming hydrogen ions. It also forms anions that are characteristic of the acid. The ionization reaction is an equilibrium. For this equilibrium, there is the usual equilibrium constant expression.

For a weak acid, HA, the ionization equation is

$$HA(aq) \rightleftharpoons H^+(aq) + A^-(aq)$$

and the equilibrium constant expression is

$$K_a = \frac{[H^+][A^-]}{[HA]}$$

For $HC_2H_3O_2$:

$$HC_2H_3O_2(aq) \rightleftharpoons H^+(aq) + C_2H_3O_2^-(aq)$$

$$K_a = \frac{[H^+][C_2H_3O_2^-]}{[HC_2H_3O_2]}$$

A weak base is a substance that, when dissolved in water, reacts with water forming hydroxide ions. It also forms cations that are characteristic of the base.

For a weak base, M, the equation and equilibrium constant expression are

$$M(aq) + H_2O(l) \rightleftharpoons MH^+(aq) + OH^-(aq)$$

$$K_b = \frac{[MH^+][OH^-]}{[M]}$$

Only the concentrations of aqueous solutes are used in K_b. The solvent, water, is not used in the K_b expression.

For NH_3:

$$NH_3(aq) + H_2O(l) \rightleftharpoons NH_4^+(aq) + OH^-(aq)$$

$$K_b = \frac{[NH_4^+][OH^-]}{[NH_3]}$$

The molecules of water themselves ionize. The self-ionization of water is represented by the equation

$$H_2O(l) \rightleftharpoons H^+(aq) + OH^-(aq)$$

Thus, all aqueous solutions contain both H^+ and OH^- ions.

The equilibrium constant expression for the self-ionization of water is

$$K_w = [H^+][OH^-]$$

In any aqueous solution at 25°C,

$$K_w = 1.0 \times 10^{-14}$$

In pure water at 25°C,

$$[H^+] = [OH^-] = 1.0 \times 10^{-7}$$

Exercise One

Write the equation for the ionization equilibrium for each of the following. Also write the equilibrium constant expression for each.

HF (a weak acid)

C_5H_5N (a weak base)

HC_6H_5O (a weak acid)

CH_3NH_2 (a weak base)

CALCULATING THE pH OF A WEAK ACID SOLUTION

The pH of a solution is an expression of its acidity.

$$pH = -\log [H^+]$$

Similarly, pOH is defined by the equation

$$pOH = -\log [OH^-]$$

In a neutral solution, $[H^+] = [OH^-]$. Therefore, in a neutral solution pH = pOH.

Example 1

Calculate the pH of a 0.050 M solution of hypochlorous acid, HOCl, at 25°C. For HOCl, $K_a = 3.5 \times 10^{-8}$ at 25°C.

$$HOCl(aq) \rightleftharpoons H^+(aq) + OCl^-(aq)$$

$$K_a = \frac{[H^+][OCl^-]}{[HOCl]} = 3.5 \times 10^{-8}$$

To find the pH, we must calculate $[H^+]$.

If no HOCl ionized, the concentrations of the solutes would be

$$[HOCl] = 0.050 \text{ M}$$

$$[H^+] = 1.0 \times 10^{-7} \text{ M}$$

$$[OCl^-] = 0$$

Because K_a is much bigger than K_w, much more H^+ will come from the ionization of HOCl than from the ionization of water. Therefore, we will simplify the calculations by ignoring H^+ from water; *i.e.*, we estimate that $[H^+] = 0$ before HOCl ionizes.

We don't know how much HOCl will ionize. Say x moles per liter ionize. Then, the concentration of HOCl decreases by x moles per liter, the concentration of H^+ increases by x moles per liter, and so does the concentration of OCl^-.

Summarize the information in a table based on the chemical equation:

	HOCl $\rightleftharpoons$	H^+ +	OCl^-
initial conc.	0.050	0	0
conc. change	$-x$	$+x$	$+x$
equil. conc.	$0.050-x$	x	x

$$\frac{(x)(x)}{(0.050-x)} = 3.5 \times 10^{-8}$$

$$x^2 = (3.5 \times 10^{-8})(0.050 - x)$$

$$x^2 = 1.75 \times 10^{-9} - 3.5 \times 10^{-8} x$$

Rearrange into an equation of the form:

$$\begin{array}{ccccccc} a\, x^2 & + & b & x & + & c & = 0 \\ x^2 & + & (3.5 \times 10^{-8}) & x & + & (-1.75 \times 10^{-9}) & = 0 \end{array}$$
$$a = 1$$
$$b = 3.5 \times 10^{-8}$$

$$c = -1.75 \times 10^{-9}$$

Apply quadratic formula:

$$x = \frac{1}{2a}\left(-b \pm \sqrt{b^2 - 4ac} \right)$$

$$x = \frac{1}{2}\left(-3.5 \times 10^{-8} \pm \sqrt{(3.5 \times 10^{-8})^2 - 4(1)(-1.75 \times 10^{-9})} \right)$$

$$= \frac{1}{2}\left(-3.5 \times 10^{-8} \pm 8.37 \times 10^{-5} \right)$$

$$= -4.2 \times 10^{-5} \quad or \quad +4.2 \times 10^{-5}$$

Because $x = [H^+]$, only the positive value is appropriate.

$$[H^+] = 4.2 \times 10^{-5} \text{ M}$$

$$pH = -\log [H^+] = -\log (4.2 \times 10^{-5} \text{ M}) = 4.38$$

A Way to Simplify the Calculations

In the calculation above, the value of x was calculated using the quadratic formula. The value obtained was 4.2×10^{-5} M.

Use this value to determine the equilibrium concentration of HOCl:

$$[HOCl] = 0.050 - x = 0.050 - (4.2 \times 10^{-5}) = 0.050$$

The value of x is insignificant compared to the initial concentration of HOCl.

We could estimate that x is insignificant before performing the calculation, that is,

$$\text{estimate } 0.050 - x = 0.050$$

Use this estimate to solve for x in Example 2.

$$[HOCl] = 0.050 - x = 0.050$$
$$[H^+] = x$$
$$[OCl^-] = x$$

$$K_a = \frac{[H^+][OCl^-]}{[HOCl]} = 3.5 \times 10^{-8}$$

$$\frac{(x)(x)}{(0.050)} = 3.5 \times 10^{-8}$$

$$x^2 = (0.050)(3.5 \times 10^{-8}) = 1.75 \times 10^{-9}$$

$$x = 4.2 \times 10^{-5}$$

This result is the same as that from the quadratic formula.

In general, we can make this simplification when the initial concentration of acid (or base) is more than 400 times the value of K_a (or K_b). That is,

if $C > 400 \times K_a$,   then estimate $C - x = C$

where C is the initial concentration of acid (or base).

In Example 1, $0.050 > 400 \times (3.5 \times 10^{-8})$, so we can estimate that $0.050 - x = 0.050$.

Exercise Two

Calculate the pH of a 0.50 M aqueous solution of ammonia, NH_3, at 25°C. For NH_3, $K_b = 1.8 \times 10^{-5}$ at 25°C.

CALCULATING THE VALUES OF K_a AND K_b

Example 2

A 0.30 M aqueous solution of aniline, $C_6H_5NH_2$, has a pH of 9.03 at 25°C. Calculate K_b for $C_6H_5NH_2$ at 25°C.

	$C_6H_5NH_2$	+	H_2O	$\rightleftharpoons$	$C_6H_5NH_3^+$	+	OH^-
initial conc.	0.30		–		0		0
conc. change	$-x$		–		$+x$		$+x$
equil. conc.	0.30–x		–		x		x

$$K_b = \frac{[C_6H_5NH_3^+][OH^-]}{[C_6H_5NH_2]}$$

The pH of the solution is 9.03. From this, the concentration of OH^- can be determined.

$$pH + pOH = 14.00$$

$$pOH = 14.00 - pH = 14.00 - 9.03 = 4.97$$

$$pOH = -\log[OH^-] = 4.97$$

$$\log[OH^-] = -4.97$$

$$[OH^-] = 10^{-4.97}$$

$$= 1.07 \times 10^{-5} \text{ M}$$

Thus, the equilibrium concentrations are:

$$[C_6H_5NH_2] = 0.30 - 1.07 \times 10^{-5} = 0.30 \text{ M}$$
$$[C_6H_5NH_3^+] = 1.07 \times 10^{-5} \text{ M}$$
$$[OH^-] = 1.07 \times 10^{-5} \text{ M}$$

$$K_b = \frac{[C_6H_5NH_3^+][OH^-]}{[C_6H_5NH_2]}$$

$$= \frac{(1.07 \times 10^{-5})(1.07 \times 10^{-5})}{0.30}$$

$$= 3.8 \times 10^{-10}$$

Exercise Three

A 0.20 M aqueous solution of hypochlorous acid, HOCl, has a pH of 4.08 at 25°C. Calculate K_a for HOCl at 25°C.

CALCULATING THE PERCENT IONIZATION

$$\textbf{percent ionization} = \frac{\text{concentration of ionized acid (or base)}}{\text{total concentration of acid (or base)}} \times 100\%$$

For a weak acid this is

$$\text{percent ionization} = \frac{\text{conc of anion}}{\text{conc of anion} + \text{conc of molecule}} \times 100\%$$

$$= \frac{[A^-]}{[A^-] + [HA]} \times 100\% = \frac{[A^-]}{C} \times 100\%$$

The total concentration of acid is $[A^-] + [HA]$, which is equal to the initial (or nominal) acid concentration, C.

Example 3

Calculate the percent ionization of nitrous acid, HNO_2, at equilibrium in a 1.0 M aqueous solution at 25°C. For HNO_2, $K_a = 4.0 \times 10^{-4}$.

	HNO_2	$\rightleftharpoons$	H^+	$+$	NO_2^-
initial conc.	1.0		0		0
conc. change	$-x$		$+x$		$+x$
equil. conc.	$1.0-x$		x		x

$$K_a = \frac{[H^+][NO_2^-]}{[HNO_2]} = 4.0 \times 10^{-4}$$

Test to see if the calculation can be simplified:

$$400 \times K_a = 400 \times (4.0 \times 10^{-4}) = 0.16$$

and 0.16 is less than the initial concentration, 1.0 M.

Therefore, estimate $1.0 - x = 1.0$.

$$K_a = \frac{[H^+][NO_2^-]}{[HNO_2]} = \frac{(x)(x)}{(1.0)} = 4.0 \times 10^{-4}$$

$$x^2 = 4.0 \times 10^{-4}$$

$$x = 2.0 \times 10^{-2}$$

$$[NO_2^-] = x = 2.0 \times 10^{-2} \text{ M}$$

Then, the percent ionization is

$$\frac{[A^-]}{C} \times 100\% = \frac{2.0 \times 10^{-2}}{1.0} \times 100\% = 2.0\%$$

This means that 2.0% of the HNO_2 molecules in a 1.0 M solution have been converted to ions.

Exercise Four

Calculate the percent ionization of methylamine, CH_3NH_2, at equilibrium in a 0.250 M aqueous solution at 25°C. For CH_3NH_2, $K_b = 4.38 \times 10^{-4}$ at 25°C.

CALCULATING THE INITIAL CONCENTRATION FROM K_a OR K_b

Example 4

Calculate the molarity of ammonia, NH_3, in an aqueous solution whose pH is 11.48 at 25°C. For NH_3, $K_b = 1.8 \times 10^{-5}$ at 25°C.

In this case, we are looking for the initial concentration of NH_3, so in the table below, we use y to represent this concentration.

	NH_3	+	H_2O	$\rightleftharpoons$	NH_4^+	+	OH^-
initial conc.	y		–		0		0
conc. change	$-x$		–		$+x$		$+x$
equil. conc.	$y - x$		–		x		x

$$K_b = \frac{[NH_4^+][OH^-]}{[NH_3]} = 1.8 \times 10^{-5}$$

The pH of the solution is 11.48.

Therefore, $pOH = 14.00 - pH = 14.00 - 11.48 = 2.52$

and $[OH^-] = 10^{-2.52} = 3.0 \times 10^{-3}$.

Therefore, $x = 3.0 \times 10^{-3}$, and the equilibrium concentrations are:

$[OH^-] = x = 3.0 \times 10^{-3}$

$[NH_4^+] = x = 3.0 \times 10^{-3}$

$[NH_3] = y - x = y - 3.0 \times 10^{-3}$

Then,

$$K_b = \frac{[NH_4^+][OH^-]}{[NH_3]} = \frac{(3.0 \times 10^{-3})(3.0 \times 10^{-3})}{(y - 3.0 \times 10^{-3})} = 1.8 \times 10^{-5}$$

$$(3.0 \times 10^{-3})^2 = (1.8 \times 10^{-5})(y - 3.0 \times 10^{-3})$$

$$9.0 \times 10^{-6} = 1.8 \times 10^{-5} y - 5.4 \times 10^{-8}$$

$$1.8 \times 10^{-5} y = 9.0 \times 10^{-6} + 5.4 \times 10^{-8} = 9.054 \times 10^{-6}$$

$$y = \frac{9.054 \times 10^{-6}}{1.8 \times 10^{-5}} = 0.50$$

Therefore, the initial concentration of NH_3 is 0.50 M.

Exercise Five

An aqueous solution of acetic acid, $HC_2H_3O_2$, has a pH of 2.87 at 25°C. Calculate the molarity of this solution. For $HC_2H_3O_2$, $K_a = 1.8 \times 10^{-5}$ at 25°C.

IONIZATION OF DIPROTIC ACIDS

Some acids are able to release more than one hydrogen ion per molecule of acid. Acids that release two hydrogen ions per molecule are called **diprotic** acids.

The ionization of diprotic acids takes place in steps. For example, carbonic acid is a diprotic acid.

step 1: $H_2CO_3(aq) \rightleftharpoons H^+(aq) + HCO_3^-(aq)$ $K_{a1} = 4.45 \times 10^{-7}$

step 2: $HCO_3^-(aq) \rightleftharpoons H^+(aq) + CO_3^{2-}(aq)$ $K_{a2} = 5.61 \times 10^{-11}$

Each step has an acid ionization constant, K_a, labeled with subscript 1 or 2 to indicate the step with which it is associated.

Example 5

What are the concentrations of H^+, HCO_3^-, and CO_3^{2-} in a 0.100 M solution of carbonic acid?

All of the HCO_3^- is formed in the first step, the CO_3^{2-} is formed only in the second step, but H^+ is formed in both steps. For a diprotic acid, deal with each step sequentially. For the first step:

	H_2CO_3	$\rightleftharpoons$	H^+	$+$	HCO_3^-
initial conc.	0.100		0		0
conc. change	$-x$		$+x$		$+x$
equil. conc.	$0.100-x$		x		x

$$K_{a1} = \frac{[H^+][HCO_3^-]}{[H_2CO_3]} = 4.45 \times 10^{-7}$$

Test to see if the calculation can be simplified:

$$400 \times K_{a1} = 400 \times (4.45 \times 10^{-7}) = 0.00018$$

and 0.00018 is less than the initial concentration, 0.100 M.

Therefore, estimate $0.100 - x = 0.100$.

$$K_{a1} = \frac{[H^+][HCO_3^-]}{[H_2CO_3]} = \frac{(x)(x)}{(0.100)} = 4.45 \times 10^{-8}$$

$$x^2 = 4.45 \times 10^{-8}$$

$$x = 2.11 \times 10^{-4}$$

$$[HCO_3^-] = x = 2.11 \times 10^{-4} \text{ M}$$

The concentration of CO_3^{2-} is determined from the second step.

$$HCO_3^- \rightleftharpoons H^+ + CO_3^{2-}$$

initial conc.

The initial concentration of HCO_3^- for step 2 is that produced in step 1. Similarly the initial $[H^+]$ is that from step 1, which is the same as the $[HCO_3^-]$ produced in step 1.

	HCO_3^-	$\rightleftharpoons$	H^+	$+$	CO_3^{2-}
initial conc.	2.11×10^{-4}		2.11×10^{-4}		0
conc. change	$-x$		$+x$		$+x$
equil. conc.	$2.11 \times 10^{-4} - x$		$2.11 \times 10^{-4} + x$		x

Test to see if the calculation can be simplified:

$$400 \times K_{a2} = 400 \times (5.61 \times 10^{-11}) = 2.2 \times 10^{-8}$$

and 2.2×10^{-8} is less than the initial concentration in step 2, namely 2.11×10^{-4} M.

Therefore, estimate $2.11 \times 10^{-4} - x = 2.11 \times 10^{-4}$.
For the same reason, $2.11 \times 10^{-4} + x = 2.11 \times 10^{-4}$.

$$K_{a2} = \frac{[H^+][CO_3^{2-}]}{[HCO_3^-]} = \frac{(2.11 \times 10^{-4})(x)}{(2.11 \times 10^{-4})} = 5.61 \times 10^{-11}$$

$$x = 5.61 \times 10^{-11}$$

$$[CO_3^{2-}] = x = 5.61 \times 10^{-11} \text{ M}$$

Because H^+ is formed in both steps, its concentration is the sum of what is produced in the two steps:

$$[H^+] = 2.11 \times 10^{-4} \text{ M} + 5.61 \times 10^{-11} \text{ M} = 2.11 \times 10^{-4} \text{ M}$$

For most diprotic acids, the second ionization constant is much smaller than the first. Because of this, virtually all H^+ is formed in the first ionization step.

Summary of results for diprotic acids:

1. The concentration of the first anion is calculated using the first ionization step.

2. The concentration of the dianion is equal to the value of K_{a2}.

3. The concentration of H^+ is equal to the concentration of the first anion.

Exercise Six

Vitamin C is ascorbic acid, $H_2C_6H_6O_6$, which is a diprotic acid.

$$H_2C_6H_6O_6(aq) \rightleftharpoons H^+(aq) + HC_6H_6O_6^-(aq) \qquad K_{a1} = 7.94 \times 10^{-5}$$

$$HC_6H_6O_6^-(aq) \rightleftharpoons H^+(aq) + C_6H_6O_6^{2-}(aq) \qquad K_{a2} = 1.62 \times 10^{-12}$$

What are the concentrations of H^+, $HC_6H_6O_6^-$, and $C_6H_6O_6^{2-}$ in a 0.085 M solution of Vitamin C?

PROBLEMS

1. Arsenious acid, H_3AsO_3, ionizes according to the equation

$$H_3AsO_3(aq) \rightleftharpoons H^+(aq) + H_2AsO_3^-(aq)$$

The value of the ionization constant for this equilibrium 6.0×10^{-10} at 25°C. For a 0.60 M solution of H_3AsO_3, calculate the pH and the percent ionization.

2. A 0.100 M aqueous solution of propanoic acid, $HC_3H_5O_2$ has a pOH of 12.065 at 25°C. Calculate the value of K_a for $HC_3H_5O_2$ at 25°C.

3. Hydrazoic acid, HN_3, ionizes according to the equation

$$HN_3(aq) \rightleftharpoons H^+(aq) + N_3^-(aq)$$

An aqueous solution of HN_3 has a pH of 3.80 at 25°C. Calculate the molarity of this solution. For HN_3, $K_a = 2.6 \times 10^{-5}$ at 25°C.

4. A 0.010 M aqueous solution of iodic acid, HIO_3, is 95% ionized at 25°C. Calculate the value of K_a for HIO_3 at 25°C.

5. Two solutions both have a pH of 3.47. One solution is HCl(aq). The other solution is formic acid, $HCHO_2$(aq), which is a weak acid with $K_a = 1.77 \times 10^{-4}$. What is the molarity of the HCl solution and of the formic acid solution?

6. Oxalic acid is a diprotic acid for which $K_{a1} = 5.90 \times 10^{-2}$ and $K_{a2} = 6.40 \times 10^{-5}$. What is the pH of a 0.10 M solution of oxalic acid, $H_2C_2O_4$? What are the concentrations of $HC_2O_4^-$ and $C_2O_4^{2-}$ in this solution?

NOTE

This is an explanation of the simplifying approximation presented in Example 1, namely, that $C - x = C$ if $C > 400 \times K$. It is unnecessary to learn this explanation in order to use the approximation.

This criterion has been selected to ensure that x is neglected only when it is less than 5% of the original concentration, C. That is, we want to neglect x only when $x < 0.05 \times C$, or $x/C < 0.05$. From the K expression, we have $x^2/C = K$, so $x = \sqrt{KC}$, and $x/C = \sqrt{K/C}$. When $x/C < 0.05$, then $\sqrt{K/C} < 0.05$, which is the same as $C > 400 \times K$.

HYDROLYSIS EQUILIBRIA

This lesson deals with:

1. Determining whether the aqueous solution of a salt will be acidic, basic, or neutral.
2. Demonstrating the relationship between the K_a and K_b of a conjugate acid-base pair.
3. Calculating the pH value of an aqueous salt solution.
4. Calculating the value of K_a or K_b from the pH value and concentration of an aqueous salt solution.
5. Calculating the amount of salt required to prepare a solution of given pH.

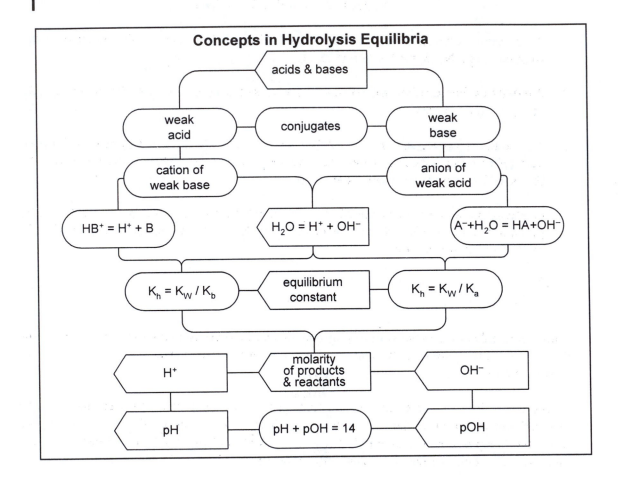

HYDROLYSIS REACTIONS

Cations of weak bases and anions of weak acids react with water. The reaction is called a **hydrolysis reaction**. (The cations of *strong* acids and anions of *strong* bases do *not* undergo hydrolysis reactions.)

Hydrolysis reaction of the anion of the weak acid HA:

$$A^- + H_2O \rightleftharpoons HA + OH^-$$

OH^- produces alkaline solution: pH > 7

Therefore, A^- is a weak base.
A^- is the conjugate base of the weak acid HA.
HA is the conjugate acid of the weak base A^-.
HA and A^- are a conjugate acid-base pair.

Hydrolysis reaction of the cation of the weak base M:

$$MH^+ + H_2O \rightleftharpoons M + H_3O^+$$

sometimes abbreviated as $MH^+ \rightleftharpoons M + H^+$

H^+ produces acidic solution: pH < 7

MH^+ is a weak acid.
MH^+ is the conjugate acid of the weak base M.
M is the conjugate base of the weak acid MH^+.
MH^+ and M are a conjugate acid-base pair.

Example 1

A. sodium propanoate, $NaC_3H_5O_2$

   $NaC_3H_5O_2$ contains Na^+ ions and $C_3H_5O_2^-$ ions.
      The base from which Na^+ ions come is NaOH. It is *a strong* base.
      The acid from which $C_3H_5O_2^-$ ions come is $HC_3H_5O_2$. It is a *weak* acid.

$NaC_3H_5O_2$ contains	Na^+	and	$C_3H_5O_2^-$
	cation of		anion of
	strong base		weak acid
	NaOH		$HC_3H_5O_2$

The anion of a weak acid undergoes a hydrolysis reaction with water, forming molecules of the weak acid and hydroxide ions. The hydroxide ions make a solution of $NaC_3H_5O_2$ basic.

$C_3H_5O_2^-$	+	H_2O	$\rightleftharpoons$	$HC_3H_5O_2$	+	OH^-
conjugate				conjugate		basic
base				acid		pH > 7

B. anilinium chloride, $C_6H_5NH_3Cl$

$C_6H_5NH_3Cl$ contains $C_6H_5NH_3^+$ and Cl^-

	cation of	anion of
	weak base	strong acid
	$C_6H_5NH_2$	HCl

$$C_6H_5NH_3^+ \ + \ H_2O \ \rightleftharpoons \ C_6H_5NH_2 \ + \ H_3O^+$$

conjugate		conjugate	acidic
acid		base	pH < 7

C. calcium nitrate, $Ca(NO_3)_2$

$Ca(NO_3)_2$ is Ca^{2+} + $2\,NO_3^-$

	cation of	anion of
	strong base	strong acid
	$Ca(OH)_2$	HNO_3

No weak acid or base. No hydrolysis. pH = 7

Exercise One

Indicate whether an aqueous solution of each of the following salts has a pH greater than, less than, or equal to 7. Also write equations for the hydrolysis reactions (if any), and identify the conjugate acid and conjugate base.

A. sodium fluoride, NaF

B. pyridinium nitrate, $(C_5H_5NH)NO_3$

C. sodium perchlorate, $NaClO_4$

D. potassium benzoate, $KC_7H_5O_2$

THE EQUILIBRIUM CONSTANT OF A HYDROLYSIS REACTION

The reaction of the anion of the weak acid with water is an equilibrium.

$$A^- + H_2O \rightleftharpoons HA + OH^-$$

It has an equilibrium constant expression.

$$K_b = \frac{[HA][OH^-]}{[A^-]}$$

(It's called K_b because A^- is a weak base.)

The K_b of A^- is related to the K_a of its conjugate acid HA. To see the relationship, multiply the K_b expression by $[H^+]/[H^+]$. (This has no effect on the value of K_b because $[H^+]/[H^+]$ is equal to 1.)

$$K_b = \frac{[HA][OH^-]}{[A^-]} \times \frac{[H^+]}{[H^+]}$$

$$= \frac{[HA]}{[H^+][A^-]} \times [H^+][OH^-]$$

$$= \frac{1}{K_a} \times K_w$$

Therefore, $K_b = \dfrac{K_w}{K_a}$.

Exercise Two

Show that $K_a = K_w/K_b$ for the hydrolysis reaction of the cation of a weak base:

$$MH^+ \rightleftharpoons M + H^+$$

pH CALCULATIONS

Example 2

Calculate the pH of a 0.50 M aqueous solution of NH_4Cl. For NH_3, $K_b = 1.8 \times 10^{-5}$.

1. Write the equation for the equilibrium reaction and its K expression.

$$NH_4^+(aq) + H_2O(l) \rightleftharpoons H_3O^+(aq) + NH_3(aq)$$

$$K_a = \frac{[H_3O^+][NH_3]}{[NH_4^+]}$$

2. Calculate the value of K_a from K_b and K_w.

$$K_a = \frac{K_w}{K_b} = \frac{1.0 \times 10^{-14}}{1.8 \times 10^{-5}} = 5.6 \times 10^{-10}$$

3. Summarize the information in a table based on the chemical equation.

	NH_4^+	+	H_2O	$\rightleftharpoons$	H_3O^+	+	NH_3
initial conc	0.50				0		0
conc change	$-x$				$+x$		$+x$
equil conc	$0.50-x$				x		x

4. Test to see if the calculation can be simplified:

$$400 \times K_a = 400 \times (5.6 \times 10^{-10}) = 2.2 \times 10^{-7}$$

This is much less than the initial concentration, 0.50 M.

Therefore, assume $0.50 - x = 0.50$.

5. Find the value of x by substituting equil conc into K expression.

$$K_a = \frac{[H_3O^+][NH_3]}{[NH_4^+]} = \frac{(x)(x)}{(0.50)} = 5.6 \times 10^{-10}$$

$$x^2 = (0.50)(5.6 \times 10^{-10})$$

$$x = 1.7 \times 10^{-5}$$

$$[H_3O^+] = x = 1.7 \times 10^{-5} \text{ M}$$

$$pH = -\log(1.7 \times 10^{-5}) = 4.77$$

Exercise Three

Calculate the pH of a 0.10 M aqueous solution of sodium acetate, $NaC_2H_3O_2$. (For acetic acid, $K_a = 1.8 \times 10^{-5}$.)

CALCULATING THE VALUE OF K

Example 3

A 0.10 M aqueous solution of pyridinium chloride, $(C_5H_5NH)Cl$, has a pH of 3.08. Calculate K_b for pyridine, C_5H_5N.

1. Write the equation for the equilibrium reaction and its K expression.

$$C_5H_5NH^+(aq) + H_2O \rightleftharpoons H_3O^+(aq) + C_5H_5N(aq)$$

$$K_a = \frac{[H_3O^+][C_5H_5N]}{[C_5H_5NH^+]}$$

2. Derive the expression for K_b from K_a and K_w.

$$K_b = \frac{K_w}{K_a}$$

$$= K_w \times \frac{[C_5H_5NH^+]}{[H_3O^+][C_5H_5N]}$$

Therefore, we need to find the equil conc of H_3O^+, C_5H_5N, and $C_5H_5NH^+$.

3. Summarize the information in a table based on the chemical equation.

	$C_5H_5NH^+$	$+$	H_2O	$\rightleftharpoons$	H_3O^+	$+$	C_5H_5N
initial conc	0.10				0		0
conc change	$-x$				$+x$		$+x$
equil conc	$0.10-x$				x		x

4. Find the value of x.

The pH = 3.08. Therefore, $[H_3O^+] = 10^{-3.08} = 8.3 \times 10^{-4}$ M.
Because $x = [H_3O^+]$, $x = 8.3 \times 10^{-4}$.

Therefore, $[C_5H_5NH^+] = 0.10 - 8.3 \times 10^{-4} = 0.10$
and $[C_5H_5N] = 8.3 \times 10^{-4}$

5. Put equil conc values into expression for K_b (step 2).

$$K_b = K_w \times \frac{[C_5H_5NH^+]}{[H_3O^+][C_5H_5N]}$$

$$= (1.0 \times 10^{-14}) \times \frac{(0.10)}{(8.4 \times 10^{-4})(8.4 \times 10^{-4})}$$

$$= 1.4 \times 10^{-9}$$

Exercise Four

A 0.045 M aqueous solution of KNO_2 has a pH of 8.0. Calculate the value of K_a for HNO_2.

CALCULATING THE CONCENTRATION OF A SOLUTION

Example 4

An aqueous solution of NaClO has a pH of 10.00. Calculate the number of moles of NaClO contained in 0.500 L of this solution. (For HClO, $K_a = 3.2 \times 10^{-8}$.)

1. Write the equation for the equilibrium reaction and its K expression.

$$ClO^-(aq) + H_2O(l) \rightleftharpoons HClO(aq) + OH^-(aq)$$

$$K_b = \frac{[HClO][OH^-]}{[ClO^-]}$$

2. Calculate the value of K_b from K_a and K_w.

$$K_b = \frac{K_w}{K_a} = \frac{1.0 \times 10^{-14}}{3.2 \times 10^{-8}} = 3.1 \times 10^{-7}$$

3. Summarize the information in a table based on the chemical equation. We do not know the initial concentration of NaClO, so we'll represent it with y.

	ClO^-	+	H_2O	$\rightleftharpoons$	$HClO$	+	OH^-
initial conc	y				0		0
conc change	$-x$				$+x$		$+x$
equil conc	$y-x$				x		x

4. Find the value of x.

The pH = 10.00. From this, we can find pOH and [OH⁻].
pOH = 14.00 − pH = 14.00 − 10.00 = 4.00
$[OH^-] = 10^{-4.00} = 1.0 \times 10^{-4}$ M.
Because $x = [OH^-]$, $x = 1.0 \times 10^{-4}$.

Therefore, $[HClO] = 1.0 \times 10^{-4}$
and $[ClO^-] = y - 1.0 \times 10^{-4}$

5. Put equil conc values into expression for K_b and solve for y.

$$K_b = \frac{[HClO][OH^-]}{[ClO^-]} = \frac{(1.0 \times 10^{-4})(1.0 \times 10^{-4})}{(y - 1.0 \times 10^{-4})} = 3.1 \times 10^{-7}$$

$$\frac{(1.0 \times 10^{-4})^2}{3.1 \times 10^{-7}} = y - 1.0 \times 10^{-4}$$

$$0.032 = y - 1.0 \times 10^{-4}$$

$$y = 0.032$$

Therefore, the initial concentration of NaClO is 0.032 M. Because we have 0.500 L of the solution, the number of moles of NaClO in the solution is

$$(0.032 \text{ mol/L})(0.500 \text{ L}) = 0.016 \text{ mol NaClO}$$

Exercise Five

Calculate the number of moles of NH_4NO_3 needed to prepare 0.500 liter of solution having a pH of 5.00. For NH_3, $K_b = 1.8 \times 10^{-5}$.

PROBLEMS

1. What is the pH of a 0.50 M solution of sodium nitrite, $NaNO_2$? For nitrous acid, $K_a = 4.5 \times 10^{-4}$.

2. A solution of 0.010 M sodium phenolate, NaC_6H_5O, has a pH of 11.00. Calculate the dissociation constant for the weak acid phenol, HC_6H_5O.

3. Calculate the pH of a 0.30 M NH_4Cl solution. For NH_3, $K_b = 1.8 \times 10^{-5}$.

4. How many grams of sodium acetate, $NaC_2H_3O_2$, must be used to make 250.0 mL of an aqueous solution having a pH of 8.50? For $HC_2H_3O_2$, $K_a = 1.8 \times 10^{-5}$.

5. What is the pH of a 0.150 M solution of potassium cyanide, KCN? Hydrocyanic acid, HCN, is a weak acid for which $K_a = 4.93 \times 10^{-10}$.

6. Sodium benzoate is the salt of benzoic acid, a weak acid with $K_a = 6.6 \times 10^{-5}$. What is the pH of a 0.25 M solution of sodium benzoate?

7. What is the molarity of a potassium acetate solution that has a pH of 9.05? (See question 4 for additional information.)

8. Potassium sorbate is used as a preservative in some foods, including cheese. It is a salt of sorbic acid, $HC_6H_7O_2$, a weak acid for which $K_a = 1.7 \times 10^{-5}$. What is the pH of a solution formed by dissolving 11.25 g of potassium sorbate in 1.75 L of solution?

BUFFER SOLUTIONS

This lesson deals with:

1. Writing equations for the buffering action that occurs when a strong acid or base is added to a buffer solution.
2. Calculating the pH value of a buffer solution.
3. Calculating the resulting pH value after a strong base is added to a weak acid buffer solution.
4. Calculating the resulting pH value after a strong acid is added to a weak acid buffer solution.
5. Calculating the resulting pH when a strong acid or base is added to a weak base buffer solution.

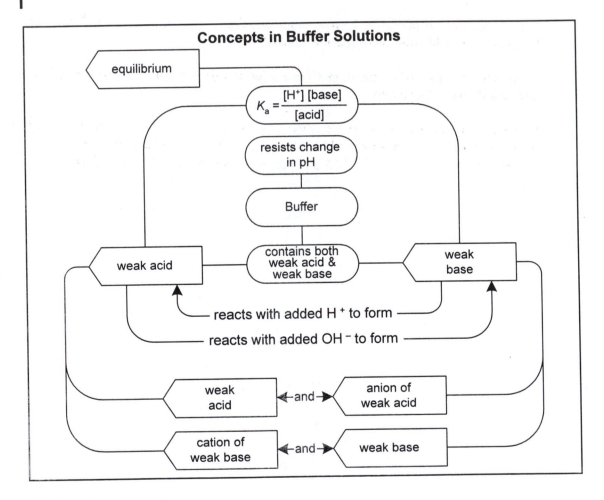

BUFFER SOLUTIONS

A buffer solution is a solution that resists changes in its pH when an acid or a base is added to it.

Two types of mixtures form buffer solutions:
a weak acid and a salt of the weak acid
a weak base and a salt of the weak base

BUFFERING ACTION

For a buffer containing a weak acid and a salt of the weak acid.
weak acid: HA
salt of weak acid: NaA (A^-)

This buffer solution contains a reaction at equilibrium.
$$HA \rightleftharpoons H^+ + A^-$$
This buffer solution contains significant amounts of both HA and A^-.

What happens if H^+ is added (e.g., from HCl)?
A^- reacts with the added H^+.
$$H^+ + A^- \longrightarrow HA$$
This is buffering action. Most of the added H^+ is consumed, so the pH changes only slightly.

What happens if OH^- is added (e.g., from NaOH)?
HA reacts with the added OH^-.
$$OH^- + HA \longrightarrow H_2O + A^-$$
This is buffering action. Most of the added OH^- is consumed, so the pH changes only slightly.

For a buffer containing a weak base and a salt of the weak base.
weak base: M
salt of weak base: (MH)Cl (MH^+)

This buffer solution contains a reaction at equilibrium.
$$M + H_2O \rightleftharpoons MH^+ + OH^-$$
This buffer solution contains significant amounts of both M and MH^+.

What happens if H^+ is added (e.g., from HCl)?
M reacts with the added H^+.
$$H^+ + M \longrightarrow MH^+$$
This is buffering action. Most of the added H^+ is consumed, so the pH changes only slightly.

What happens if OH^- is added (e.g., from NaOH)?
MH^+ reacts with the added OH^-.
$$OH^- + MH^+ \longrightarrow H_2O + M$$
This is buffering action. Most of the added OH^- is consumed, so the pH changes only slightly.

Exercise One

A. A buffer solution contains $HC_2H_3O_2$ and $NaC_2H_3O_2$.

Write the equation for the buffering action that occurs when HCl is added to this buffer solution.

Write the equation for the buffering action that occurs when NaOH is added to this buffer solution.

B. Another buffer solution contains NH_3 and NH_4Cl.

Write the equation for the buffering action that occurs when HCl is added to this buffer solution.

Write the equation for the buffering action that occurs when NaOH is added to this buffer solution.

CALCULATING THE pH OF A BUFFER SOLUTION

Example 1

Calculate the pH of a buffer solution prepared by dissolving 0.10 mole of cyanic acid, HCNO, and 0.50 mole of sodium cyanate, NaCNO, in enough water to make 0.500 liter of solution. For HCNO, $K_a = 2.0 \times 10^{-4}$ at 25°C.

1. Write the equation for the equilibrium reaction and its K expression.

$$HCNO(aq) \rightleftharpoons H^+(aq) + CNO^-(aq)$$

$$K_a = \frac{[H^+][CNO^-]}{[HCNO]} = 2.0 \times 10^{-4}$$

2. Summarize the information in a table based on the chemical equation.

	HCNO	$\rightleftharpoons$	H^+	+	CNO^-
initial moles	0.10		0		0.50
initial conc	0.20		0		1.0
conc change	$-x$		$+x$		$+x$
equil conc	$0.20-x$		x		$1.0+x$

3. Test to see if the calculation can be simplified:

$$400 \times K_a = 400 \times (2.0 \times 10^{-4}) = 0.08$$

0.08 is less than both 0.20 and 1.0.

Therefore, assume $0.20 - x = 0.20$ and $1.0 + x = 1.0$.

4. Find the value of x by substituting equilibrium concentrations into K expression.

$$K_a = \frac{[H^+][CNO^-]}{[HCNO]} = \frac{(x)(1.0)}{(0.20)} = 2.0 \times 10^{-4}$$

$$x = \frac{0.20}{1.0} \times (2.0 \times 10^{-4})$$

$$x = 4.0 \times 10^{-5}$$

$$[H^+] = x = 4.0 \times 10^{-5} \text{ M}$$

$$pH = -\log(4.0 \times 10^{-5}) = 4.40$$

Exercise Two

A. Calculate the pH of a buffer solution prepared by dissolving 0.400 mole of formic acid, $HCHO_2$, and 0.100 mole of sodium formate, $NaCHO_2$, in 0.250 liter of solution. For formic acid, $K_a = 1.77 \times 10^{-4}$ at 25°C.

B. Calculate the pH of a buffer solution containing 0.20 mole of NH_3 and 0.15 mole of NH_4Cl in 0.500 liter of solution. For NH_3, $K_b = 1.8 \times 10^{-5}$ at 25°C.

CALCULATIONS INVOLVING BUFFERING ACTION

Adding a Base to a Weak Acid Buffer Solution

Example 2

Calculate the pH of the buffer solution in Example 1 after 0.020 mole of NaOH has been added to it.

1. Write the equation for the equilibrium reaction and its K expression.

$$HCNO(aq) \rightleftharpoons H^+(aq) + CNO^-(aq)$$

$$K_a = \frac{[H^+][CNO^-]}{[HCNO]} = 2.0 \times 10^{-4}$$

2. Summarize the information in a table based on the chemical equation.

	HCNO $\rightleftharpoons$	H^+ +	CNO^-
initial conc	0.20	0	1.0

Now take into account the buffering action that results when 0.020 mole of NaOH is added.

$$OH^- \quad + \quad HCNO \quad \longrightarrow \quad CNO^- \quad + \quad H_2O$$

0.020 mol added	0.020 mol consumed	0.020 mol formed	

Therefore, as a result of buffering action,

the concentration of HCNO decreases by $\dfrac{0.020 \text{ mol}}{0.500 \text{ L}} = 0.040$ mol/L

and

the concentration of CNO^- increases by $\dfrac{0.020 \text{ mol}}{0.500 \text{ L}} = 0.040$ mol/L.

Put these results of buffering action into the table.

	HCNO	$\rightleftharpoons$	H^+	+	CNO^-
initial conc	0.20		0		1.0
buff action	−0.040		0		+0.040
conc change	−x		+x		+x
equil conc	0.16−x		x		1.04+x

3. Simplifications can still be made.

    Assume $0.16 - x = 0.16$ and $1.04 + x = 1.04$.

4. Find the value of x by substituting equilibrium concentrations into K expression.

$$K_a = \frac{[H^+][CNO^-]}{[HCNO]} = \frac{(x)(1.04)}{(0.16)} = 2.0 \times 10^{-4}$$

$$x = \frac{0.16}{1.04} \times (2.0 \times 10^{-4})$$

$$x = 3.1 \times 10^{-5}$$

$$[H^+] = x = 3.1 \times 10^{-5} \text{ M}$$

$$pH = -\log(3.1 \times 10^{-5}) = 4.51$$

Exercise Three

A buffer solution contains 0.400 mole of formic acid, $HCHO_2$, and 0.100 mole of sodium formate, $NaCHO_2$, in 1.00 liter of solution. Calculate the pH of this buffer solution after 0.100 mole of NaOH is added. For $HCHO_2$, $K_a = 1.77 \times 10^{-4}$ at 25°C.

CALCULATIONS INVOLVING BUFFERING ACTION

Adding an Acid to a Weak Acid Buffer Solution

Example 3

Calculate the pH of the buffer solution from Example 1 after 0.030 mole of HCl is added.

Take into account the buffering action that results when 0.030 mole of HCl is added.

$$H^+ \quad + \quad CNO^- \quad \longrightarrow \quad HCNO$$

0.030 mol	0.030 mol	0.030 mol
added	consumed	formed

Therefore, as a result of buffering action,

the concentration of CNO^- decreases by $\dfrac{0.030 \text{ mol}}{0.500 \text{ L}} = 0.060$ mol/L

and

the concentration of HCNO increases by $\dfrac{0.030 \text{ mol}}{0.500 \text{ L}} = 0.060$ mol/L.

Put these results of buffering action into the table.

	HCNO	$\rightleftharpoons$	H^+	+	CNO^-
initial conc	0.20		0		1.0
buff action	+0.060		0		−0.060
conc change	−x		+x		+x
equil conc	0.26−x		x		0.94+x

Make simplifications:

Assume $0.26 - x = 0.26$ and $0.94 + x = 0.94$.

Find the value of x by substituting equilibrium concentrations into K expression.

$$K_a = \frac{[H^+][CNO^-]}{[HCNO]} = \frac{(x)(0.94)}{(0.26)} = 2.0 \times 10^{-4}$$

$$x = \frac{0.26}{0.94} \times (2.0 \times 10^{-4})$$

$$x = 5.5 \times 10^{-5}$$

$$pH = -\log(5.5 \times 10^{-5}) = 4.26$$

Exercise Four

A buffer solution contains 0.400 mole of formic acid, $HCHO_2$, and 0.100 mole of sodium formate, $NaCHO_2$, in 1.00 liter of solution. Calculate the pH of this solution after 0.050 mole of HCl is added. For $HCHO_2$, $K_a = 1.77 \times 10^{-4}$ at 25°C.

CALCULATIONS INVOLVING BUFFERING ACTION

Buffering Action in a Weak Base Buffer Solution

Example 4

A buffer solution is prepared by dissolving 0.100 mole of methylamine, CH_3NH_2, and 0.0600 mole of methylammonium chloride, CH_3NH_3Cl, in 0.250 liter of solution. Calculate the pH of this solution after 0.0200 mole of HCl has been added. For CH_3NH_2, $K_b = 4.38 \times 10^{-4}$.

1. Write the equation for the equilibrium reaction and its K expression.

$$CH_3NH_2(aq) + H_2O(l) \rightleftharpoons CH_3NH_3^+(aq) + OH^-(aq)$$

$$K_b = \frac{[CH_3NH_3^+][OH^-]}{[CH_3NH_2]} = 4.38 \times 10^{-4}$$

2. Summarize the information in a table based on the chemical equation. The volume of the solution is 0.250 L.

	CH_3NH_2	+ H_2O $\rightleftharpoons$	$CH_3NH_3^+$	+ OH^-
initial moles	0.100		0.0600	
initial conc	0.400		0.240	

Take into account the buffering action. The added H^+ is consumed by CH_3NH_2.

	H^+	+ CH_3NH_2	$\longrightarrow$ $CH_3NH_3^+$
	0.020 mol	0.020 mol	0.020 mol
	added	consumed	formed

Therefore, as a result of buffering action,

the concentration of CH_3NH_2 decreases by $\dfrac{0.0200 \text{ mol}}{0.250 \text{ L}} = 0.0800$ mol/L

and

the concentration of $CH_3NH_3^+$ increases by $\dfrac{0.0200 \text{ mol}}{0.250 \text{ L}} = 0.0800$ mol/L.

	CH_3NH_2	+ H_2O $\rightleftharpoons$	$CH_3NH_3^+$	+ OH^-
initial conc	0.400		0.240	0
buff action	−0.0800		+0.0800	
conc change	−x		+x	+x
at equil	0.320−x		0.320+x	x

3. Simplifications:

$$[CH_3NH_2] = 0.320 - x = 0.320$$
$$[CH_3NH_3^+] = 0.320 + x = 0.320$$
$$[OH^-] = x$$

4. Solve for x and find pH.

$$K_b = \frac{[CH_3NH_3^+][OH^-]}{[CH_3NH_2]} = \frac{(0.32)(x)}{(0.32)} = 4.38 \times 10^{-4}$$

$$x = \frac{0.32}{0.32} \times (4.38 \times 10^{-4})$$

$$x = 4.38 \times 10^{-4} = [OH^-]$$

$$pOH = -\log(4.38 \times 10^{-4}) = 3.358$$

$$pH = 14.000 - pOH = 14.000 - 3.358 = 10.642$$

Exercise Five

A. A buffer solution contains 0.20 mole NH_3 and 0.60 mole NH_4Cl in 750 mL of solution. Calculate the pH of this buffer solution after 0.10 mole of NaOH is added to it. For NH_3, $K_b = 1.8 \times 10^{-5}$ at 25°C.

B. One liter of a buffer solution is prepared so that it contains 0.10 mole of NH_3 and 0.10 mole of NH_4Cl. When one mole of NaOH is added to this solution, the pH changes drastically. Why did the buffer solution fail to maintain its pH?

PROBLEMS

1. a. Calculate the pH of a solution made by adding 0.50 mole of nitrous acid, HNO_2, and 0.10 mole of potassium nitrite, KNO_2, to enough water to make 1.0 liter of solution. The K_a for HNO_2 is 4.5×10^{-4} at 25°C.
   b. Calculate the pH of this solution after 0.010 mole of HCl is added to it.

2. Ethyl amine, $C_2H_5NH_2$, ionizes in aqueous solution according to the equation

$$C_2H_5NH_2 + H_2O \rightleftharpoons C_2H_5NH_3^+ + OH^-$$

   The value of K_b for this reaction is 5.6×10^{-4} at 25°C.
   a. Calculate the pH of a solution made by adding 0.20 mole of $C_2H_5NH_2$ and 0.10 mole of ethylammonium chloride, $(C_2H_5NH_3)Cl$, to enough water to make 0.50 liter of solution.
   b. Calculate the pH of this solution after 0.010 mole of HCl has been added to it.

SOLUBILITY EQUILIBRIA

This lesson deals with:

1. Writing the solubility product expression for the dissolution of a sparingly soluble salt.
2. Calculating the K_{sp} value from the molar solubility of a salt.
3. Calculating the molar solubility of a salt from its K_{sp}.
4. Calculating the solubility of a salt in a solution containing a common ion.
5. Determining whether a precipitate will form when two solutions are mixed.

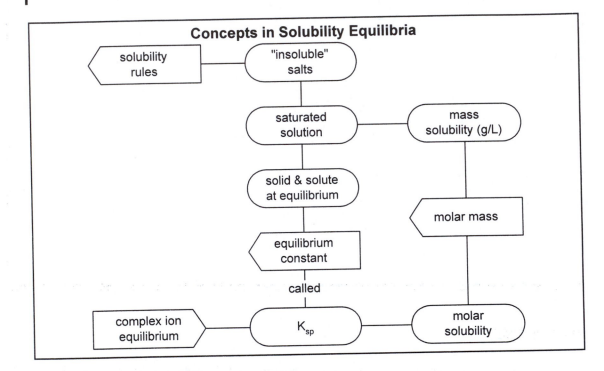

THE SOLUBILITY PRODUCT (K_{sp}) EXPRESSION

When a salt dissolves in water, its ions separate into aqueous ions.

$$AgCl(s) \rightleftharpoons Ag^+(aq) + Cl^-(aq)$$

$$Cu_3(PO_4)_2(s) \rightleftharpoons 3\,Cu^{2+}(aq) + 2\,PO_4^{3-}(aq)$$

Frequently, such an equation is written without the state indications.

$$Cu_3(PO_4)_2 \rightleftharpoons 3\,Cu^{2+} + 2\,PO_4^{3-}$$

When solid is added to water in an amount greater than can dissolve, the excess remains undissolved in the mixture. An equilibrium is established between excess solid and ions in solution.

As for any chemical equilibrium, we may write an equilibrium constant expression in the usual manner.

For AgCl, $K_{sp} = [Ag^+][Cl^-]$

For $Cu_3(PO_4)_2$, $K_{sp} = [Cu^{2+}]^3 [PO_4^{3-}]^2$

Remember that solids are not included in an equilibrium constant expression.
The equilibrium constant for a solubility equilibrium is called a **solubility product**, and it is represented by K_{sp}.

Exercise One

Write the chemical equation for the solubility equilibrium of each of the following salts, and write the solubility product expression for each.

A. $MgCO_3$

B. $Cu(OH)_2$

C. Cu_2S

D. Bi_2S_3

CALCULATING THE VALUE OF K_{sp}

Example 1

At 25°C, only 1.0×10^{-5} mole of $BaSO_4$ dissolves in one liter of water. What is the value of K_{sp} for $BaSO_4$?

Write the solubility equilibrium equation and its K_{sp} expression.

$$BaSO_4 \rightleftharpoons Ba^{2+} + SO_4^{2-}$$

$$K_{sp} = [Ba^{2+}][SO_4^{2-}]$$

When 1.0×10^{-5} mol of $BaSO_4$ dissolve in 1.0 liter of water, 1.0×10^{-5} mol of Ba^{2+} ions and 1.0×10^{-5} mol of SO_4^{2-} ions are formed in solution. Then, the concentrations of the ions are

$$[Ba^{2+}] = 1.0 \times 10^{-5} \text{ M}$$

$$[SO_4^{2-}] = 1.0 \times 10^{-5} \text{ M}$$

Then,

$$K_{sp} = [Ba^{2+}] [SO_4^{2-}] = (1.0 \times 10^{-5})(1.0 \times 10^{-5}) = 1.0 \times 10^{-10}$$

In calculating the K_{sp} follow these steps:

1. Write the balanced equation for the dissolution of the salt.
2. Write the equilibrium constant expression.
3. Find the concentration in moles per liter of each ion in the equilibrium constant expression.
4. Substitute the concentrations into the equilibrium constant expression and calculate the K_{sp}.

Example 2

At a certain temperature, only 0.022 gram of Ag_2CrO_4 will dissolve in 0.500 L of water. Calculate the K_{sp} for Ag_2CrO_4.

1.
$$Ag_2CrO_4 \rightleftharpoons 2 \, Ag^+ + CrO_4^{2-}$$

2.
$$K_{sp} = [Ag^+]^2 [CrO_4^{2-}]$$

3. The solubility of Ag_2CrO_4 is 0.022 gram per one half liter. Convert this to molarity.

$$0.022 \text{ g } Ag_2CrO_4 \left(\frac{1 \text{ mol } Ag_2CrO_4}{332 \text{ g } Ag_2CrO_4} \right) = 6.6 \times 10^{-5} \text{ mol } Ag_2CrO_4$$

$$\frac{6.6 \times 10^{-5} \text{ mol } Ag_2CrO_4}{0.500 \text{ L}} = 1.3 \times 10^{-4} \text{ M } Ag_2CrO_4$$

Each mole of Ag_2CrO_4 produces two moles of Ag^+ and one mole of CrO_4^{2-}. Therefore,

$$[Ag^+] = 2 \times (1.3 \times 10^{-4}) = 2.6 \times 10^{-4}$$

$$[CrO_4^{2-}] = 1.3 \times 10^{-4}$$

4.
$$\begin{aligned} K_{sp} &= [Ag^+]^2 [CrO_4^{2-}] \\ &= (2.6 \times 10^{-4})^2 (1.3 \times 10^{-4}) \\ &= (6.8 \times 10^{-8})(1.3 \times 10^{-4}) \\ &= 8.8 \times 10^{-12} \end{aligned}$$

Exercise Two

A. Calculate the K_{sp} for SrF_2, given that only 8.9×10^{-4} moles dissolve in a liter of solution.

B. It takes 33 liters of water to dissolve 1.0 gram of Ag_2CO_3. Calculate the K_{sp} for Ag_2CO_3.

Table 1. Solubility Product Constants at 25°C

Salt	K_{sp}
Barium fluoride (BaF_2)	1.8×10^{-7}
Copper(I) sulfide (Cu_2S)	2.3×10^{-48}
Iron(III) hydroxide ($Fe(OH)_3$)	2.6×10^{-39}
Lead chloride ($PbCl_2$)	1.2×10^{-5}
Magnesium carbonate ($MgCO_3$)	6.8×10^{-6}
Silver chloride ($AgCl$)	1.8×10^{-10}
Silver phosphate (Ag_3PO_4)	8.9×10^{-17}

CALCULATING SOLUBILITY FROM THE VALUE OF K_{sp}

The **molar solubility** is the number of moles of a substance that will dissolve in one liter of water.

Example 3

Use the value of its K_{sp} from Table 1 to calculate the molar solubility of $PbCl_2$.

1.
$$PbCl_2 \rightleftharpoons Pb^{2+} + 2\ Cl^-$$

2.
$$K_{sp} = [Pb^{2+}][Cl^-]^2 = 1.2 \times 10^{-5}$$

3. Summarize the information in a table based on the chemical equation.
   Let x = molar solubility of $PbCl_2$

	$PbCl_2(s)$ $\rightleftharpoons$	Pb^{2+}	+	$2\ Cl^-$
initial conc		0		0
conc change		$+x$		$+2x$
equil conc		x		$2x$

$[Pb^{2+}] = x$
$[Cl^-] = 2x$

4. Substitute these values into the K_{sp} expression and solve for x.
$$(x)(2x)^2 = 1.2 \times 10^{-5}$$
$$4x^3 = 1.2 \times 10^{-5}$$
$$x^3 = 3.0 \times 10^{-6}$$
$$x = (3.0 \times 10^{-6})^{1/3} = 0.014$$

Therefore, the molar solubility of $PbCl_2$ is 0.014 mole/liter.

Example 4

How many grams of $PbCl_2$ will dissolve in 100.0 mL of water?

Use the molar solubility of $PbCl_2$ calculated in Example 3.

$$0.100 \text{ L} \left(\frac{0.014 \text{ mol } PbCl_2}{1 \text{ liter}} \right) = 0.0014 \text{ mol } PbCl_2$$

$$0.0014 \text{ mol } PbCl_2 \left(\frac{278 \text{ g } PbCl_2}{1 \text{ mol } PbCl_2} \right) = 0.39 \text{ g } PbCl_2$$

Exercise Three

A. Use its K_{sp} value to calculate the molar solubility of Cu_2S.

B. How many grams of $MgCO_3$ will dissolve in 20.0 liters of water?

SOLUBILITY IN SOLUTIONS — THE COMMON ION EFFECT

The solubility of a salt can be affected by other substances also in the solution.

Example 5

What is the molar solubility of AgCl in 0.10 M NaCl?

1.
$$AgCl \rightleftharpoons Ag^+ + Cl^-$$

2.
$$K_{sp} = [Ag^+][Cl^-] = 1.8 \times 10^{-10}$$

3. Let x be the molar solubility of AgCl in 0.10 M NaCl.
   The initial concentration of Cl^- is 0.10 M from NaCl.

	AgCl(s) $\rightleftharpoons$	Ag$^+$	+	Cl$^-$
initial conc		0		0.10
conc change		$+x$		$+x$
equil conc		x		$0.10+x$

4. Substitute these values into the K_{sp} expression and solve for x.
$$(x)(0.10 + x) = 1.8 \times 10^{-10}$$
$$0.10x + x^2 = 1.8 \times 10^{-10}$$
$$x^2 + 0.10x - 1.8 \times 10^{-10} = 0$$

Use the quadratic formula to solve for x.

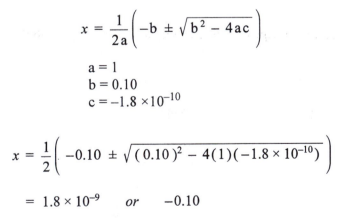

$$x = \frac{1}{2a}\left(-b \pm \sqrt{b^2 - 4ac}\right)$$

$a = 1$
$b = 0.10$
$c = -1.8 \times 10^{-10}$

$$x = \frac{1}{2}\left(-0.10 \pm \sqrt{(0.10)^2 - 4(1)(-1.8 \times 10^{-10})}\right)$$

$$= 1.8 \times 10^{-9} \quad or \quad -0.10$$

Because $x = [Ag^+]$, only the positive value is appropriate. Therefore, the molar solubility of AgCl in 0.10 M NaCl is 1.8×10^{-9} mole/liter.

Because the value of x is very small compared to 0.10, we could have ignored it in performing the calculation. We could assume that $0.10 + x = 0.10$. This would simplify the calculation:

$$(x)(0.10) = 1.8 \times 10^{-10}$$

$$x = \frac{1.8 \times 10^{-10}}{0.10}$$

$$= 1.8 \times 10^{-9}$$

This gives the same result as the quadratic formula.

We could have suspected that x would be very small, because K_{sp} is very small. In general, in solving most solubility problems, we can ignore x when it is added to or subtracted from initial concentrations, because most K_{sp} values are small.

Exercise Four

A. How many moles of $PbCl_2$ will dissolve in 1.0 liter of 0.20 M $Pb(NO_3)_2$?

B. How many grams of $MgCO_3$ will dissolve in 20.0 liters of 1.0 M Na_2CO_3?

PRECIPITATION

When two different ionic solutions are mixed, the ions in the mixture may combine to form a salt that is not very soluble. When this happens, particles of solid form in the mixture. The formation of this solid is called **precipitation**.

A solution of $Ba(NO_3)_2$ can be mixed with a solution of NaF. The salt BaF_2 is not very soluble.

$$BaF_2 \rightleftharpoons Ba^{2+} + 2\,F^-$$

$$K_{sp} = [Ba^{2+}]\,[F^-]^2 = 1.8 \times 10^{-7}$$

If the ion concentrations make the value of equilibrium constant expression greater than the K_{sp} a precipitate will form.

If the value of the equilibrium constant expression is less than the K_{sp} no precipitate will form.

Example 6

When 20.0 mL of 0.010 M $Ba(NO_3)_2$ is mixed with 30.0 mL of 0.020 M NaF, will a precipitate form?

We must find $[Ba^{2+}]$ and $[F^-]$ in the final mixture.

volume of mixture = 20.0 mL + 30.0 mL = 50.0 mL

moles Ba^{2+} = (0.010 M) (0.020 L) = 2.0×10^{-4} moles

$$[Ba^{2+}] = \frac{2.0 \times 10^{-4}\ \text{mol}}{0.050\ \text{L}} = 4.0 \times 10^{-3}\ \text{M}$$

moles F^- = (0.020 M) (0.030 L) = 6.0×10^{-4} moles

$$[F^-] = \frac{6.0 \times 10^{-4}\ \text{mol}}{0.050\ \text{L}} = 1.2 \times 10^{-2}\ \text{M}$$

Calculate the value of the K_{sp} expression for the mixture.

$$[Ba^{2+}]\,[F^-]^2 = (4.0 \times 10^{-3})\,(1.2 \times 10^{-2})^2 = 5.8 \times 10^{-7}$$

The value (5.8×10^{-7}) is greater than the value of the K_{sp} (1.8×10^{-7}). Therefore, a precipitate of BaF_2 forms in the mixture.

Exercise Five

Will a precipitate of $Fe(OH)_3$ form when one drop (0.050 mL) of 8.0 M NaOH is added to 100.0 mL of 0.20 M $FeCl_3$?

PROBLEMS

1. Milk of magnesia contains $Mg(OH)_2$. At 25°C, the K_{sp} for $Mg(OH)_2$ is 5.6×10^{-11}. What is the pH of a saturated $Mg(OH)_2$ solution?

2. The K_{sp} for AgBr is 5.4×10^{-13}. What is the molar solubility of AgBr in a 0.040 M solution of NaBr?

3. At 25°C, the molar solubility of Ag_3PO_4 is 4.3×10^{-5} moles per liter. Calculate the value of K_{sp} for Ag_3PO_4.

4. What is the molar solubility of $Cu(OH)_2$ in pure water? What is its molar solubility in a buffer solution with a pH of 10.0? For $Cu(OH)_2$, the $K_{sp} = 2.5 \times 10^{-16}$.

RATES OF CHEMICAL REACTIONS

This lesson deals with:

1. Relating the rates of disappearance and formation of the various substances in a chemical reaction.
2. Determining the general rate equation from initial rate data.
3. Determining the order of a reaction from time-concentration data.
4. Deriving equations relating the half-life to the rate constant.
5. Determining the half-life of a reaction from time-concentration data.
6. Determining the rate constant from reaction half-life.
7. Calculating the activation energy from rate constants.
8. Calculating the rate constant at one temperature from the activation energy and the rate constant at a different temperature.

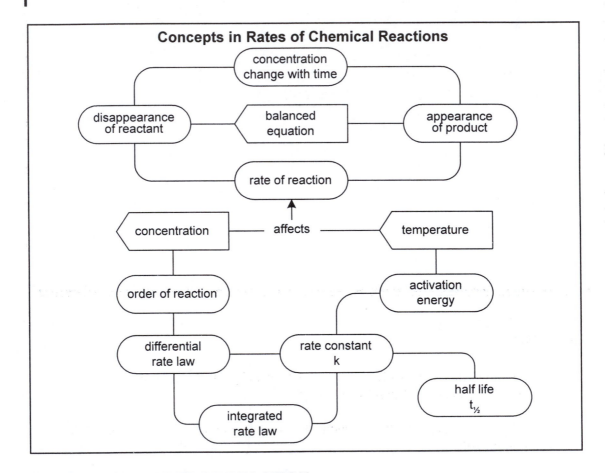

Concepts in Rates of Chemical Reactions

EXPRESSING THE RATE OF REACTION

Chemical reactions take place at various rates.

An example of a fast reaction is the one that inflates automobile safety air bags.

fast: $\qquad$ $2\,NaN_3(s) \longrightarrow 2\,Na(g) + 3\,N_2(g)$

An example of a slow reaction is the rusting of automobile fenders.

slow: $4\ Fe(s)\ +\ 3\ O_2(g) \longrightarrow 2\ Fe_2O_3(s)$

The rate of a chemical reaction is expressed as a change in concentration with time. The concentration may be of any reactant or product in the reaction.

Example 1

Express the rate of the following reaction in terms of each of its products and reactants.

$$2\ A\ +\ B \longrightarrow 3\ C\ +\ D$$

The molar concentration (molarity) of D is represented by [D].
The change in concentration is represented by d[D].
The time during which the change occurs is represented by dt.
Then,

$$\text{rate} = \frac{d[D]}{dt}$$

The rate may also be expressed in terms of C. However, according to the balanced chemical equation, for an amount of D produced in the reaction, three times as much C is produced. Therefore, C is formed 3 times as fast as D, or the rate of formation of D is one-third that of C.

$$\frac{d[D]}{dt} = \frac{1}{3}\frac{d[C]}{dt}$$

The reactant B is consumed at the same rate as D is formed, because their coefficients are the same in the balanced chemical equation. However, the concentration of B is decreasing while that of D is increasing. Therefore, the rates expressed in terms of B and D have opposite signs.

$$\frac{d[D]}{dt} = -\frac{d[B]}{dt}$$

The reactant A is consumed at twice the rate that D is formed, or D is formed at half the rate that A is consumed. This is expressed as

$$\frac{d[D]}{dt} = -\frac{1}{2}\frac{d[A]}{dt}$$

Thus,

$$\text{rate} = \frac{d[D]}{dt} = \frac{1}{3}\frac{d[C]}{dt} = -\frac{d[B]}{dt} = -\frac{1}{2}\frac{d[A]}{dt}$$

The rate is expressed as the change in concentration of reactant or product with time. The change in concentration is divided by the coefficient of the reactant or product in the balanced chemical equation. Rates expressed in terms of reactants have a minus sign before them, because the concentration change of a reactant is negative (it decreases), and by convention, the rate of a reaction is expressed as a positive number.

Exercise One

A. Express the rate of each reaction by supplying the proper coefficient and sign in each blank in each equation.

1. $2 \, Fe^{3+} + 2 \, I^- \longrightarrow 2 \, Fe^{2+} + I_2$

$$\frac{d[I_2]}{dt} = \underline{\hspace{1cm}} \frac{d[Fe^{2+}]}{dt} = \underline{\hspace{1cm}} \frac{d[I^-]}{dt} = \underline{\hspace{1cm}} \frac{d[Fe^{3+}]}{dt}$$

2. $Zn^{2+} + 4 \, NH_3 \longrightarrow Zn(NH_3)_4^{2+}$

$$-\frac{d[NH_3]}{dt} = \underline{\hspace{1cm}} \frac{d[Zn(NH_3)_4^{2+}]}{dt} = \underline{\hspace{1cm}} \frac{d[Zn^{2+}]}{dt}$$

3. $N_2 + 3 \, H_2 \longrightarrow 2 \, NH_3$

$$-\frac{d[N_2]}{dt} = \underline{\hspace{1cm}} \frac{d[NH_3]}{dt} = \underline{\hspace{1cm}} \frac{d[H_2]}{dt}$$

B. In the reaction below, suppose $d[H_2O]/dt$ is 0.42 mole/liter-min.

$$4 \, NH_3 + 7 \, O_2 \longrightarrow 4 \, NO + 6 \, H_2O$$

What are the values of the following rates?

$$\frac{d[NO]}{dt}$$

$$\frac{d[O_2]}{dt}$$

THE GENERAL RATE EQUATION

The rate of a chemical reaction depends on the concentrations of the substances in the reaction mixture. For a general chemical reaction represented by

$$A + B \longrightarrow C + D$$

the dependence of the rate on concentration can be expressed by the equation

$$\text{rate} = k[A]^p[B]^q.$$

This equation is called a general rate equation. In this equation, k is the rate constant, p is the order of A, and q is the order of B. The overall order of the reaction is $p + q$. All three, k, p, and q, must be determined experimentally.

Example 2

Using the data in the table determine the values of k, p, and q in the rate equation for the reaction below.

$$2\,NO + 2\,H_2 \longrightarrow N_2 + 2\,H_2O$$

$$\frac{d[N_2]}{dt} = k\,[NO]^p\,[H_2]^q$$

trial	initial [NO]	initial [H$_2$]	initial $d[N_2]/dt$
(1)	0.20	0.10	0.024 M min^{-1}
(2)	0.40	0.10	0.096 M min^{-1}
(3)	0.40	0.20	0.19 M min^{-1}

In trial (2) the concentration of NO was doubled from that in trial (1), while the concentration of H_2 remained constant. Doubling [NO] increased the rate by a factor of

$$\frac{0.096}{0.024} = 4$$

Because $(2)^p = (2)^2 = 4$, this indicates that the value of p is 2.

In trial (3) the concentration of H_2 was doubled from that in trial (2), while the concentration of NO remained constant. Doubling [H$_2$] increased the rate by a factor of

$$\frac{0.19}{0.096} = 2$$

Because $(2)^q = (2)^1 = 2$, this indicates that the value of q is 1.

Now, using the values of p and q, we use data from the table to determine the value of k.

$$\frac{d[N_2]}{dt} = k[NO]^p[H_2]^q$$

$$0.024 = k(0.20)^2(0.10)^1$$

$$k = \frac{0.024}{(0.20)^2(0.10)} = 6.0$$

The complete rate equation, then, is

$$\frac{d[N_2]}{dt} = 6.0[NO]^2[H_2]$$

To find the rate equation from concentration and initial rate data:
1. Write the general form of the rate equation, e.g.,

$$\frac{d[N_2]}{dt} = k[NO]^p[H_2]^q$$

2. Deduce the values of the exponents from the given data.
3. Find the value of the rate constant.

Exercise Two

Use the data in the table to find the complete rate equation for the reaction

$$C_2H_4 + HI \longrightarrow C_2H_5I$$

trial	initial $[C_2H_4]$	initial $[HI]$	initial $d[C_2H_4]/dt$
(1)	0.20	0.10	4.0 M min^{-1}
(2)	0.10	0.10	2.0 M min^{-1}
(3)	0.20	0.20	8.0 M min^{-1}

A. Write the general form of the rate equation.

$$\frac{d[C_2H_5I]}{dt} =$$

B. 1. In which two sets of data is $[C_2H_4]$ doubled while $[HI]$ is not changed?

When $[C_2H_4]$ is doubled, by what factor does the rate increase?

To what exponent must 2 be raised to give this factor? _____

Therefore, the value of the exponent of $[C_2H_4]$ is _____.

2. In which two sets of data is [HI] doubled while [C_2H_4] remains constant?

   When [HI] is doubled, by what factor does the rate increase?

   Therefore, the value of the exponent of HI is _____.

C. Because the values of the exponents p and q are known, the value of the rate constant can be determined. Write the rate equation using the proper values of the exponents.

Choose a set of data from the table and substitute it into this rate equation and solve for the value of the rate constant k.

THE INTEGRATED RATE EQUATION

It is easier to measure changes in concentrations with time than to measure the initial rate of a reaction. The integrated form of the rate equation is more suitable for use with concentration versus time data.

Consider the general reaction $X \longrightarrow Y + Z$

for which the rate equation is $\dfrac{d[Y]}{dt} = k[X]^p$

There are several cases for this general reaction.

First-Order Reaction ($p = 1$)

The rate equation:

$$\frac{d[Y]}{dt} = k[X]$$

The integrated first-order rate equation:

$$\ln[X]_t = -kt + \ln[X]_0$$

      where t is the time

               $[X]_t$ is the concentration of X at time t

      and $[X]_0$ is the initial concentration of X.

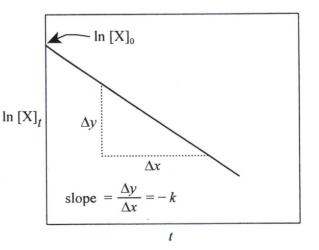

This has the form of a linear equation

$$y = m\,x + b$$

If y is plotted versus x, the result is a straight line with slope m and y-intercept b.

For the integrated first-order rate equation, if $\ln[X]_t$ is plotted versus t, the result is a straight line with slope equal to $-k$ and y-intercept equal to $\ln[X]_0$.

Second-Order Reaction ($p = 2$)

The rate equation:

$$\frac{d[Y]}{dt} = k[X]^2$$

The integrated second-order rate equation:

$$\frac{1}{[X]_t} = kt + \frac{1}{[X]_0}$$

This has the form of a linear equation.

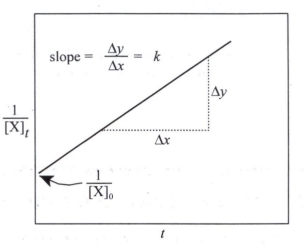

For the integrated second-order rate equation, if $1/[X]_t$ is plotted versus t, the result is a straight line with slope equal to k and y-intercept equal to $1/[X]_0$.

Zero-Order Reaction ($p = 0$)

The rate equation:

$$\frac{d[Y]}{dt} = k[X]^0 = k$$

The integrated rate law:

$$[X]_t = -kt + [X]_0$$

For the integrated zero-order rate equation, if $[X]_t$ is plotted versus t, the result is a straight line with slope equal to $-k$ and y-intercept equal to $[X]_0$.

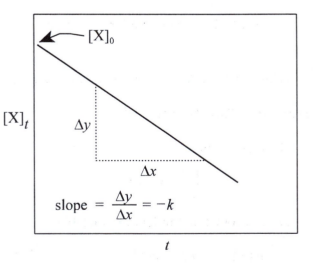

Summary

We can determine the order of a reaction by determining which plot gives a straight line. From the slope of this straight line we can determine the value of the rate constant. Table 1 contains a summary of these relationships.

Table 1. Summary of kinetic relationships for orders 0, 1, and 2.

Order	Rate Law	Integrated Form	Units of k
0	$R = k$	$[A] = -kt + [A]_0$	$M \cdot s^{-1}$
1	$R = k[A]$	$\ln[A] = -kt + \ln[A]_0$	s^{-1}
2	$R = k[A]^2$	$\dfrac{1}{[A]} = kt + \dfrac{1}{[A]_0}$	$M^{-1} \cdot s^{-1}$

Exercise Three

Fill in the blanks with the appropriate word(s). The rate of the reaction
A ⟶ B + D was determined by measuring the change in concentration of A with time.

A. If the reaction is zero order in A, a plot of _____ versus time will give a

   straight line. The rate constant may be evaluated from the slope of this line, because

   slope=_____ .

B. If the reaction is first order in A, a straight line will be obtained when _____ is

   plotted versus time. The slope of this line will determine the rate constant, because

   the slope equals _____ .

C. If the reaction is second order in A, a plot of _____ versus time gives a straight
   line. The rate constant may be evaluated from the slope, because the
   slope = _____ .

D. The following graph was obtained from the data.

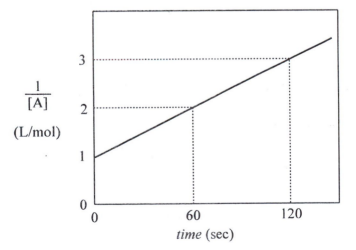

This graph indicates that the reaction is _____ order in X. The value of the slope of the line is _____ . Therefore the value of the rate constant for the reaction is _____ . The initial concentration of A in this experiment was $[A]_0 =$ _____ .

HALF-LIFE OF A FIRST-ORDER REACTION

The **half-life** is the time required for half of the initial reactant to be consumed. For a first-order reaction:

$$\ln [X]_t = -kt + \ln [X]_0$$

According to the definition of half-life, when $t = t_{1/2}$, then $[X]_t = \frac{1}{2} [X]_0$.

Therefore,

$$\ln\left(\frac{1}{2} [X_0] \right) = -kt_{1/2} + \ln [X_0]$$

$$kt_{1/2} = \ln [X_0] - \ln\left(\frac{1}{2} [X_0] \right)$$

$$kt_{1/2} = \ln\left(\frac{[X_0]}{\frac{1}{2} [X_0]} \right)$$

$$kt_{1/2} = \ln (2)$$

$$kt_{1/2} = 0.693$$

Example 3

The half-life of a certain first-order reaction is 5.0 minutes. Find the rate constant.

For a first-order reaction, $kt_{1/2} = 0.693$.

$$k (5.0 \text{ min}) = 0.693$$

$$k = \frac{0.693}{5.0 \text{ min}} = 0.14 \text{ min}^{-1}$$

Exercise Four

The rate constant for a certain first-order reaction is 0.030 min^{-1}. What is the half-life of this reaction? How long will it take for the concentration of the reactant to drop from 0.40 M to 0.20 M? From 0.20 M to 0.05 M?

THE EFFECT OF TEMPERATURE

An increase in temperature increases the rate of a chemical reaction.
A change in temperature changes the value of the rate constant.

The **Arrhenius equation**:

$$\ln k = -\frac{E_a}{R}\frac{1}{T} + \ln A$$

where k is the rate constant of a chemical reaction
E_a is the activation energy of the reaction (in units of J/mol)
R is the molar gas constant (8.314 J/mol·K)
T is the absolute temperature (Kelvin)
A is the frequency factor of the reaction

The Arrhenius equation is an equation for a straight line.
Plot $\ln k$ versus $1/T$:
the slope is $-E_a/R$ and the y-intercept is $\ln A$.

Another form of the Arrhenius equation relates E_a to the values of the rate constants at two different temperatures.

Suppose at temperature T_1 the rate constant is k_1 and at temperature T_2 the rate constant is k_2. Then,

$$\ln \frac{k_1}{k_2} = \frac{E_a}{R} \left(\frac{1}{T_2} - \frac{1}{T_1} \right)$$

Example 4

The rate constant for the reaction below was determined at two temperatures.

$$2\ N_2O_5 \longrightarrow 4\ NO_2 + O_2$$

Temp (°C)	Rate constant k	ln k	Temp (K)	$1/T$
25	4.2×10^{-5}	-10.08	298	3.356×10^{-3}
45	6.2×10^{-4}	-7.38	318	3.145×10^{-3}

Determine the value of the activation energy of this reaction.

$$\ln \frac{k_1}{k_2} = \frac{E_a}{R} \left(\frac{1}{T_2} - \frac{1}{T_1} \right)$$

Let T_1 be 25°C and T_2 be 45°C. Then, k_1 is 4.2×10^{-5} and k_2 is 6.2×10^{-4}.

We need $1/T_1$ and $1/T_2$ in Kelvin.

$$1/T_1 = 1/(273+25) = 3.356 \times 10^{-3}\ K^{-1}$$
$$1/T_2 = 1/(273+45) = 3.145 \times 10^{-3}\ K^{-1}$$

We need $\ln (k_1/k_2)$.

$$\ln (4.2 \times 10^{-5})/(6.2 \times 10^{-4}) = \ln (0.068) = -2.69$$

Then,

$$-2.69 = \frac{E_a}{R} \left(3.145 \times 10^{-3}\ K^{-1} - 3.356 \times 10^{-3}\ K^{-1} \right)$$

$$-2.69 = \frac{E_a}{R} \left(-2.11 \times 10^{-4}\ K^{-1} \right)$$

$$\frac{E_a}{R} = \frac{-2.69}{-2.11 \times 10^{-4}\ K^{-1}} = 1.27 \times 10^4\ K$$

Because R = 8.314 J/mol-K,

$$E_a = (1.27 \times 10^4\ K)\,(8.314\ J/mol\text{-}K) = 1.1 \times 10^5\ J/mol$$

Exercise Five

The rate constant for the decomposition of HBr is 4.77×10^{-8} L/mol-sec at 225°C and 1.66×10^{-4} L/mol-sec at 616°C. Find the value of the activation energy for this reaction.

CALCULATING A RATE CONSTANT FROM E_a

Example 5

The activation energy for a certain reaction is 114 kJ/mol, and the rate constant at 787 K is 9.1×10^5 L/mol·sec. What is the value of the rate constant at 650 K?

$$\ln \frac{k_1}{k_2} = \frac{E_a}{R} \left(\frac{1}{T_2} - \frac{1}{T_1} \right)$$

$R = 8.314$ J/mol-K
$E_a = 114000$ J/mol
At $T_1 = 787$ K, $k_1 = 9.1 \times 10^5$ L/mol-sec
At $T_2 = 650$ K, $k_2 = ?$

$$\ln \frac{9.1 \times 10^5 \text{ L/mol-s}}{k_2} = \frac{114000 \text{ J/mol}}{8.314 \text{ J/mol-K}} \left(\frac{1}{650 \text{ K}} - \frac{1}{787 \text{ K}} \right)$$

$$\ln \frac{9.1 \times 10^5 \text{ L/mol-s}}{k_2} = 3.67$$

$$\frac{9.1 \times 10^5 \text{ L/mol-s}}{k_2} = e^{3.67} = 39.25$$

$$k_2 = \frac{9.1 \times 10^5 \text{ L/mol-s}}{39.25} = 2.3 \times 10^4 \text{ L/mol-s}$$

Exercise Six

For the reaction

$$2 \text{ NOCl} \longrightarrow 2 \text{ NO} + \text{Cl}_2$$

the rate constant is 1.5×10^{-5} L/mol-s at 300 K. The activation energy is 90.3 kJ/mol. Calculate the value of the rate constant at 400 K.

PROBLEMS

1. For the reaction $2 \text{ NOCl} \longrightarrow 2 \text{ NO} + \text{Cl}_2$, the following data were measured at 100°C.

time (sec)	[NOCl] (mol/L)
0	0.500
10	0.357
20	0.278
30	0.227

a. What is the order of the reaction?

b. Calculate the rate constant and indicate its units.

2. In a certain first-order reaction, the concentration of reactant decreased from 1.50 M to 0.50 M in 12 minutes. Calculate the rate constant.

3. For the decomposition reaction

$$2 N_2O_5 \longrightarrow 2 N_2O_4 + O_2$$

the rate was found to be first order in $[N_2O_5]$. From the following data determine E_a of this reaction.

temp (°C)	$k \ (s^{-1})$
27	4.1×10^{-5}
37	1.6×10^{-4}
47	6.2×10^{-4}
57	2.1×10^{-3}

4. A reaction that follows a first-order rate law is found to be 50% complete after 15 minutes at 300 K and 50% complete after 3 minutes at 320 K.

    (a) What is the rate constant at each temperature?
    (b) What is the energy of activation?

5. A reaction with $E_a = 61.9$ kJ/mol has a rate constant of 0.080 $M^{-1} s^{-1}$ at 100°C. What is the value of the rate constant at 37°C?

6. Determine the order and find the value of the rate constant for the following reaction.

$$PCl_5 \longrightarrow PCl_3 + Cl_2$$

time (min)	$[PCl_5]$ (M)
0	1.00
30	0.63
60	0.40
90	0.25
120	0.16

7. Dinitrogen pentoxide, N_2O_5, decomposes by first-order process with a rate constant of $3.7 \times 10^{-5} s^{-1}$ at 298 K.

    (a) What is the half-life in hours for the decomposition of N_2O_5 at 298 K?
    (b) If the concentration of N_2O_5 starts at 2.33×10^{-2} M, what will be its concentration after 2.0 hours?
    (c) How many minutes will it take for the concentration of N_2O_5 to decrease from 0.0233 M to 0.0176 M?

8. Use the integrated form of the second-order rate equation and follow the method described on page 274 to derive the equation for the half life of a second-order reaction. Suggest a reason that this half life is less useful than that of a first-order reaction.

MECHANISMS OF CHEMICAL REACTIONS

This lesson deals with:

1. Identifying the molecularity of elementary reactions.
2. Writing the rate law equation for elementary reactions.
3. Adding the steps of a mechanism to obtain the equation for the overall reaction.
4. Identifying intermediates and catalysts from the mechanism of a reaction.
5. Writing the rate law equation for a mechanism in which the first step is the rate-determining step.
6. Writing the rate law equation for a mechanism in which the rate-determining step is not the first step.
7. Assessing a mechanism by comparing the rate law equation derived from the mechanism to the experimentally determined rate law equation for the reaction.

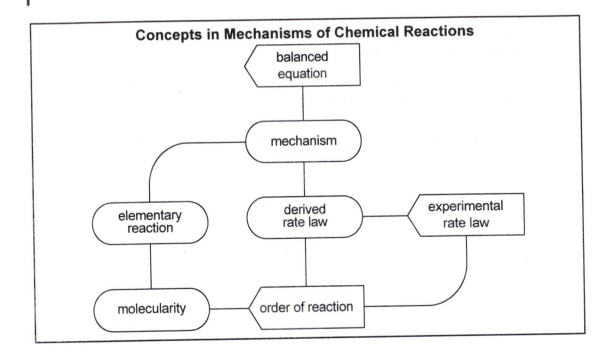

MECHANISMS AND ELEMENTARY REACTIONS

Most chemical reactions do not occur in a single step. Instead, most reactions involve a series of steps. A mechanism describes step-by-step how the atoms in the reactants are rearranged into the products. Each step in a mechanism is called an *elementary reaction*. An elementary reaction describes which particles are directly involved in each step. Chemical equations are used to represent elementary reactions. Chemical equations are also used to represent overall reactions, which indicate the products and reactants, but do not imply how the transformation of reactants into products takes place.

Recognizing elementary reactions

Whether a chemical equation represents an elementary reaction or an overall reaction depends on its context. In many cases they look exactly alike. The way to know that

$$CO + H_2O \longrightarrow CO_2 + H_2$$

represents an elementary reaction and not an overall reaction is by how it is used.

On the other hand, an equation that indicates states, such as,

$$H_3O^+(aq) + CO_3^{2-}(aq) \longrightarrow HCO_3^-(aq) + H_2O(l)$$

always represents an overall reaction. An elementary reaction indicates explicitly the particles involved in a reaction, where $H_3O^+(aq)$ represents the H_3O^+ ion and the water molecules surrounding it.

Rate laws and elementary reactions

Because an elementary reaction indicates exactly which particles are involved in the reaction, the rate law for an elementary reaction can be determined from its chemical equation.

Elementary reactions are classified as unimolecular, bimolecular, or termolecular by the number of particles involved as reactants.

Unimolecular reactions

Unimolecular reactions are elementary reactions that involve only one particle (molecule or ion) as reactant.

$$A \longrightarrow products$$

The rate of a unimolecular reaction is proportional to the concentration of A and does not depend on the concentration of the products.

$$-\frac{d[A]}{dt} = k[A]$$

This is an example of a unimolecular reaction and its rate law equation:

$$NOBr_2 \longrightarrow NO + Br_2 \qquad -\frac{d[NOBr_2]}{dt} = k[NOBr_2]$$

Bimolecular reactions

Bimolecular reactions are elementary reactions that involve two reactant particles.

$$A + B \longrightarrow products$$

The rate of a bimolecular reaction is directly proportional to the concentrations of both reactant particles and does not depend on the concentrations of the products.

$$-\frac{d[A]}{dt} = k[A][B]$$

These are examples of bimolecular reactions and their rate law equations:

$$\text{NO} + \text{O}_3 \longrightarrow \text{NO}_2 + \text{O}_2 \qquad -\frac{d\,[\text{NO}]}{d\,t} = k\,[\text{NO}]\,[\text{O}_3]$$

$$2\,\text{NO}_2 \longrightarrow \text{N}_2\text{O}_4 \qquad -\frac{d\,[\text{NO}_2]}{d\,t} = k\,[\text{NO}_2]^2$$

Termolecular reactions

Termolecular reactions are elementary reactions that involve three reactant particles.

$$\text{A} + \text{B} + \text{C} \longrightarrow \text{products}$$

The rate of a termolecular reaction is proportional to the concentrations of all three of its reactant particles.

$$-\frac{d\,[\text{A}]}{d\,t} = k\,[\text{A}]\,[\text{B}]\,[\text{C}]$$

This is an example of a termolecular elementary reaction and its rate law equation:

$$2\,\text{NO} + \text{O}_2 \longrightarrow 2\,\text{NO}_2 \qquad -\frac{d\,[\text{NO}]}{d\,t} = k\,[\text{NO}]^2\,[\text{O}_2]$$

The likelihood of three particles simultaneously coming together is much lower than that for two or a single particle reacting. Therefore, termolecular elementary reactions are seldom encountered. Elementary reactions involving four particles are even less likely.

Example 1

Indicate the molecularity and write the rate law equation for the elementary reaction whose equation is below. What is the order of the reaction?

$$2\,\text{NO}_2 + \text{Ar} \longrightarrow 2\,\text{NO} + \text{O}_2 + \text{Ar}$$

This elementary reaction involves three particles as reactants, 2 NO_2 and Ar. The equation could be written as

$$\text{NO}_2 + \text{NO}_2 + \text{Ar} \longrightarrow 2\,\text{NO} + \text{O}_2 + \text{Ar}$$

Therefore, it is a termolecular reaction.

The rate law equation will involving two NO_2 and one Ar.

$$-\frac{d\,[\text{NO}_2]}{d\,t} = k\,[\text{NO}_2]\,[\text{NO}_2]\,[\text{Ar}]$$

This can be written as

$$-\frac{d\,[\text{NO}_2]}{d\,t} = k\,[\text{NO}_2]^2\,[\text{Ar}]$$

It is a termolecular reaction, and its rate law indicates that it is third order overall.

Exercise 1

For each of the elementary reactions below, indicate its molecularity (uni-, bi-, or ter-) and write a rate law equation for it.

	molecularity	rate law equation

(a) $2\,Cl \longrightarrow Cl_2$ _____   _____

(b) $O_3 \longrightarrow O_2 + O$ _____   _____

(c) $2\,HI + O \longrightarrow I_2 + H_2O$ _____   _____

REACTION MECHANISMS

A reaction mechanism is the sequence of elementary reactions that a chemical reaction follows in converting reactants to products. The sum of all of the elementary reactions is the equation for the overall chemical reaction.

Example 2

A certain reaction is believed to take place by the following 3-step mechanism.

Step 1:	$Cl_2 \longrightarrow Cl + Cl$
Step 2:	$Cl + CO \longrightarrow COCl$
Step 3:	$COCl + Cl \longrightarrow COCl_2$

What is the overall reaction?

The overall reaction is the sum of the individual steps in the mechanism. To find it, add all the reactants on the left and all the products on the right.

$$Cl_2 + Cl + CO + COCl + Cl \longrightarrow Cl + Cl + COCl + COCl_2$$

Then delete the particles that appear on both sides.

$$Cl_2 + \cancel{Cl} + CO + \cancel{COCl} + \cancel{Cl} \longrightarrow \cancel{Cl} + \cancel{Cl} + \cancel{COCl} + COCl_2$$

$$Cl_2 + CO \longrightarrow COCl_2$$

This is the equation for the overall reaction that follows the mechanism above.

EXERCISE TWO

For each of the mechanisms below, find the overall reaction that corresponds to it.

(a)
$$NO + Cl_2 \longrightarrow NOCl_2$$
$$NOCl_2 + NO \longrightarrow NOCl + NOCl$$

(b)
$$OCl^- + H_2O \longrightarrow HOCl + OH^-$$
$$I^- + HOCl \longrightarrow HOI + Cl^-$$

(c)
$$NO + NO \longrightarrow N_2O_2$$
$$N_2O_2 + H_2 \longrightarrow N_2O + H_2O$$
$$N_2O + H_2 \longrightarrow N_2 + H_2O$$

(d)
$$H_2O + CO \longrightarrow H_2 + CO_2$$

INTERMEDIATES AND CATALYSTS

The mechanism for a reaction may contain particles that do not appear in the equation for the overall chemical reaction. There are two kinds of such particles: intermediates and catalysts.

Intermediates are particles that are generated and consumed during the course of a reaction.

Catalysts are particles that are present at the start and at the end of a reaction.

Example 3

Below is a proposed mechanism for a certain reaction. Identify any intermediates or catalysts in the mechanism.

Step 1: $ClO_2 + SO_2 \longrightarrow ClO + SO_3$
Step 2: $ClO + O_2 \longrightarrow ClO_2 + O$
Step 3: $O + SO_2 \longrightarrow SO_3$

To determine if there are intermediates or catalysts, the equation for the overall reaction is needed. Adding the steps together gives the overall equation. The equation for the overall chemical reaction is

$$2\,SO_2 + O_2 \longrightarrow 2\,SO_3$$

The particles in the mechanism that do not appear in the overall chemical equation are ClO_2, ClO, and O. These are either intermediates or catalysts.

Intermediates are generated during the reaction. Catalysts are present at the start.

If a particle appears in the mechanism for the first time on the left side of a step, that means it is present at the start of the reaction and, therefore, is a catalyst.

If it appears for the first time in the mechanism on the right side of a step, then it is generated during reaction and is an intermediate.

In this mechanism, ClO_2 appears for the first time on the left side of step 1. Therefore, it is a catalyst.

The particle ClO appears for the first time in the mechanism on the right side of step 1, therefore it is an intermediate.

The particle O appears for the first time on the right side of step 2, so it, too, is an intermediate.

Conclusion: ClO_2 and O are intermediates
                ClO_2 is a catalyst

EXERCISE THREE

Below is a proposed mechanism for a certain reaction. Identify any intermediates or catalysts in the mechanism.

Step 1: $OCl^- + H_2O \longrightarrow HOCl + OH^-$

Step 2: $I^- + HOCl \longrightarrow HOI + Cl^-$

Step 3: $HOI + OH^- \longrightarrow OI^- + H_2O$

RATE LAWS AND MECHANISMS

The rate law for an overall chemical reaction cannot be determined from the chemical equation for that reaction. However, the rate law can be derived from the mechanism of the reaction.

The rate of an overall chemical reaction is determined by the rate of the slowest step in its mechanism.

When the slowest step is the first step in the mechanism, then the rate law for that step is the rate law for the overall reaction.

Consider the reaction for which the overall equation is

$$NO_2(g) + CO(g) \longrightarrow NO(g) + CO_2(g)$$

This reaction may take place by the following mechanism.

Step 1:	$NO_2 + NO_2 \longrightarrow NO_3 + NO$
Step 2:	$NO_3 + CO \longrightarrow NO_2 + CO_2$

First step slow

The two steps of this mechanism can take place at different rates. Step 1 may be much slower than step 2.

Step 1:	$NO_2 + NO_2 \longrightarrow NO_3 + NO$	slow
Step 2:	$NO_3 + CO \longrightarrow NO_2 + CO_2$	fast

The "slow" and "fast" labels indicate the relative rates of the two steps. Both steps may be very fast, but the second is much faster than the first.

The rate law for the slow step, step 1, is

$$-\frac{d[NO_2]}{dt} = k_1 [NO_2]^2$$

In this equation, the rate constant is given a subscript 1 to indicate that it is the rate constant for step 1. Because the slow step determines the rate of the reaction, the rate equation above for the slow step is also the rate equation for the overall reaction.

Later step slow

It is also possible that the second step is slower than the first.

Step 1:	$NO_2 + NO_2 \longrightarrow NO_3 + NO$	fast
Step 2:	$NO_3 + CO \longrightarrow NO_2 + CO_2$	slow

In this case, step 2 determines the rate of the overall reaction. The rate law for step 2 is

$$-\frac{d[CO]}{dt} = k_2 [NO_3] [CO]$$

This, however, is not the rate equation for the overall reaction. This rate equation contains the concentration of an intermediate, $[NO_3]$. Because the concentration of intermediates is difficult to determine or control, it is not used in the rate equation for the overall reaction.

The concentration of intermediate can be eliminated by considering what happens in the fast first step. This fast step produces the intermediate NO_3 more rapidly than it is consumed in the slower step 2. Therefore, the concentrations of the products of the first step, NO_3 and NO, build up. As their concentrations build, the reverse reaction of step 1 begins to take place at a significant rate.

Reverse of step 1: $NO_3 + NO \longrightarrow NO_2 + NO_2$

Eventually, the rate of the reverse of step 1 increases until NO_3 is consumed at the same rate at which it is formed. The rate for the formation of NO_3 is the rate of step 1:

$$\frac{d\,[NO_3]}{dt} = k_1\,[NO_2]^2$$

The rate at which NO_3 is consumed by the reverse of step 1 is

$$-\frac{d\,[NO_3]}{dt} = k_{-1}\,[NO_3]\,[NO]$$

In this equation, k_{-1} is the rate constant for the reverse of step 1. Because these two rates are equal,

$$k_1\,[NO_2]^2 = k_{-1}\,[NO_3]\,[NO]$$

This equation can be solved for the concentration of the intermediate NO_3.

$$[NO_3] = \frac{k_1}{k_{-1}}\frac{[NO_2]^2}{[NO]}$$

This expression for $[NO_3]$ can be substituted into the rate equation for the slow step, step 2.

$$-\frac{d\,[CO]}{dt} = k_2\frac{k_1}{k_{-1}}\frac{[NO_2]^2}{[NO]}\,[CO]$$

This is a rate equation that is based on the rate equation for the slow, rate-determining step, but which does not contain intermediates, only reactant and products. The three rate constants in this equation are often combined into one, the rate constant of the overall reaction: $k = k_2k_1/k_{-1}$.

$$-\frac{d\,[CO]}{dt} = k\frac{[NO_2]^2\,[CO]}{[NO]}$$

Example 4

What is the rate equation for the overall reaction whose mechanism is below?

$$NO + Br_2 \longrightarrow NOBr_2 \qquad \text{fast}$$
$$NOBr_2 + NO \longrightarrow NOBr + NOBr \qquad \text{slow}$$

The slow step, the second step, in the mechanism determines the rate. The rate equation for the second step is

$$\frac{d\,[NOBr]}{dt} = k_2\,[NOBr_2]\,[NO]$$

To see if either $NOBr_2$ or NO is an intermediate, compare them to the overall reaction equation. The overall reaction is

$$2\,NO + Br_2 \longrightarrow 2\,NOBr$$

NO is a reactant, so its concentration may be used in the rate law.

$NOBr_2$ does not appear in the overall equation, and it first appears in the mechanism on the right side of step 1. Therefore, $NOBr_2$ is an intermediate. The rate law for the overall reaction should not be expressed in terms of $[NOBr_2]$.

While the fast first step waits on the slow second step, the reverse rate of the first step becomes equal to the forward rate.

$$k_{-1}\,[NOBr_2] = k_1\,[NO]\,[Br_2]$$

Solve this for $[NOBr_2]$.

$$[NOBr_2] = \frac{k_1}{k_{-1}}\,[NO]\,[Br_2]$$

Substitute this expression for $[NOBr_2]$ into the rate law for the slow step.

$$\frac{d\,[NOBr]}{dt} = k_2\,\frac{k_1}{k_{-1}}\,[NO]\,[Br_2]\,[NO]$$

Simplifying by consolidating the three mechanism rate constants into one: $k = k_2 k_1 / k_{-1}$.

$$\frac{d\,[NOBr]}{dt} = k\,[NO]^2\,[Br_2]$$

This is the rate law for the overall reaction described by the given mechanism.

EXERCISE FOUR

Derive the rate law for the overall reaction from each of the following mechanisms.

(a)   Step 1 $Cl_2 \longrightarrow Cl + Cl$ fast
      Step 2 $CHCl_3 + Cl \longrightarrow CCl_3 + HCl$ slow
      Step 3 $CCl_3 + Cl \longrightarrow CCl_4$ fast

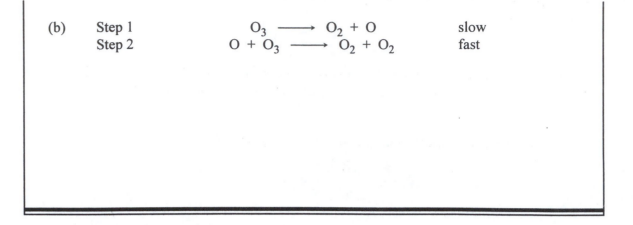

(b)   Step 1 $O_3 \longrightarrow O_2 + O$ slow
       Step 2 $O + O_3 \longrightarrow O_2 + O_2$ fast

ASSESSING THE VALIDITY OF A MECHANISM

In order to be valid, a mechanism for a reaction must yield a rate law equation that agrees with the experimentally determined rate law. If the proposed mechanism does not agree with the experimental rate law, then it is certain that the reaction does not follow the proposed mechanism. If, on the other hand, the mechanism does agree with the experimental rate equation, then the mechanism is valid, but not proven. Agreement between a rate law derived from a proposed mechanism and an experimental rate law does not prove that the reaction follows the mechanism.

Example 5

The reaction between hydrogen and iodine

$$H_2(g) + I_2(g) \longrightarrow 2\ HI(g)$$

has an experimentally-determined rate law which indicates that

$$-\frac{d[I_2]}{dt} = k[H_2][I_2]$$

Below are three proposed mechanisms. Which of these is a valid mechanism for the reaction?

Mechanism A
       Step 1 $I_2 \longrightarrow I + I$ slow
       Step 2 $I + H_2 \longrightarrow HI + H$ fast
       Step 3 $H + I \longrightarrow HI$ fast

Mechanism B
       Step 1 $H_2 + I_2 \longrightarrow 2\ HI$

Mechanism C
       Step 1 $I_2 \longrightarrow I + I$ fast
       Step 2 $I + H_2 \longrightarrow H_2I$ fast
       Step 3 $H_2I + I \longrightarrow 2\ HI$ slow

In Mechanism A, the rate determining step is the first step, so the rate law equation for this mechanism is

$$-\frac{d[I_2]}{dt} = k[I_2]$$

This does **not** agree with the experimentally determined rate law, so this mechanism is not a valid for this reaction; the reaction does not take place the way this mechanism describes.

In Mechanism B, there is only one step, and its rate equation is

$$-\frac{d[I_2]}{dt} = k[H_2][I_2]$$

This agrees with the experimental rate law, so this mechanism is valid. The reaction **may** take place in the way this mechanism describes. However, this does not prove that the reaction occurs this way.

In Mechanism C, the rate-determining step is the third step. The rate law for this step is

$$\frac{d[HI]}{dt} = k_3[H_2I][I]$$

Both H_2I and I are intermediates, so their concentrations must be eliminated from the equation before it can be compared to the experimental rate law.

From step 2 of this mechanism,

$$k_2[I][H_2] = k_{-2}[H_2I]$$

so

$$[H_2I] = \frac{k_2}{k_{-2}}[I][H_2]$$

Substituting this into the rate law for step 3 gives

$$\frac{d[HI]}{dt} = k_3\frac{k_2}{k_{-2}}[I][H_2][I]$$

or

$$\frac{d[HI]}{dt} = k_3\frac{k_2}{k_{-2}}[I]^2[H_2] \qquad \text{(equation 1)}$$

From step 1 of this mechanism,

$$k_1[I_2] = k_{-1}[I]^2$$

so

$$[I] = \sqrt{\frac{k_1}{k_{-1}}[I_2]}$$

Substituting this into the equation 1 above gives

$$\frac{d\,[\text{HI}]}{d\,t} = k_3 \frac{k_2}{k_{-2}} \left(\sqrt{\frac{k_1}{k_{-1}}\,[\text{I}_2]} \right)^2 [\text{H}_2]$$

or

$$\frac{d\,[\text{HI}]}{d\,t} = k_3 \frac{k_2}{k_{-2}} \frac{k_1}{k_{-1}}\,[\text{I}_2]\,[\text{H}_2]$$

which can be simplified by combining the rate constants into one:

$$\frac{d\,[\text{HI}]}{d\,t} = k\,[\text{I}_2]\,[\text{H}_2]$$

Therefore, Mechanism C agrees with the experimentally determined rate law equation. Mechanism C is valid for the reaction, but not proven.

Both Mechanism B and Mechanism C agree with the experimental rate equation. To determine which of these is more appropriate would require additional experimental evidence. For example, suppose iodine atoms were detected while the reaction is occurring. Mechanism B cannot explain this, while Mechanism C does. Such an observation would rule out Mechanism B for this reaction.

EXERCISE FIVE

Ozone is decomposed by reaction with nitrogen dioxide.

$$\text{O}_3(g) + 2\,\text{NO}_2(g) \longrightarrow \text{N}_2\text{O}_5(g) + \text{O}_2(g)$$

The experimental rate law for this reaction is

$$-\frac{d\,[\text{O}_3]}{d\,t} = k\,[\text{O}_3]\,[\text{NO}_2]$$

Is either of the following two proposed mechanisms consistent with the experimental rate law?

Mechanism A

Step 1	$\text{NO}_2 + \text{NO}_2 \longrightarrow \text{N}_2\text{O}_4$	fast
Step 2	$\text{N}_2\text{O}_4 + \text{O}_3 \longrightarrow \text{N}_2\text{O}_5 + \text{O}_2$	slow

Mechanism B

Step 1	$\text{NO}_2 + \text{O}_3 \longrightarrow \text{NO}_3 + \text{O}_2$	slow
Step 2	$\text{NO}_3 + \text{NO}_2 \longrightarrow \text{N}_2\text{O}_5$	fast

PROBLEMS

1. For each of the following elementary reactions, indicate the molecularity and write its rate law.

    (a) $NO + NO \longrightarrow N_2O_4$
    (b) $Cl_2 \longrightarrow Cl + Cl$
    (c) $2 NO_2 \longrightarrow NO + NO_3$
    (d) $OH + NO_2 + N_2 \longrightarrow HNO_3 + N_2$
    (e) $ClO^- + H_2O \longrightarrow HClO + OH^-$

2. The following mechanism has been proposed for a process which contributes to the destruction of the stratospheric ozone layer.

Step 1	$O_3 + NO \longrightarrow NO_2 + O_2$
Step 2	$NO_2 + O \longrightarrow NO + O_2$

    (a) What is the equation for the overall reaction?
    (b) What is the rate law equation for each step.
    (c) Is there a catalyst in the reaction? If so, identify it.
    (d) Is there an intermediate in the reaction? If so, identify it.

3. The rate law equation for the reaction
    $$Cl_2(aq) + H_2S(aq) \longrightarrow S(s) + 2 HCl(aq)$$
    is determined by experiment to be
    $$-\frac{d[Cl_2]}{dt} = k[Cl_2][H_2S]$$
    Which of the following mechanisms are consistent with the experimental rate law?

    Mechanism A
Step 1	$Cl_2 \longrightarrow Cl^+ + Cl^-$	slow
Step 2	$Cl^- + H_2S \longrightarrow HCl + HS^-$	fast
Step 3	$Cl^+ + HS^- \longrightarrow HCl + S$	fast

    Mechanism B
Step 1	$Cl_2 + H_2S \longrightarrow HCl + Cl^+ + HS^-$	slow
Step 2	$Cl^+ + HS^- \longrightarrow HCl + S$	fast

    Mechanism C
Step 1	$Cl_2 \longrightarrow Cl + Cl$	fast
Step 2	$Cl + H_2S \longrightarrow HCl + HS$	fast
Step 3	$HS + Cl \longrightarrow HCl + S$	slow

COORDINATION COMPOUNDS: COMPOSITION AND STRUCTURE

This lesson deals with:

1. Representing the composition of a coordination compound by its chemical formula.
2. Identifying the central atom, its oxidation number, coordination number and ligands.
3. Relating some properties of coordination compounds to their composition and formulas.
4. Drawing geometric structures of complex ions.
5. Naming coordination compounds.

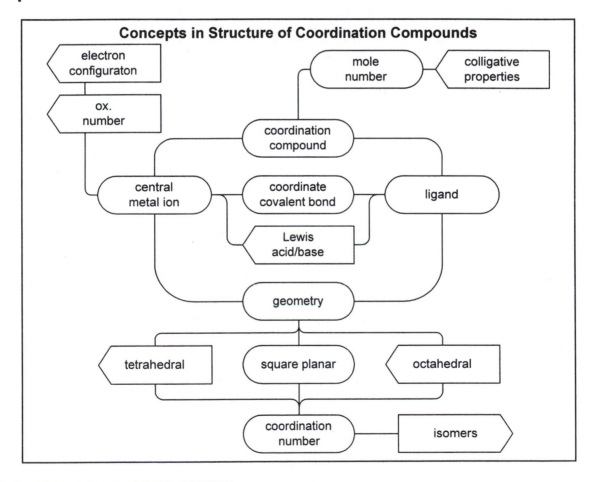

COORDINATION COMPOUNDS

A **coordination compound** contains a transition metal or transition-metal ion joined to neutral molecules or ions by coordinate covalent bonds.

A **coordinate covalent bond** forms when an atom accepts a pair of electrons from another atom.

Example 1

$[Cu(NH_3)_4]SO_4$ is a coordination compound; it contains the $[Cu(NH_3)_4]^{2+}$ ion.

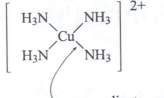

— a coordinate covalent bond

A coordinate covalent bond is formed when both electrons of the bond come from one atom.

$$Cu^{2+} \longleftarrow :N-H$$

Here, both electrons in the bond between Cu and N come from N.
Cu^{2+} has accepted a pair of electrons from N.

A **Lewis base** is a molecule or ion that donates a pair of electrons.
A **Lewis acid** is a molecule or ion that accepts a pair of electrons.

Each NH_3 is a Lewis base.
The Cu^{2+} is a Lewis acid.

Terminology

The **central atom** of a coordination compound is the transition metal, which acts as the Lewis acid.

The Lewis bases which donate electron pairs to the central atom are called **ligands**. Ligands are covalently bonded to the central atom.

In the compound of Example 1,
Cu^{2+} is the central atom.
Each of the four NH_3 is a ligand.

The central atom, together with its ligands, is called a complex.

$[Cu(NH_3)_4]^{2+}$ is a complex ion.

Counter ions balance the charge of the complex ion. Counter ions are not covalently bonded to the central atom.

In Example 1, the counter ion is SO_4^{2-}.

$[Cu(NH_3)_4] SO_4$

complex ion   counter ion

coordination compound

In writing the formula for a coordination compound, the complex is usually written inside square brackets.

Example 2

Typical examples of central atoms:
$$Ag^+, Cu^{2+}, Ni^{2+}, Co^{2+}, Fe^{3+}, Cr^{3+}, Pt^{4+}.$$

Typical examples of ligands:
$$NH_3, H_2O, OH^-, Cl^-, Br^-, I^-, CN^-.$$

A. Write the formula for the complex ion containing Cr^{3+} as the central atom with five OH^- and one H_2O as ligands.

1. The central atom in square brackets.
$$[Cr \qquad]$$

2. Add the ligands
$$[Cr(OH)_5 (H_2O)]$$

3. Determine the charge of the complex ion.
$$(Cr^{3+}) + 5(OH^-) + (H_2O) = (+3) + 5(-1) + (0) = -2$$

The formula for the complex ion is $[Cr(OH)_5(H_2O)]^{2-}$

The formula for a coordination compound that contains this complex ion is, for example, $Na_2[Cr(OH)_5(H_2O)]$.

B. Write the formula of the coordination compound containing

central atom: Pt^{4+}
ligands: four NH_3 and two Cl^-
counter ions: NO_3^-

$$[Pt(NH_3)_4Cl_2]$$

Charge on complex ion: $(+4) + 4(0) + 2(-1) = +2$
The complex ion is $[Pt(NH_3)_4Cl_2]^{2+}$.
Two NO_3^- are needed to balance the charge.

The coordination compound is $[Pt(NH_3)_4Cl_2](NO_3)_2$

Exercise One

Write the formula for the coordination compounds containing the indicated central atom, ligands, and counter ions.

A. central atom: Fe^{3+} ligands: six CN^- counter ion: K^+

formula of complex ion: _____

formula of coordination compound: _____

B. central atom: Ag^+ ligands: two OH^- counter ion: Na^+

   formula of complex ion: _____

   formula of coordination compound: _____

C. central atom: Ni^{2+} ligands: four NH_3, two H_2O counter ion: SO_4^{2-}

   formula of complex ion: _____

   formula of coordination compound: _____

D. central atom: Pt^{4+} ligands: four H_2O, two Cl^- counter ion: Cl^-

   formula of complex ion: _____

   formula of coordination compound: _____

ADDITIONAL TERMINOLOGY

Coordination Number

The **coordination number** of the central metal is the number of electron pairs accepted by the central metal from the ligands. Each ligand in Example 2 donates *one* pair of electrons and thus the coordination number is equal to the number of ligands.

Example 3

What is the coordination number of the central atoms in each of the following?

A. $[Cr(H_2O)_2(NH_3)_2Br_2]^+$

   There are six ligands: two H_2O, two NH_3, and two Br_2.
   Each ligand donates one pair of electrons.
   The coordination number of chromium is 6.

B. $[Zn(H_2O)_2(OH)_2]$

   There are four ligands, and each ligand donates one pair of electrons.
   The coordination number of zinc is 4.

The most common coordination numbers are 2, 4, and 6.

Oxidation Number of the Central Atom

To determine the oxidation number or charge of a central metal use the relationship:

(oxidation no. of metal) + (total charge of ligands) = charge of complex

Example 4

Find the oxidation number of the central atom in $K_2[Co(H_2O)_2Br_4]$.

The complex ion is $[Co(H_2O)_2Br_4]^{2-}$

(ox. no. of Co) + (charge of ligands)   = -2

There are four Br^- ligands.

(ox. no. of Co) + 4 (−1) =   −2
ox. no. of Co   =   +2

The central ion in this compound is Co^{2+}.

Exercise Two

Find the coordination number and oxidation number of the central atom in each coordination compound below.

	central atom	oxidation number	coordination number
A. $K_2[PtCl_6]$	_____	_____	_____
B. $[Fe(H_2O)_6](NO_3)_2$	_____	_____	_____
C. $Na_2[Ni(CN)_4]$	_____	_____	_____
D. $[Zn(H_2O)_2(OH)_2]$	_____	_____	_____

COMPOSITION FROM EXPERIMENTAL DATA

Information about the composition of a coordination compound can be obtained from its mole number and from the moles of free anions per mole of compound.

The mole number (i) of a compound is the moles of particles (ions and molecules) per mole of compound. The mole number can be determined by measuring the conductivity of a solution of a compound.

The mole numbers of some sodium compounds:

$$NaCl(s) \longrightarrow Na^+(aq) + Cl^-(aq)$$
Total moles of ions is 2. Therefore, $i = 2$.

$$Na_2SO_4(s) \longrightarrow 2\,Na^+(aq) + SO_4^{2-}(aq)$$
Total moles of ions is 3. Therefore, $i = 3$.

$$Na_3PO_4(s) \longrightarrow 3\,Na^+(aq) + PO_4^{3-}(aq)$$
Total moles of ions is 4. Therefore, $i = 4$.

The moles of free chloride ions per mole of compound can be determined by mixing a solution of the compound with a solution of $AgNO_3$. Only free Cl^- ions form a precipitate of AgCl. The Cl^- ions that are bound as ligands are not free to form AgCl.

NaCl contains 1 mole of free Cl^- per mole of compound.
$$NaCl(aq) + AgNO_3(aq) \longrightarrow NaNO_3(aq) + AgCl(s)$$

$CaCl_2$ contains 2 moles of free Cl^- per mole of compound.
$$CaCl_2(aq) + 2\,AgNO_3(aq) \longrightarrow Ca(NO_3)_2(aq) + 2\,AgCl(s)$$

Table 1 contains experimental results for four coordination compounds of cobalt.

Table 1. Properties of Coordination Compounds of Cobalt

	Composition of Compound	Color of Compound	Mole Number i	Moles of free Cl^- per Mole of Compound
A.	$CoCl_3(NH_3)_6$	yellow	four	three
B.	$CoCl_3(NH_3)_5$	purple	three	two
C.	$CoCl_3(NH_3)_4$	green	two	one
D.	$CoCl_3(NH_3)_4$	violet	two	one

Example 5

Identify the complex ion in each coordination compound in Table 1.

A. $CoCl_3(NH_3)_6$

The mole number is four, therefore there are four ions in the coordination compound. These are the complex ion and three counter ions.

The three free moles of Cl^- are the three counter ions.

Therefore, the complex ion contains Co and six NH_3: $[Co(NH_3)_6]$

Because the total charge of the counter ions is –3, the charge of the complex ion must be +3.

The formula of the complex ion is $[Co(NH_3)_6]^{3+}$ and the formula of the coordination compound is $[Co(NH_3)_6]Cl_3$.

B. $CoCl_3(NH_3)_5$

The mole number is three, therefore there are three ions in the coordination compound. These are the complex ion and two counter ions.

The two free moles of Cl^- are the two counter ions.

Therefore, the complex ion contains Co, five NH_3, and the other Cl^-:
   $[Co(NH_3)_5Cl]$

Because the total charge of the counter ions is –2, the charge of the complex ion must be +2.

The formula of the complex ion is $[Co(NH_3)_5Cl]^{2+}$ and the formula of the coordination compound is $[Co(NH_3)_5Cl]Cl_2$.

C. and D. $CoCl_3(NH_3)_4$

Both compounds have the same mole number and number of free Cl^-. Therefore, the composition of their complex ions are the same, although their colors show they are different in some other way. (The way they differ will be described later.)

The mole number is two. Therefore, there are two ions in the coordination compound. These are the complex ion and one counter ion.

The one free mole of Cl^- is the counter ion.

Therefore, the complex ion contains Co, four NH_3, and the other two Cl^-:
   $[Co(NH_3)_4Cl_2]$

Because the total charge of the counter ion is –1, the charge of the complex ion must be +1.

The formula of the complex ion is $[Co(NH_3)_4Cl_2]^+$ and the formula of the coordination compound is $[Co(NH_3)_4Cl_2]Cl$.

Exercise Three

A. In the three blanks give the formula indicating the composition of the complex and the counter ions for the coordination compounds below.

composition	moles Cl^- per mole compound	mole number	formula of coordination compound
$PtCl_4(NH_3)_6$	four	five	$[Pt(NH_3)_6]Cl_4$
$PtCl_4(NH_3)_4$	two	three	_____
$PtCl_4(NH_3)_2$	none	one	_____
$PtCl_5(NH_3)K$	none	two	_____

B. Complete the table below.

formula of coordination compound	formula of complex ion	formula of counter ions	total number of ions
$[Cr(H_2O)_6]Cl_3$	_____	_____	_____
$[Cr(H_2O)_5Cl]Cl_2$	_____	_____	_____
$[Cr(H_2O)_4Cl_2]Cl$	_____	_____	_____
$NH_4[Cr(H_2O)_2Cl_4]$	_____	_____	_____

GEOMETRIC STRUCTURE OF COMPLEXES

The geometric structure of complexes describes the arrangement of the bonded ligand atoms around the central atom. The terminology is similar to that used for simple molecules, as described in Lesson 13.

Example 6

Describe the geometry of each of the following complexes.

A. $[Ag(NH_3)_2]^+$

The coordination number of Ag^+ is 2. Therefore, the geometry of this complex indicates the arrangement of the three atoms, Ag^+ and the two ligand atoms bonded to it, namely the N atoms.

In this complex, these atoms are arranged in a straight line.

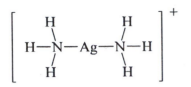

linear geometry

B. [Ni(CN)$_4$]$^{2-}$

The coordination number of Ni^{2+} is 4. Therefore, the geometry of this complex indicates the arrangement of five atoms, Ni^{2+} and the four ligand atoms bonded to it, namely the four C atoms.

In this complex, the four C atoms are arranged in a square around the Ni^{2+} ion.

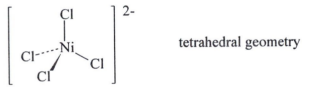   square planar geometry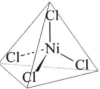

C. [NiCl$_4$]$^{2-}$

The coordination number of Ni^{2+} is 4.

In this complex, the four Cl$^-$ ions are arranged at the corners of a tetrahedron surrounding the Ni^{2+} ion.

   tetrahedral geometry

In these structures, a dashed line represents a bond to an atom behind the plane of the page, while the wedge represents a bond to an atom in front of the page.

D. [FeF$_6$]$^{3-}$

The coordination number of Fe^{3+} is 6.

The six F$^-$ ions are arranged at the corners of an octahedron surrounding the Fe^{3+} ion.

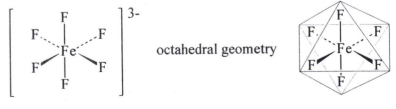   octahedral geometry

The octahedral geometry of complex ions is often represented by the shorthand notation illustrated at the right.

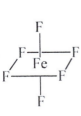

Example 7

In Example 5, there were two complex ions with the same composition but different colors.

$$[Co(NH_3)_4Cl_2]^+ \qquad \text{green}$$
$$[Co(NH_3)_4Cl_2]^+ \qquad \text{violet}$$

Draw structures to show how they are different.

In both, the coordination number of Co^{3+} is 6. Both complexes are octahedral.

In placing two Cl^- ligands at the corners of an octahedron, we discover that there are two different ways they can be placed.

They can be placed   They can be placed
at adjacent corners.   at opposite corners.

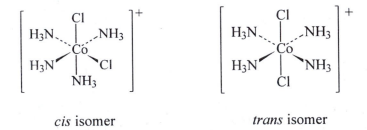

This produces two different complex ions with the same composition.

cis isomer   *trans* isomer

Compounds having the same composition but different geometric structures are called **geometric isomers**.

The ligands at adjacent corners of the octahedron are said to be *cis* to each other, and those at opposite corners are said to be *trans* to each other. The two geometric isomers shown above are labeled *cis* and *trans*, because the chloride ligands are *cis* and *trans* to each other respectively.

We must be careful when labeling *cis* and *trans* isomers, because it is the relative positions of the ligands that determines which isomer it is, not the absolute position. The structure of the *cis* isomer can be drawn in a variety of orientations.

All of the following structures represent the *cis* isomer.

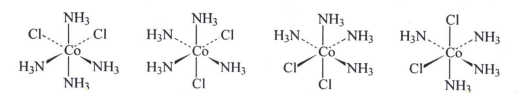

Linear and tetrahedral complexes do not have geometric isomers, but square complexes can.

Example 8

Draw structures for the two geometric isomers of the square planar complex $[Pt(NH_3)_2Cl_2]$.

The two NH_3 ligands can be placed either at adjacent corners of the square or at the opposite corners of the square.

The structure at the left is the *cis* isomer, the one on the right is the *trans* isomer.

Exercise Four

A. For each complex below, give the coordination number of the central atom, and the expected geometry of the complex.

	coordination number of central atom	geometry of complex
$[Fe(H_2O)_6]^{3+}$	_____	_____
$[Ni(NH_3)_2Cl_2]$ (has only one isomer)	_____	_____
$[CuCl_2]^-$	_____	_____
$[Ni(H_2O)_3(NH_3)_3]^{2+}$	_____	_____
$[Au(CN)_2Cl_2]^-$ (has two isomers)	_____	_____

B. Draw structures for the two isomers of $[Ni(NH_3)_3(H_2O)_3]^{2+}$.

C. Draw structures for the two isomers of $[Pt(H_2O)_2(NH_3)Cl]^+$.

POLYDENTATE LIGANDS

Polydentate ligands donate more than one pair of electrons to the central atom. Each pair of electrons is from a different atom in the ligand molecule.

Bidentate ligands donate two pairs of electrons.

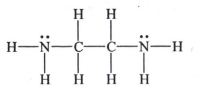

ethylenediamine

Ethylenediamine has two lone pairs on two different nitrogen atoms. Each lone pair can form a coordinate covalent bond to a metal.

The complex below has two ethylenediamine ligands and two chloride ion ligands.
$$[Co(NH_2CH_2CH_2NH_2)_2Cl_2]^+$$

The coordination number of the central atom is the number of electron pairs it has accepted. Because each ethylenediamine donates two pairs of electrons and each chloride ion only one, the coordination number of Co^{3+} in this complex is 6.

The geometry of the complex is octahedral.

The two chloride ligands can be either *cis* or *trans*, so there are two geometric isomers.

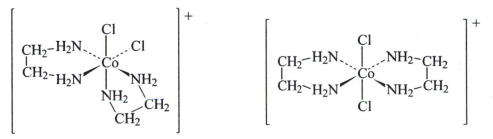

       cis isomer   *trans* isomer

$NH_2CH_2CH_2NH_2$ is often abbreviated as "en".

$$[Co(en)_2Cl_2]^+ \text{ is } [Co(NH_2CH_2CH_2NH_2)_2Cl_2]^+$$

Tridentate ligands donate *three* pairs of electrons.

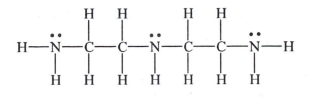

diethylenetriamine (dien)

Hexadentate ligands donate *six* pairs of electrons.

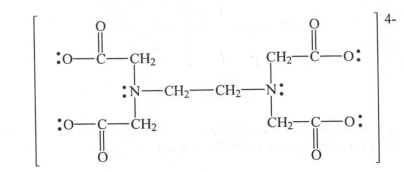

anion of ethylenediaminetetraacetic acid (EDTA)

Table 2 contains the formulas and names of some common ligands of a variety of types.

Table 2. Some Common Ligands (abbreviations in parentheses)

A. Neutral Monodentate

:OH_2	aqua	:CO	carbonyl
:NH_3	ammine	:NC_5H_5	pyridine (py)

B. Charged Monodentate

F^-	fluoro	I^-	iodo
Cl^-	chloro	:OH^-	hydroxo
Br^-	bromo	:CN^-	cyano
:OSO_3^{2-}	sulfato	:ONO^-	nitrito

C. Neutral Polydentate

bidentate:

$NH_2CH_2CH_2NH_2$ ethylenediamine (en)

tridentate:

$NH_2CH_2CH_2NHCH_2CH_2NH_2$ diethylenetriamine (dien)

D. Charged Polydentate

bidentate:

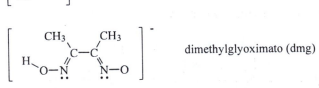

ethylenediaminetetraacetato   (EDTA)   [structure at top of page]

Exercise Five

A. Give the oxidation number and coordination number of the central atom in each complex below.

complex	oxidation number of central atom	coordination number of central atom
$[Ni(en)_3]^{2+}$	_____	_____
$[Co(CO_3)_3]^{3-}$	_____	_____
$[Pt(C_2O_4)_2]^{2-}$	_____	_____
$[Ni(dmg)_2]$	_____	_____

B. Draw structures for the two isomers of $[Cr(C_2O_4)_2(H_2O)_2]^-$.

NAMING COORDINATION COMPOUNDS

Coordination compounds are named in a systematic fashion.

1. The cation is named first, then the anion.

2. Naming the complex.

   a. The ligands are named first, in alphabetical order.

   b. Prefixes are used before the name of the ligand to indicate the number of ligands in the complex.
   For simple ligands, the prefixes are:

no. of ligands	prefix
one	(none)
two	di
three	tri
four	tetra
five	penta
six	hexa

If the name of the ligand contains one of these prefixes, e.g., ethylenediamine, then the name of the ligand is enclosed in parentheses and one of the following prefixes is used.

no. of ligands	prefix
two	bis
three	tris
four	tetrakis

c.   After the name of the ligands comes the name of the central metal.

d.   A Roman numeral in parentheses indicates the oxidation state of the metal.

Example 9

Name the following complex ions

A. $[Ag(NH_3)_2]^+$

The complex contains ammonia ligand, called ammine.
    There are two of these ligands: diammine.
The metal is Ag^+: silver(I).

The name of the complex is diamminesilver(I) ion.

B. $[Co(en)_3]^{3+}$

The complex contains ethylenediamine.
    There are three of these ligands, and their name includes "di."
        Therefore, the part of the name for the ligands is
            tris(ethylenediamine)
The metal is Co^{3+}: cobalt(III)

The name of the complex is tris(ethylenediamine)cobalt(III) ion.

C. $[Cr(NH_3)_4Cl_2]^+$

There are four ammonia ligands: tetraammine.

There are two chloride ligands: dichloro
The ligands are named in alphabetical order, ammine preceding chloro:
    tetraamminedichloro
The metal is Cr^{3+}: chromium(III)

The name of the complex is tetraamminedichlorochromium(III) ion.

D. $[Fe(CN)_6]^{3-}$

When the complex has a negative charge, the suffix "ate" is added to the name of the metal.

"cobalt" becomes "cobaltate"
"chromium" becomes "chromate"
"iron" becomes "ferrate"
"copper" becomes "cuprate"

The name of $[Fe(CN)_6]^{3-}$ is hexacyanoferrate(III) ion.

Example 10

Name the coordination compound $K_3[Co(CO_3)_3]$.

This compound contains K^+ counter ions and the $[Co(CO_3)_3]^{3-}$

The cations is named first: potassium

The anion is the complex ion.
    There are three carbonate ligands: tricarbonato
    The metal is Co^{3+} in an anionic complex: cobaltate(III)

The name of the coordination compound is
                potassium tricarbonatocobaltate(III)

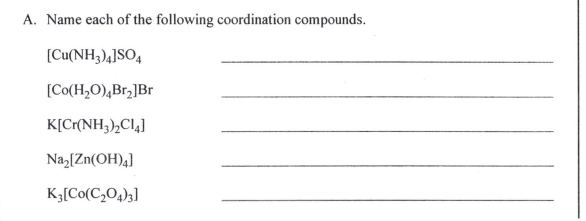

Exercise Six

A. Name each of the following coordination compounds.

$[Cu(NH_3)_4]SO_4$ _____

$[Co(H_2O)_4Br_2]Br$ _____

$K[Cr(NH_3)_2Cl_4]$ _____

$Na_2[Zn(OH)_4]$ _____

$K_3[Co(C_2O_4)_3]$ _____

B. Write the formula for each of the following coordination compounds.

hexaamminenickel(II) sulfate   _____

potassium amminepentachloroplatinate(IV)

_____

dichlorobis(ethylenediamine)cobalt(III) chloride

_____

PROBLEMS

1. The following questions are in reference to the coordination compound $[M(NH_3)_5Cl]Cl_3$.

   a. What is the oxidation number of the central metal M?
   b. If the atomic number of M is 28, how many d-electrons does the central metal have?
   c. The electric conductivity of an aqueous solution of this coordination compound will be similar to that of a solution of (choose one):
      1. CH_4
      2. $LaCl_3$
      3. Na_2SO_4
      4. NaH_2PO_4

2. Chemical analysis shows that five different coordination compounds all have the empirical formula $K_2CoCl_2I_2(NH_3)_2$. Electrical conductivity studies of aqueous solutions of the five compounds indicated properties similar to those of Na_2SO_4.

   a. Based on the above information, write a structural formula for each of the five different complex ions. The structural formulas should indicate the differences between the five complex ions.
   b. What is the oxidation number of the central atom?

COORDINATION COMPOUNDS:
COLOR AND MAGNETIC PROPERTIES

This lesson deals with:

1. Writing electron configurations for central atoms.
2. Drawing d-orbital diagrams for central atoms.
3. Writing crystal field splitting energy diagrams for octahedral, tetrahedral and square planar geometries.
4. Relating the magnitude of the crystal field splitting of energy levels to the color and magnetic properties of a complex.

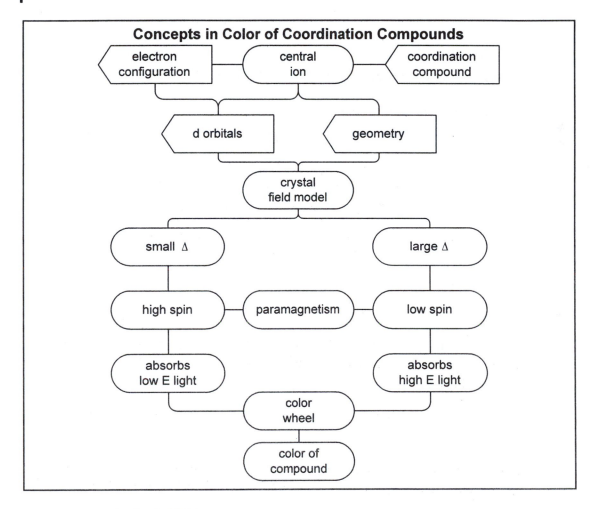

CRYSTAL FIELD THEORY

The color and magnetic properties of complex ions can be explained using crystal field theory.

The color of a complex ion depends on the identity of the ligands.

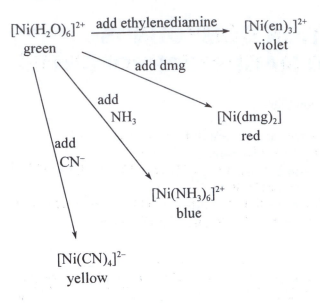

The color of a complex is caused by the absorption of light by the complex.

White light is composed of all colors of the visible spectrum.

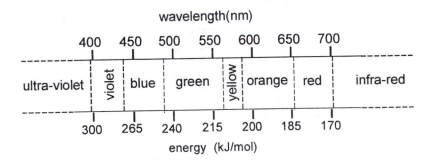

Figure 1. Visible region of the electromagnetic spectrum.

Figure 1 is a diagram of the visible region of the electromagnetic spectrum. It shows the relationship between wavelength of radiation and the color we see when radiation of that wavelength enters our eyes.

For example, when radiation with a wavelength of 550 nm enters our eyes, we see green.

Figure 1 also shows the relationship between wavelength of visible radiation and energy. It shows that green light with a wavelength of 550 nm corresponds to an energy of 215 kJ/mol.

When a complex absorbs only a portion of the visible spectrum, some of the colors are not absorbed. These are the colors we see.

The relationship between what is absorbed and the color we see can be diagramed with a complementary color wheel.

The complementary color wheel shows the relationship between complementary colors. Complementary colors are opposite each other in the wheel. For example, red is the complement of green and yellow is the complement of violet.

When a substance absorbs light of a certain color, we perceive the substance to have a color complementary to the absorbed color. If a substance absorbs in the red region of the spectrum, we see the substance as the complement of red, namely green.

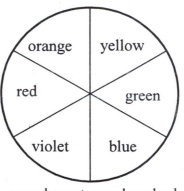
complementary color wheel

Example 1

Adding ammonia to a green solution containing $Ni(H_2O)_6^{2+}$ turns the solution blue, as a result of the formation of $Ni(NH_3)_6^{2+}$. Which complex absorbs light of higher energy?

Use the complementary color wheel to determine the color of the light absorbed by each complex.

$Ni(H_2O)_6^{2+}$ is green, therefore it absorbs red light.

$Ni(NH_3)_6^{2+}$ is blue, therefore it absorbs orange light.

According to Figure 1, red light has energy of about 180 kJ/mol, and orange light has energy of about 190 kJ/mol. Orange light is higher energy than red light.

Therefore, light absorbed by $Ni(NH_3)_6^{2+}$ is higher in energy than that absorbed by $Ni(H_2O)_6^{2+}$.

Exercise One

A. What is the color of a substance that absorbs yellow light?

B. The substance that gives leaves their green color is chlorophyll. Approximately what wavelength of light does chlorophyll absorb?

C. In each pair of substances below, choose the one that absorbs light of higher energy. The observed color of each substance is indicated.

Absorbs higher energy?

1. substance A (yellow), substance B (green)   A   or   B

2. substance B (green), substance C (violet)   B   or   C

3. substance D (red), substance E (blue)   D   or   E

4. substance C (violet), substance F (orange)   C   or   F

COLOR, MAGNETISM, AND ELECTRONS

Color and magnetism of complexes are a result of the behavior of electrons.

The energy from absorbed light moves an electron from a low-energy orbital to a higher-energy orbital.

Magnetism results from unpaired electrons in a substance.

Crystal Field Theory explains how the color and magnetism of transition-metal complexes are related to their electron configurations.

According to Crystal Field Theory, the energy of the d orbitals of a transition metal in a complex is affected by the electrons of the ligands.

In an octahedral complex, some of the d orbitals are closer to the ligand's electrons than other d orbitals.

The $d_{x^2-y^2}$ and the d_{z^2} orbitals are closer to the ligands than are the d_{xy}, d_{xz}, and d_{yz}. The figures below show the $d_{x^2-y^2}$ and the d_{z^2} orbitals with the electron pairs from six ligands arranged in an octahedron around them.

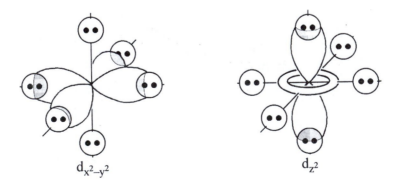

$d_{x^2-y^2}$ d_{z^2}

Below are the d_{xy}, d_{xz}, and d_{yz} orbitals with the electron pairs from six ligands arranged in an octahedron around them. Electrons in these d orbitals are not as close to the ligand electrons as are electrons in the $d_{x^2-y^2}$ and d_{z^2} orbitals.

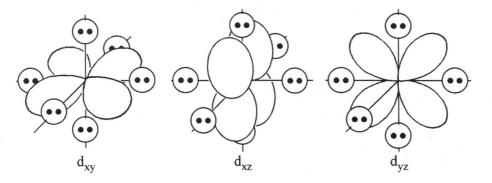

$$d_{xy} \qquad d_{xz} \qquad d_{yz}$$

Because electrons repel each other, and because electrons in the d_{xy}, d_{xz}, and d_{yz} orbitals are farther from the ligand electrons than are electrons in the $d_{x^2-y^2}$ and d_{z^2} orbitals, the electrons in the metal atom will occupy the d_{xy}, d_{xz}, and d_{yz} orbitals in preference to the $d_{x^2-y^2}$ and d_{z^2} orbitals.

In other words, in an octahedral complex, the d_{xy}, d_{xz}, and d_{yz} orbitals are lower in energy than the $d_{x^2-y^2}$ and d_{z^2} orbitals.

This difference in energy can be represented with an orbital-energy diagram, such as the one in Figure 2.

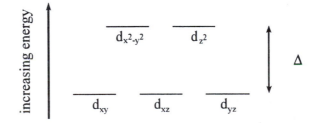

Figure 2.   Orbital-energy diagram for an
octahedral complex.

In this figure, the higher-energy orbitals are written above the lower-energy orbitals. The vertical separation between them corresponds to the energy difference between them. This difference is labeled Δ, and called the crystal-field splitting energy.

The magnitude of Δ depends upon both the identity of the ligands and the identity of the central atom.

The magnitude of Δ is equal to the energy of the light absorbed by the complex. When a complex absorbs light, an electron moves from one of the lower-energy d orbitals to one of the higher-energy d orbitals.

Example 2

Explain the colors of the complexes below using the crystal-field splitting of the d orbitals of the central atom. Estimate the value of Δ for each complex.

$$[Cr(H_2O)_6]^{3+}$$
violet

$$[Cr(NH_3)_6]^{3+}$$
orange

The water complex is violet. Therefore, it absorbs yellow light.
The ammonia complex is orange. Therefore, it absorbs blue light.

Blue light is higher in energy than is yellow light. Therefore, the crystal-field splitting is greater in the ammonia complex than in the water complex.

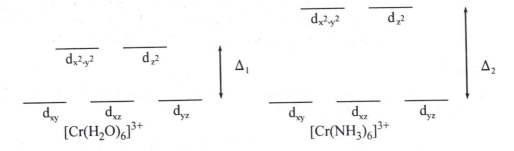

NH_3 ligands cause a larger splitting in the energies of the d orbitals of Cr^{3+} than do H_2O ligands.

Because $[Cr(H_2O)_6]^{3+}$ absorbs yellow light, the value of its Δ is about 210 kJ/mol, as indicated by Figure 1. Similarly, Δ for $[Cr(NH_3)_6]^{3+}$ is about 260 kJ/mol.

The energy absorbed by these complexes when they absorb visible light causes an electron to move from the lower-energy d orbitals to the higher-energy ones.

Cr^{3+} has 3 electrons in its d orbitals. That is, its configuration is $[Ar]\,3d^3$.
When a Cr^{3+} complex absorbs light, the arrangement of its electrons changes, as represented by the diagram below.

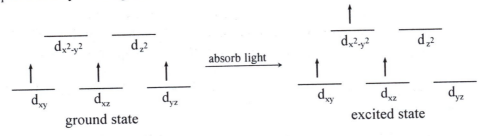

Exercise Two

A. Draw the crystal field splitting diagrams for $[Ni(H_2O)_6]^{2+}$, which is green, and $[Ni(NH_3)_6]^{2+}$, which is blue. Label the d-orbitals and place the nickel(II) valence electrons into the orbitals. (Show only the ground-state configuration.)

B. Which complex has the larger Δ?

C. Refer to Figure 1 and indicate the approximate value of Δ in kJ/mol for $[Ni(H_2O)_6]^{2+}$.

MAGNETIC PROPERTIES

The valence electron configuration of Fe^{2+} is [Ar] $3d^6$.

Place the six electrons into the d orbitals of an octahedral Fe^{2+} complex. There are two possibilities.

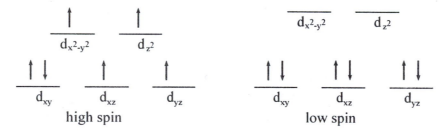

In the case on the left, the electrons spread out filling all d orbitals before pairing. This leads to 4 unpaired electrons.
In the case on the right, the electrons occupy only the low-energy d orbitals. This leads to no unpaired electrons.
The case on the left is called high spin because there are more unpaired electrons than in the case on the right, which is called low spin.

Because unpaired electron spin is the origin of paramagnetism, a complex with the electron arrangement on the left will be paramagnetic (attracted to a magnet). Because there are no unpaired electrons in the arrangement on the right, a complex with this arrangement will be diamagnetic (not attracted to a magnet).

Which arrangement occurs in a particular complex depends on the value of Δ. It takes a small amount of energy to get 2 electrons to occupy the same orbital. This is why electrons spread out in orbitals of the same energy, occupying the orbitals one electron per orbital, before pairing up. The amount of energy required to pair two electrons into the same orbital is called the **pairing energy**.

If Δ is greater than the pairing energy, the electrons will pair up before filling all d orbitals. This leads to a low-spin complex, like the one on the right above.

If Δ is smaller than the pairing energy, the electrons will spread out over the d orbitals before pairing up. This leads to a high-spin complex, like the one on the left above.

If Δ > pairing energy, then low-spin complex.
If Δ < pairing energy, then high-spin complex.

Example 3

$[Fe(CN)_6]^{4-}$ is yellow. $[Fe(H_2O)_6]^{2+}$ is pale green. One of these complexes is high-spin, the other is low-spin. How many unpaired electrons does each have?

Electron configuration of Fe^{2+}: $[Ar]\,3d^6$

$[Fe(CN)_6]^{4-}$ is yellow. It absorbs violet light.
$[Fe(H_2O)_6]^{2+}$ is green. It absorbs red light.
   Red light is of lower energy than violet light.
   Therefore, the value of Δ for $[Fe(CN)_6]^{4-}$ is greater than Δ for $[Fe(H_2O)_6]^{2+}$.

Draw orbital-energy diagrams for the two complexes:

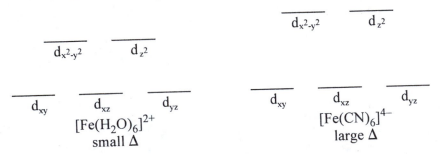

Because one complex is high spin and one is low spin, for one complex, Δ is smaller than the electron pairing energy, and for the other complex, Δ is larger than the pairing energy.

Put the 6 electrons into the orbital-energy diagrams:

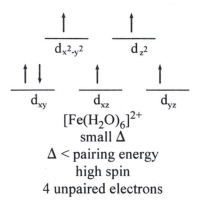

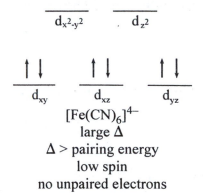

$[Fe(H_2O)_6]^{2+}$
small Δ
Δ < pairing energy
high spin
4 unpaired electrons

$[Fe(CN)_6]^{4-}$
large Δ
Δ > pairing energy
low spin
no unpaired electrons

Exercise Three

A. Consider the two complex ions $[Mn(H_2O)_6]^{2+}$ and $[Mn(CN)_6]^{4-}$. The first is pale red and is a high-spin complex. The second is a low-spin complex.

1. Sketch the crystal field splitting energy diagram for $[Mn(H_2O)_6]^{2+}$. Label each orbital and put in the electrons.

2. Do the same for $[Mn(CN)_6]^{4-}$.

3. What is the color of the light absorbed by $[Mn(H_2O)_6]^{2+}$?

4. Based on the magnitude of its Δ compared to that of the H_2O complex, what color of light might be absorbed by $[Mn(CN)_6]^{4-}$?

5. What color is a solution of $[Mn(CN)_6]^{4-}$ likely to be?

B. The magnitude of splitting of orbital energies of a transition-metal ion depends upon the identity of the ligand. Based on the color of the complexes given below, arrange the ligands in order of the magnitude of splitting they cause, beginning with the ligand causing the smallest Δ.

Least _____ < _____ < _____ < _____ Greatest

Complex	Color	Complex	Color
$[Co(H_2O)_6]^{3+}$	pale red	$[Co(CN)_6]^{3-}$	yellow
$[Co(NH_3)_6]^{3+}$	pale orange	$[Co(CO_3)_3]^{3-}$	green

TETRAHEDRAL AND SQUARE CRYSTAL FIELD SPLITTING

The ligands in tetrahedral and square complexes also cause splitting of d-orbital energies.

For tetrahedral complexes, Δ is smaller than the pairing energy. Therefore, tetrahedral complexes are high spin.

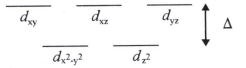

tetrahedral complex

For square complexes, the splitting between the three lower levels is smaller than the pairing energy, so electrons spread out in these four orbitals before pairing up. However, the splitting between the top two levels is larger than the pairing energy. Therefore, electrons enter the top level ($d_{x^2-y^2}$) only after the lower levels are completely filled with 8 electrons.

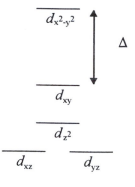

square complex

Example 4

A. Draw an orbital-energy diagram for the blue tetrahedral complex $[CoCl_4]^{2-}$ and determine the number of unpaired electrons in the complex.

Co^{2+}: $[Ar]\,3d^7$

Tetrahedral complexes have small Δ and are high spin. Therefore, the electrons spread out in the d orbitals before pairing.

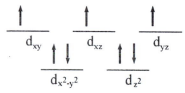

tetrahedral $[CoCl_4]^{2-}$

There are 3 unpaired electrons in $[CoCl_4]^{2-}$.

B. Draw an orbital-energy diagram for the red square complex $[Ni(dmg)_2]$ and determine the number of unpaired electrons in the complex.

In a square complex, the first four electrons spread out in the lowest four orbitals before pairing up. Then, only after the lowest four orbitals are filled with 8 electrons, do any electrons enter the top level.

Ni^{2+}: $[Ar]\,3d^8$

$[Ni(dmg)_2]$ has no unpaired electrons.

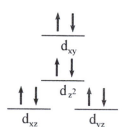

square $[Ni(dmg)_2]$

Exercise Four

$Ni(CN)_4^{2-}$ is yellow and has no unpaired electrons. $NiCl_4^{2-}$ is deep blue and has two unpaired electrons.

A. Which complex is high spin and which low spin?

high spin   _____

low spin   _____

B. Indicate the correct geometry for each complex ion:

$NiCl_4^{2-}$ _____

$Ni(CN)_4^{2-}$ _____

C. Sketch the correct splitting energy diagram for each complex ion in the designated spaces. Label each orbital and fill in the electrons.

$NiCl_4^{2-}$ $Ni(CN)_4^{2-}$

PROBLEMS

1. The magnitude of splitting caused by various ligands has been determined for cobalt complexes. The order is
    large splitting : $CN^- > NH_3 > H_2O > F^- > Cl^-$: small splitting
    Of the two complexes below, one is high spin and the other low spin. One absorbs at 590 nm and the other at 480 nm. Identify the absorption and spin state of each complex.

    $$[CoF_6]^{3-} \qquad\qquad [Co(CN)_6]^{3-}$$

2. Air humidity indicators that are pink in humid conditions and blue in dry conditions contain cobalt(II) and chloride ions. The blue color is that of a tetrahedral complex of cobalt(II) and chloride ions. The pink color is that of an octahedral complex of cobalt(II) and water.

    a.   Write the formula for the blue complex.
    b.   Write the formula for the pink complex.
    c.   How many d electrons does cobalt(II) have?
    d.   Draw the splitting energy diagram for the tetrahedral complex ion which has 3 unpaired electrons.
    e.   Draw the splitting energy diagram for the pink complex ion which is in a high-spin state.
    f.   Write a balanced equation representing the formation of the pink complex ion from the blue complex ion and water (from humid air).

3. Rubies are red and emeralds are green. Both of these owe their color to the presence of a small amount of Cr^{3+} ions in the crystal. In rubies, the Cr^{3+} ions are surrounded by oxide ions in an octahedral arrangement. In emeralds, the Cr^{3+} ions are surrounded by silicate ions, also octahedrally arranged. Draw crystal field splitting diagrams for the d orbitals of the Cr^{3+} ions in rubies and for those in emeralds. Be sure to show which has the larger crystal field splitting energy. Label the orbitals, and place the d electrons in the appropriate orbitals.

4. The hexaaquacobalt(III) complex $[Co(H_2O)_6]^{3+}$ is red, and the hexamminecobalt(III) complex $[Co(NH_3)_6]^{3+}$ is orange. One of these two is high spin and the other is low spin.
    a.   How many d electrons does Co^{3+} have?
    b.   How many unpaired electrons does each Co^{3+} in the high-spin complex contain?
    c.   How many unpaired electrons does each Co^{3+} in the low-spin complex contain?
    d.   Which complex is high spin?
    e.   The carbonate ion, CO_3^{2-}, can act as a bidentate ligand. It forms an octahedral complex with Co^{3+}, $[Co(CO_3)_3]^{3-}$, and this complex is green. Is the $[Co(CO_3)_3]^{3-}$ complex high spin or low spin?

COORDINATION COMPOUNDS:
COMPLEX ION EQUILIBRIA

This lesson deals with:

1. Calculating the equilibrium concentrations of the species involved in the dissociation of a complex ion.
2. Calculating the value of K_{diss} for a complex ion from concentration data.
3. Relating the value of K_{diss} to the value of K_f.
4. Calculating the concentration of ligand required to form a complex with a given fraction of metal ion in solution.
5. Calculating the molar solubility of a water-insoluble salt in a solution of ligand.
6. Calculating the amount of ligand required to dissolve a given amount of a water-insoluble salt.

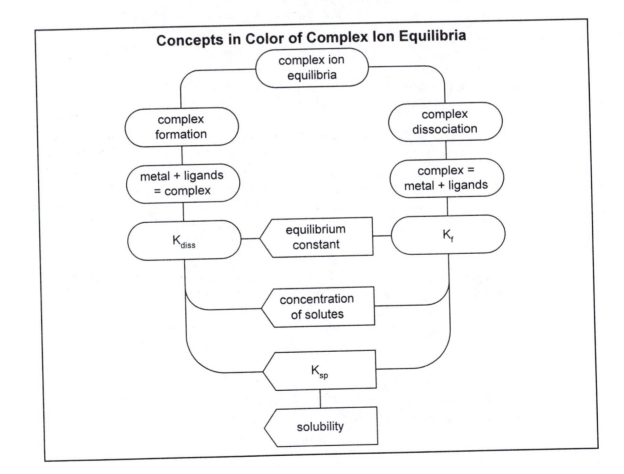

Concepts in Color of Complex Ion Equilibria

DISSOCIATION OF A COMPLEX ION

Complex Ion: Metal Ion and Ligands

In solution, a complex ion dissociates to some extent to its constituent metal ion and ligands. The dissociation reaction reaches equilibrium.

$$CuCl_2^-(aq) \rightleftharpoons Cu^+(aq) + 2\,Cl^-(aq)$$

As for any equilibrium, we can write an equilibrium constant expression. The equilibrium constant for the dissociation of a complex is called the dissociation constant.

$$K_{diss} = \frac{[Cu^+][Cl^-]^2}{[CuCl_2^-]} = 3.0 \times 10^{-6}$$

Example 1

Calculate the concentrations of the ions in a 0.50 M solution of $CuCl_2^-$.

1. Summarize the information in a table based on the chemical equation:

	$CuCl_2^-$	$\rightleftharpoons$	Cu^+	+	$2\,Cl^-$
initial conc	0.50		0		0
conc change	$-x$		$+x$		$+2x$
equil conc	$0.50-x$		x		$2x$

2. Assume that x will be small compared to 0.50, so $0.50 - x = 0.50$.

3. Find the value of x by substituting equil conc into K expression.

$$K_{diss} = \frac{[Cu^+][Cl^-]^2}{[CuCl_2^-]} = \frac{(x)(2x)^2}{(0.50)} = 3.0 \times 10^{-6}$$

$$4x^3 = (0.50)(3.0 \times 10^{-6})$$

$$x^3 = 3.8 \times 10^{-7}$$

$$x = \sqrt[3]{3.8 \times 10^{-7}} = 7.2 \times 10^{-3}$$

Therefore, the concentrations are

$[Cu^+] = x = 7.2 \times 10^{-3}$ M

$[Cl^-] = 2x = 1.4 \times 10^{-2}$ M

$[CuCl_2^-] = 0.50 - x = 0.50$ M

Note that the last of these shows our assumption to be correct.

Exercise One

Calculate the concentrations of the ions in a 0.080 M solution of $Fe(CN)_6^{3-}$. For $Fe(CN)_6^{3-}$, the $K_{diss} = 1.0 \times 10^{-31}$.

$$Fe(CN)_6^{3-} \rightleftharpoons Fe^{3+} + 6\,CN^-$$

CALCULATING THE VALUE OF A DISSOCIATION CONSTANT

Example 2

A solution is prepared by dissolving 0.10 mole of $ZnSO_4$ in enough 6.0 M $NH_3(aq)$ to make 1.0 liter of solution. In this solution, the concentration of uncomplexed Zn^{2+} ions is 8.1×10^{-14} M. Calculate the value of the dissociation constant of $Zn(NH_3)_4^{2+}$.

1. Write the equation for the equilibrium reaction and its K expression.

$$Zn(NH_3)_4^{2+} \rightleftharpoons Zn^{2+} + 4\,NH_3 \qquad\qquad K_{diss} = \frac{[Zn^{2+}]\,[NH_3]^4}{[Zn(NH_3)_4^{2+}]}$$

2. Summarize the information in a table based on the chemical equation.

	$Zn(NH_3)_4^{2+}$ $\rightleftharpoons$	Zn^{2+} +	$4\,NH_3$
initial conc	0	0.10	6.0
conc change	$+x$	$-x$	$-4x$
equil conc	x	$0.10-x$	$6.0-4x$

3. Find the value of x and the equilibrium concentrations.

$$[Zn^{2+}] = 8.1 \times 10^{-14} = 0.10 - x$$
$$x = 0.10 - 8.1 \times 10^{-14} = 0.10$$
$$[Zn(NH_3)_4^{2+}] = x = 0.10$$
$$[NH_3] = 6.0 - 4x = 6.0 - 4(0.10) = 5.6$$

4. Put equil conc values into expression for K_{diss}.

$$K_{diss} = \frac{[Zn^{2+}]\,[NH_3]^4}{[Zn(NH_3)_4^{2+}]}$$

$$= \frac{(8.1 \times 10^{-14})\,(5.6)^4}{(0.10)}$$

$$= 8.0 \times 10^{-10}$$

Sometimes the equilibrium reaction is written as a formation reaction.

$$Zn^{2+} + 4\,NH_3 \rightleftharpoons Zn(NH_3)_4^{2+}$$

For this reaction, the equilibrium constant is called a formation constant, K_f.

$$K_f = \frac{[Zn(NH_3)_4^{2+}]}{[Zn^{2+}]\,[NH_3]^4}$$

The formation constant is the inverse of the dissociation constant.

$$K_f = \frac{1}{K_{diss}}$$

For the $Zn(NH_3)_4^{2+}$ complex in Example 2,

$$K_f = \frac{1}{8.0 \times 10^{-10}} = 1.3 \times 10^9$$

Exercise Two

A. When 0.020 mole of $Ni(NO_3)_2$ is dissolved in enough 0.200 M NH_3 to make 1.0 liter solution, the concentration of uncomplexed Ni^{2+} is 1.5×10^{-4} M. Calculate the value of K_{diss} for $Ni(NH_3)_6^{2+}$.

B. What is the value of K_f for $Ni(NH_3)_6^{2+}$?

COMPLEX FORMATION

Example 3

How many moles of HCl must be added to 1.0 liter of 0.20 M $CdCl_2$ to produce a complex with 95% of the cadmium in the form of $CdCl_4^{2-}$? For $CdCl_4^{2-}$, the $K_{diss} = 4.0 \times 10^{-3}$.

	$CdCl_4^{2-}$	$\rightleftharpoons$	Cd^{2+}	$+$	$4\ Cl^-$
initial conc	0		0.20		0.40
add Cl^- conc					$+y$
conc change	$+x$		$-x$		$-4x$
equil conc	x		$0.20-x$		$0.40+y-4x$

We want 95% of the Cd in the solution to be in the form of $CdCl_4^{2-}$.

Therefore,

$$[CdCl_4^{2-}] = (0.95)(0.20) = 0.19 = x$$

Then,

$$[Cd^{2+}] = 0.20 - x = 0.20 - 0.19 = 0.01$$

$$[Cl^-] = 0.40 + y - 4x = 0.40 + y - 4(0.19) = y - 0.36$$

Substitute these values into the K_{diss} expression and solve for y.

$$K_{diss} = \frac{[Cd^{2+}][Cl^-]^4}{[CdCl_4^-]} = \frac{(0.01)(y-0.36)^4}{(0.19)} = 4.0 \times 10^{-3}$$

$$(y-0.36)^4 = \frac{0.19}{0.01}(4.0 \times 10^{-3})$$

$$(y-0.36)^4 = 0.076$$

$$y-0.36 = (0.076)^{1/4}$$

$$y-0.36 = 0.52$$

$$y = 0.88$$

Therefore, it is necessary to add 0.88 mol HCl per liter to complex 95% of the Cd. Because the volume of the solution is 1.0 liter, the number of moles of HCl needed is 0.88 mol.

Exercise Three

How many moles of NaSCN must he added to 1.0 liter of a 0.0500 M solution of $AgNO_3$ to form a complex with 99% of the silver as $Ag(SCN)_2^-$? For $Ag(SCN)_2^-$, the $K_{diss} = 1.0 \times 10^{-10}$.

ENHANCED SOLUBILITY THROUGH COMPLEX FORMATION

Many salts that are only slightly soluble in water are very soluble in solutions containing substances with which they form a complex. For example, silver chloride is only slightly soluble in water.

$$AgCl(s) \rightleftharpoons Ag^+(aq) + Cl^-(aq)$$

$$K_{sp} = [Ag^+][Cl^-] = 1.6 \times 10^{-10}$$

However, silver ions in solution form a complex with ammonia.

$$Ag^+(aq) + 2\,NH_3(aq) \rightleftharpoons Ag(NH_3)_2^+(aq)$$

$$K_f = \frac{[Ag(NH_3)_2^+]}{[Ag^+][NH_3]^2} = 2.5 \times 10^7$$

When the complex forms, the concentration of free Ag^+ in the solution diminishes, and more AgCl dissolves to replace it. The net effect of the dissolving of AgCl and the formation of the $Ag(NH_3)_2^+$ complex is represented by the following equation.

$$AgCl(s) + 2\,NH_3(aq) \rightleftharpoons Ag(NH_3)_2^+(aq) + Cl^-(aq)$$

This reaction reaches an equilibrium and has an equilibrium constant.

$$K = \frac{[Ag(NH_3)_2^+][Cl^-]}{[NH_3]^2}$$

The value of this equilibrium constant is related to the K_{sp} of AgCl and to the K_f of $Ag(NH_3)_2^+$.
The relationship is

$$K = \frac{[Ag(NH_3)_2^+][Cl^-]}{[NH_3]^2} = [Ag^+][Cl^-] \times \frac{[Ag(NH_3)_2^+]}{[Ag^+][NH_3]^2} = K_{sp} \times K_f$$

Therefore, for the solubility of AgCl in $NH_3(aq)$,

$$K = (1.6 \times 10^{-10}) \times (2.5 \times 10^7) = 4.0 \times 10^{-3}$$

Because $K_f = 1 / K_{diss}$, the value of the K is also equal to K_{sp} / K_{diss}.

Example 4

Calculate the molar solubility of AgCl in 6.0 M aqueous ammonia.

	AgCl(s)	+	$2\,NH_3$	$\rightleftharpoons$	$Ag(NH_3)_2^+$	+	Cl^-
initial conc			6.0		0		0
conc change			$-2x$		$+x$		$+x$
equil conc			$6.0-2x$		x		x

$$K = \frac{[Ag(NH_3)_2^+][Cl^-]}{[NH_3]^2} = 4.0 \times 10^{-3}$$

$$\frac{(x)(x)}{(6.0-2x)^2} = 4.0 \times 10^{-3}$$

$$\frac{x}{(6.0-2x)} = \sqrt{4.0 \times 10^{-3}}$$

$$x = (0.0632)(6.0 - 2x)$$

$$x = 0.379 - 0.126\,x$$

$$1.126\,x = 0.379$$

$$x = \frac{0.379}{1.126} = 0.34$$

The molar solubility of AgCl in 6.0 M NH_3(aq) is 0.34 mole/liter. This is much greater than the molar solubility of AgCl in pure water, which is 1.3×10^{-4} mole/liter.

Exercise Four

Calculate the molar solubility of AgBr in a 0.10 M solution of $Na_2S_2O_3$. For AgBr, $K_{sp} = 5.0 \times 10^{-13}$. For $Ag(S_2O_3)_2{}^{3-}$, the $K_{diss} = 1.0 \times 10^{-13}$. (Remember, $K_f = 1/K_{diss}$.)

DISSOLVING A SOLID BY COMPLEX FORMATION

Example 5

A 0.10-mole sample of insoluble $Zn(OH)_2$ is mixed with 0.500 liter of water. Calculate the amount of NaOH(s) that must be added in order to dissolve the $Zn(OH)_2$. For $Zn(OH)_2$, $K_{sp} = 5.0 \times 10^{-17}$. For $Zn(OH)_4^{2-}$, $K_{diss} = 3.0 \times 10^{-16}$.

We want to find the initial concentration of OH^- required to dissolve 0.10 mol $Zn(OH)_2$ in 0.500 L. Let y be the necessary OH^- concentration.

$$Zn(OH)_2(s) \quad + \quad 2\,OH^- \quad \rightleftharpoons \quad Zn(OH)_4^{2-}$$

initial conc	y	0
conc change	$-2x$	$+x$
equil conc	$y{-}2x$	x

$$K = \frac{[Zn(OH)_4^{2-}]}{[OH^-]^2} = \frac{K_{sp}}{K_{diss}} = \frac{5.0 \times 10^{-17}}{3.0 \times 10^{-16}} = 0.17$$

We want to dissolve 0.10 mol of $Zn(OH)_2$ in 0.500 L. Therefore, the equilibrium concentration of $Zn(OH)_4^{2-}$ will be 0.10 mol/0.500 L = 0.20 M.

$$[Zn(OH)_4^{2-}] = 0.20 = x$$

Therefore,

$$[OH^-] = y - 2x = y - 2(0.20) = y - 0.40$$

Then,

$$\frac{0.20}{(y-0.40)^2} = 0.17$$

$$\frac{0.20}{0.17} = (y-0.40)^2$$

$$\sqrt{1.18} = y - 0.40$$

$$y = 1.5$$

The needed initial concentration of OH^- is 1.5 M. For 0.500 L, the amount of NaOH needed is

$$(1.5 \text{ mol/L}) \times (0.500 \text{ L}) = 0.75 \text{ mol NaOH}$$

Exercise Five

How much NaCN must be used to dissolve 0.20 mole of ZnS in 1.0 liter of solution? For ZnS, the $K_{sp} = 1.0 \times 10^{-21}$. For $Zn(CN)_4^{2-}$, the $K_{diss} = 1.0 \times 10^{-17}$.

PROBLEMS

1. Calculate $[Ag^+]$, $[Ag(NH_3)_2^+]$, and $[NH_3]$ when 1.0 gram of $AgNO_3$ is dissolved in 500.0 mL of 1.0 M NH_3. For $[Ag(NH_3)_2^+]$, the $K_{diss} = 4.0 \times 10^{-8}$.

2. What is the concentration of NH_3 when $[Cu^{2+}] = [Cu(NH_3)_4^{2+}]$ in the equilibrium
    $$Cu(NH_3)_4^{2+} \rightleftharpoons Cu^{2+} + 4\,NH_3$$
    For $Cu(NH_3)_4^{2+}$, the $K_{diss} = 2.0 \times 10^{-13}$.

3. What is the molar solubility of CuS in 1.0 M NH_3? For CuS, the $K_{sp} = 9.0 \times 10^{-36}$. For $Cu(NH_3)_4^{2+}$, the $K_{diss} = 2.0 \times 10^{-13}$.

4. What is the concentration of uncomplexed silver ions in a solution prepared by dissolving 0.010 mol of $AgNO_3$ in 0.20 M NH_3 to make 1.0 L of solution. For $Ag(NH_3)_2^+$, $K_{diss} = 5.9 \times 10^{-8}$.

5. Challenge Problem
    How much KCN is needed to prevent precipitation of $ZnCO_3$ in 1.00 liter of 3.0×10^{-3} M Zn^{2+}, when the concentration of CO_3^{2-} is 2.0×10^{-3} M? For $ZnCO_3$, the $K_{sp} = 1.4 \times 10^{-11}$. For $Zn(CN)_4^{2-}$, the $K_{diss} = 1.0 \times 10^{-17}$.

BALANCING OXIDATION-REDUCTION EQUATIONS

This lesson deals with:

1. Identifying the oxidized and reduced atoms in a chemical equation.
2. Separating a redox equation into its half-reactions.
3. Balancing redox equations for acidic and basic aqueous solutions by the ion-electron method.

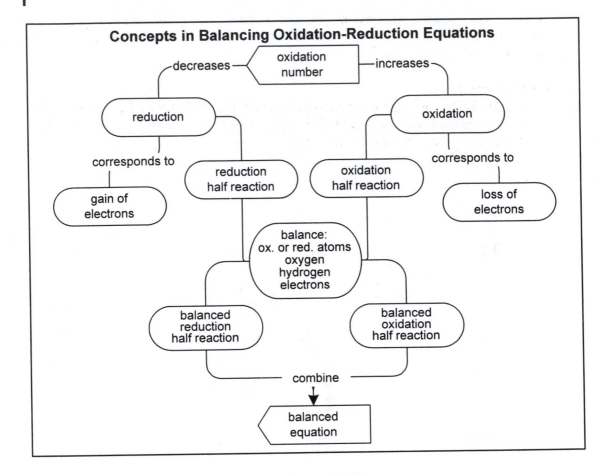

OXIDATION-REDUCTION (REDOX) REACTIONS

Definitions:

oxidation: increase in oxidation number

reduction: decrease in oxidation number

Example:

In an acid solution, hypochlorous acid reacts with dichromate ions:

$$HClO(aq) + Cr_2O_7^{2-}(aq) \longrightarrow ClO_3^-(aq) + Cr^{3+}(aq) + H_2O(l)$$

Assign oxidation numbers to all of the elements.

$$\overset{+1\qquad\ +6}{HClO(aq) + Cr_2O_7^{2-}(aq)} \longrightarrow \overset{+5}{ClO_3^-(aq)} + \overset{+3}{Cr^{3+}(aq)} + \overset{-2}{H_2O(l)}$$
$$\underset{+1\ -2\qquad\quad -2\qquad\qquad\ -2\qquad\qquad\qquad\ +1}{}$$

The oxidation number of hydrogen does not change.
The oxidation number of oxygen does not change.
The oxidation number of chlorine increases.

$$\overset{+1}{HClO} \longrightarrow \overset{+5}{ClO_3^-}$$

$HClO$ is oxidized to ClO_3^-

Although it is the oxidation number of chlorine that increases, we say that HClO (hypochlorous acid) is oxidized, because it is that material that provides the chlorine.

The oxidation number of chromium decreases.

$$\overset{+6}{Cr_2O_7^{2-}} \longrightarrow \overset{+3}{Cr^{3+}}$$

$Cr_2O_7^{2-}$ is reduced to Cr^{3+}

The oxidation-reduction (redox) reaction.

Oxidation and reduction always occur together; it is not possible to have one without the other.

Exercise One

Indicate whether or not each of the following unbalanced reactions is a redox reaction. Also, indicate which element is oxidized and which is reduced in each redox reaction.

A. $CO_2 + H_2 \longrightarrow CO + H_2O$

B. $PCl_3 + H_2O \longrightarrow H_3PO_3 + HCl$

C. $H_2S + KClO_3 \longrightarrow H_2SO_3 + KCl$

TEN STEPS FOR BALANCING REDOX EQUATIONS

Here is a summary of the steps involved in balancing a redox chemical equation. Each step will be examined in detail in the rest of the lesson.

1. Assign oxidation numbers to the elements in the reactants and products and identify which element is oxidized and which element is reduced.
2. Separate the equation into the oxidation process and the reduction process (*i.e.*, write two half-reactions).
3. Balance the number of oxidized and reduced atoms in the half-reactions.
4. Balance the elements other than oxygen and hydrogen that are neither oxidized nor reduced. (This step is used in only a few reactions.)
5. Balance the number of oxygen atoms in each half-reaction by adding H_2O molecules as needed.
6. Balance the number of hydrogen atoms in each half-reaction by adding H^+ ions as needed.
7. Balance each half-reaction with respect to charge by adding electrons so that the total charges on each side are equal.
8. Multiply the two half-reactions by appropriate integers to equate the number of electrons involved in each.
9. Add the two half-reactions. If the reaction takes place in basic solution, add to both sides of the equation a number of hydroxide ions equal to the number of hydrogen ions in the equation.
10. Check the numbers of atoms of all elements and the total charge on both sides of the equation to be sure the equation is balanced.

Step one was described in the previous section, so we will start with step two.

STEP TWO

Separate the equation into half-reactions.

In the equation from the first page of this lesson,

$$HClO(aq) + Cr_2O_7^{2-}(aq) \longrightarrow ClO_3^-(aq) + Cr^{3+}(aq)$$

we determined that $HClO$ is oxidized to ClO_3^-, and $Cr_2O_7^{2-}$ is reduced to Cr^{3+}.

For the oxidation half-reaction, put the material that is oxidized on the left of a reaction arrow, and its product on the right.

Oxidation half-reaction: $HClO \longrightarrow ClO_3^-$

Perform the analogous procedure for the reduction half-reaction.

Reduction half-reaction: $Cr_2O_7^{2-} \longrightarrow Cr^{3+}$

Exercise Two

The reactions represented by the equations below occur in acid solution.
Write the oxidation and reduction half-reactions for these equations.

A. $MnO_4^-(aq) + C_2O_4^{2-}(aq) \longrightarrow Mn^{2+}(aq) + CO_2(g)$

oxidation half-reaction:

reduction half-reaction:

B. $H_2S(aq) + NO_3^-(aq) \longrightarrow S(s) + NO(g)$

oxidation half-reaction:

reduction half-reaction:

STEP THREE

Balance the number of oxidized and reduced atoms in the half-reaction.

oxidation half-reaction: $HClO \longrightarrow ClO_3^-$ Cl balanced

There is one chlorine atom on each side, so this is already balanced.

reduction half-reaction: $Cr_2O_7^{2-} \longrightarrow Cr^{3+}$ Cr not balanced

There are two chromium atoms on the left, but only one on the right.
Balance the chromium atoms by placing a 2 in front of Cr^{3+} on the right.

$$Cr_2O_7^{2-} \longrightarrow 2\,Cr^{3+} \quad \text{Cr balanced}$$

Exercise Three

Balance the oxidized and reduced atoms in the half-reactions from Exercise Two.

A. oxidation half-reaction:

reduction half-reaction:

B. oxidation half-reaction:

reduction half-reaction:

STEP FIVE

(We are skipping step four now, because it does not apply to our example reaction. The example reaction does not involve elements other than oxygen and hydrogen that are neither oxidized nor reduced. We will consider later an example using step four.)

Balance the number of oxygen atoms in each half-reaction by adding water molecules as needed.

$$\text{oxidation half-reaction: } HClO \longrightarrow ClO_3^- \qquad \text{oxygen unbalanced}$$

There is one oxygen atom on the left, but three on the right. Two more are needed on the left. Therefore, add two water molecules on the left to provide the needed oxygen atoms.

$$2\,H_2O + HClO \longrightarrow ClO_3^- \qquad \text{oxygen balanced}$$

$$\text{reduction half-reaction: } Cr_2O_7^{2-} \longrightarrow 2\,Cr^{3+} \qquad \text{oxygen unbalanced}$$

There are seven oxygen atoms on the left, but none on the right. Add seven water molecules on the right to provide the needed oxygen atoms.

$$Cr_2O_7^{2-} \longrightarrow 2\,Cr^{3+} + 7\,H_2O \qquad \text{oxygen balanced}$$

Exercise Four

Balance the number of oxygen atoms in the half-reactions from Exercise Three.

A. oxidation half-reaction:

reduction half-reaction:

B. oxidation half-reaction:

reduction half-reaction:

STEP SIX

Balance the number of hydrogen atoms in each half-reaction by adding hydrogen ions (H^+) as needed.

$$\text{oxidation half-reaction: } 2\,H_2O + HClO \longrightarrow ClO_3^- \quad \text{H unbalanced}$$

There are five hydrogen atoms on the left, but none on the right. Supply the needed five atoms on the right by adding five H^+ ions.

$$2\ H_2O + HClO \longrightarrow ClO_3^- + 5\ H^+ \quad \text{H balanced}$$

reduction half-reaction: $Cr_2O_7^{2-} \longrightarrow 2\ Cr^{3+} + 7\ H_2O \qquad \text{H unbalanced}$

There are no hydrogen atoms on the left, but fourteen on the right. Supply the needed fourteen atoms on the left by adding 14 H^+ ions.

$$14\ H^+ + Cr_2O_7^{2-} \longrightarrow 2\ Cr^{3+} + 7\ H_2O \qquad\qquad \text{H balanced}$$

Exercise Five

Balance the number of hydrogen atoms in the half-reactions from Exercise Four.

A. oxidation half-reaction:

reduction half-reaction:

B. oxidation half-reaction:

reduction half-reaction:

STEP SEVEN

Balance each half-reaction with respect to charge by adding electrons so that the total charge on the left side of each half-reaction equals the total charge on the right side.

oxidation: $\underbrace{2\ H_2O + HClO}_{\text{total charge} = 0} \longrightarrow \underbrace{ClO_3^- + 5\ H^+}_{\text{total charge} = -1 + 5 = +4}$ charges unequal

The total charge on the left side of the oxidation half-reaction is 0, but on the right, the total charge is +4. Adding electrons will reduce the total charge. So reduce +4 on the right to 0 by adding 4 electrons.

$$2\ H_2O + HClO \longrightarrow ClO_3^- + 5\ H^+ + 4\ e^- \quad \text{charges equal}$$

reduction: $\underbrace{14\ H^+ + Cr_2O_7^{2-}}_{+14 - 2 = +12} \longrightarrow \underbrace{2\ Cr^{3+} + 7\ H_2O}_{2(+3) = +6}$ charges unequal

Add 6 electrons to the left to balance the charges.

$$6\ e^- + 14\ H^+ + Cr_2O_7^{2-} \longrightarrow 2\ Cr^{3+} + 7\ H_2O \quad \text{charges equal}$$

Exercise Six

Balance the charge in each of the half-reactions from Exercise Five.

A. oxidation half-reaction:

   reduction half-reaction:

B. oxidation half-reaction:

   reduction half-reaction:

STEP EIGHT

Multiply each half-reaction by an appropriate integer so that both half-reactions involve the same number of electrons.

$$\text{oxidation:} \quad 2\,H_2O + HClO \longrightarrow ClO_3^- + 5\,H^+ + 4\,e^-$$

$$\text{reduction:} \quad 6\,e^- + 14\,H^+ + Cr_2O_7^{2-} \longrightarrow 2\,Cr^{3+} + 7\,H_2O$$

The oxidation half-reaction involves 4 electrons, but the reduction half-reaction involves 6 electrons.

The least common multiple of 4 and 6 is 12.
If the oxidation half-reaction is multiplied by 3 and the reduction half-reaction is multiplied by 2, then both half-reactions will involve 12 electrons.

$$3 \times 4\,e^- = 12\,e^-$$
$$2 \times 6\,e^- = 12\,e^-$$

$$\text{oxidation:} \quad 6\,H_2O + 3\,HClO \longrightarrow 3\,ClO_3^- + 15\,H^+ + 12\,e^-$$

$$\text{reduction:} \quad 12\,e^- + 28\,H^+ + 2\,Cr_2O_7^{2-} \longrightarrow 4\,Cr^{3+} + 14\,H_2O$$

STEP NINE

Add the two half-reactions together, and add OH^- if the reaction takes place in basic solution.

$$6\,H_2O + 3\,HClO + 12\,e^- + 28\,H^+ + 2\,Cr_2O_7^{2-} \longrightarrow$$
$$3\,ClO_3^- + 15\,H^+ + 12\,e^- + 4\,Cr^{3+} + 14\,H_2O$$

Because the reaction takes place in acid solution, we do not need to add OH^- ions.

Always simplify the equation.

Delete the electrons and subtract excess H_2O and H^+
    Subtract 6 H_2O from each side.
    Subtract 15 H^+ from each side.

$$3\ HClO + 13\ H^+ + 2\ Cr_2O_7{}^{2-} \longrightarrow 3\ ClO_3{}^- + 4\ Cr^{3+} + 8\ H_2O$$

STEP TEN

Check to see that the equation is balanced.

We must check that:
    (a) there are the same number of atoms of each element on each side of the equation,
and
    (b) the total charge is the same on both sides of the equation.

In our equation the numbers of atoms of each element are the same on both sides, and the total charge on the left is the same as that on the right, namely +9.

Exercise Seven

Complete the balancing of the equations from Exercise Six.

A.

B.

SUMMARY OF THE STEPS

1. Identify oxidized and reduced reactants.
2. Separate the equation into oxidation and reduction half-reactions
3. Balance the oxidized and reduced atoms.
4. Balance non-redox atoms other than H and O. (We will see this case in a later example.)
5. Balance the oxygen atoms using H_2O.
6. Balance the hydrogen atoms using H^+

7. Balance the total charges by adding electrons.
8. Multiply the half-reactions by the proper integers so that both half-reactions involve the same number of electrons.
9. Add the half-reactions, eliminate the electrons, and remove excess water and excess hydrogen ions. Add OH^- if the reaction occurs in basic solution. (We will see an example of a basic solution in the next example.)
10. Check to see the equation is balanced.

Exercise Eight

Balance the following equations for reactions that take place in acidic solution.

A. $Fe^{2+}(aq) + MnO_4^-(aq) \longrightarrow Fe^{3+}(aq) + Mn^{2+}(aq)$

B. $NO_3^-(aq) + I_2(s) \longrightarrow IO_3^-(aq) + NO_2(g)$

C. $H_2O_2(aq) + I^-(aq) \longrightarrow H_2O(l) + I_2(s)$

REACTIONS THAT OCCUR IN BASIC SOLUTIONS

An additional procedure in step 9.

Convert all H^+ ions to H_2O molecules by adding the necessary number of OH^- ions to both sides of the balanced equation.

Example: $Cl_2 \longrightarrow Cl^- + ClO_3^-$ in basic solution

Step 1. Cl_2 oxidation number 0
        Cl^- oxidation number -1
        ClO_3^- oxidation number of Cl is $+5$
        Cl_2 is reduced to Cl^- and Cl_2 is oxidized to ClO_3^-

Step 2. $Cl_2 \longrightarrow ClO_3^-$ oxidation
        $Cl_2 \longrightarrow Cl^-$ reduction

Step 3. $Cl_2 \longrightarrow 2\ ClO_3^-$
        $Cl_2 \longrightarrow 2\ Cl^-$

Step 4. Doesn't apply.

Step 5. $6\ H_2O + Cl_2 \longrightarrow 2\ ClO_3^-$
        $Cl_2 \longrightarrow 2\ Cl^-$

Step 6. $6\ H_2O + Cl_2 \longrightarrow 2\ ClO_3^- + 12\ H^+$
        $Cl_2 \longrightarrow 2\ Cl^-$

Step 7. $6\ H_2O + Cl_2 \longrightarrow 2\ ClO_3^- + 12\ H^+ + 10\ e^-$
        $2\ e^- + Cl_2 \longrightarrow 2\ Cl^-$

Step 8. $6\ H_2O + Cl_2 \longrightarrow 2\ ClO_3^- + 12\ H^+ + 10\ e^-$
        $10\ e^- + 5\ Cl_2 \longrightarrow 10\ Cl^-$

Step 9. $6\ H_2O + 6\ Cl_2 \longrightarrow 2\ ClO_3^- + 12\ H^+ + 10\ Cl^-$

> The modification for basic solution:
>   Add to both sides of the equation the number of OH^- ions equal to the number of H^+ ions in the equation.

> This equation has 12 H^+ on the right, so add 12 OH^- to both sides.

> $12\ OH^- + 6\ H_2O + 6\ Cl_2 \longrightarrow 2\ ClO_3^- + 12\ H^+ + 12\ OH^- + 10\ Cl^-$

> Combine the OH^- and H^+ into H_2O molecules.

> $12\ OH^- + 6\ H_2O + 6\ Cl_2 \longrightarrow 2\ ClO_3^- + 12\ H_2O + 10\ Cl^-$

Subtract excess H_2O (6 molecules) from both sides.

$$12\ OH^- + 6\ Cl_2 \longrightarrow 2\ ClO_3^- + 6\ H_2O + 10\ Cl^-$$

Step 10. Same number of atoms on both sides; same charge (-12) on both sides.

In this example, the same substance was both oxidized and reduced. This type of reaction is called a **disproportionation** reaction.

Exercise Nine.

Balance the following equations for reactions that occur in basic solutions.

A. $S^{2-} + NO_3^- \longrightarrow NO_2^- + S(s)$

B. $CrO_4^{2-}(aq) + Fe(s) \longrightarrow Cr(OH)_3(s) + Fe(OH)_3(s)$

C. $N_2H_4 + ClO^- \longrightarrow N_2(g) + Cl^-$

STEP FOUR

Some redox reactions involve elements other than H and O that are neither oxidized nor reduced. These elements are balanced in step four.

Example: $UF_6^- + MnO_4^- \longrightarrow UO_2^{2+} + HF + Mn^{2+}$ (in acidic solution)

Step 1. $\begin{array}{cccccc} +5 & +7 & & +6 & +1 & +2 \\ UF_6^- & + MnO_4^- & \longrightarrow & UO_2^{2+} & + HF & + Mn^{2+} \\ -1 & -2 & & -2 & -1 & \end{array}$

Step 2. oxidation half-reaction: $UF_6^- \longrightarrow UO_2^{2+}$
reduction half-reaction: $MnO_4^- \longrightarrow Mn^{2+}$

Step 3. oxidation half-reaction: $UF_6^- \longrightarrow UO_2^{2+}$
reduction half-reaction: $MnO_4^- \longrightarrow Mn^{2+}$

Step 4. When a half-reaction involves atoms (other than O or H) that are neither oxidized nor reduced, they should be balanced in step 4.

Here, the oxidation half-reaction involves fluorine with a -1 oxidation number. It should be balanced using HF, which is given in the original equation.

oxidation half-reaction: $UF_6^- \longrightarrow UO_2^{2+} + 6\ HF$

Exercise Ten

Complete the balancing of the previous example. Also, balance equation B below.

A. $UF_6^- + MnO_4^- \longrightarrow UO_2^{2+} + HF + Mn^{2+}$

B. $CuCl_4^{2-} + SnCl_4^{2-} \longrightarrow CuCl(s) + SnCl_6^{2-} + Cl^-$

A FINAL NOTE

The steps outlined here work for balancing nearly any oxidation-reduction equation that can be written. This does not mean that the reaction actually occurs. Simply because a balanced equation may be written, we should not conclude that the reaction is an experimental fact.

PROBLEMS

Balance the following redox equations.

1. $C_2H_4(g) + MnO_4^- \longrightarrow Mn^{2+} + CO_2$ (acidic solution)

2. $PbO_2(s) + Cl^- \longrightarrow Pb^{2+} + Cl_2$ (acidic solution)

3. $As_2S_3 + NO_3^- \longrightarrow H_3AsO_4 + HSO_4^- + NO$ (acidic solution)

4. $H_2O_2 + ClO_2 \longrightarrow ClO_2^- + O_2$ (basic solution)

5. $HgS(s) + Cl^- + NO_3^- \longrightarrow HgCl_4^{2-} + NO_2 + S(s)$ (acidic solution)

6. $Cu(s) + NH_3 + O_2 \longrightarrow Cu(NH_3)_4^{2+}$ (basic solution)

7. Unlike the above reactions, which actually occur, the following equation does not represent an actual reaction but is only an exercise in balancing oxidation-reduction equations.

   $(NH_4)_3[Co(SCN)_6] + MnO_4^- \longrightarrow NO_3^- + SO_4^{2-} + CO_2 + Co^{2+} + Mn^{2+}$

   (acidic solution)

VOLTAIC CELLS

This lesson deals with:

1. Defining a voltaic cell.
2. Defining standard state conditions.
3. Finding the standard potential of a half-reaction by using a table of standard reduction potentials (SRP's).
4. Determining relative oxidizing and reducing agent strengths.
5. Calculating the standard potential of a redox reaction by using a table of SRP's.
6. Determining whether a redox reaction is spontaneous.
7. Determining the standard potential cell reaction of a voltaic cell, given its components.
8. Calculating the potential of a voltaic cell at non-standard state conditions using the Nernst equation.
9. Calculating the concentration of an ion in a voltaic cell, given the cell potential.
10. Calculating the equilibrium constant of a reaction using standard reduction potentials.

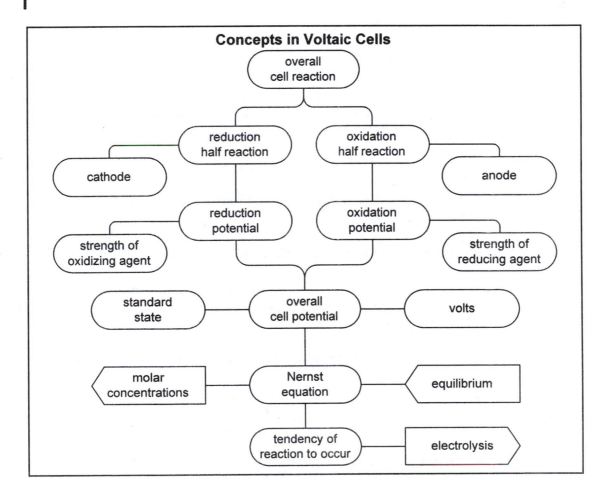

ELECTROCHEMICAL CELLS AND POTENTIALS

An oxidation-reduction reaction is one in which electrons are transferred from one reactant to another. Oxidation corresponds to the loss of electrons, and reduction corresponds to the gain of electrons. An **electrochemical cell** is a device in which an oxidation-reduction reaction takes place, such that the oxidation process occurs in a different location than the reduction process. The oxidation takes place at the surface of an electrically conductive material, called the **anode**. The reduction takes place at another electrically conductive surface called the **cathode**. Together, the anode and cathode are called **electrodes**. The electrons released at the anode travel through an **external circuit** to the cathode. The cell itself contains an electrically conductive medium called the **electrolyte**.

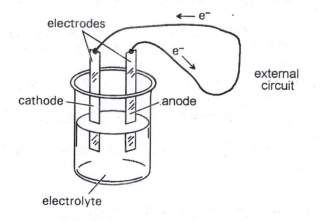

There are two kinds of electrochemical cells:
1. A **voltaic cell** uses a spontaneous oxidation-reduction reaction to produce electrical energy. Automobile batteries and AA alkaline cells are voltaic cells.
2. An **electrolytic cell** uses electrical energy to produce a non-spontaneous oxidation-reduction reaction.

The amount of energy available from a voltaic cell or needed for an electrolytic cell is expressed by the **cell potential** or **potential** (E).

Cell potential indicates the tendency of the cell reaction to occur.

Cell potential is expressed in volts.

A voltaic cell has a positive cell potential ($E > 0$).

An electrolytic cell has a negative cell potential ($E < 0$).

Voltage depends on temperature, pressure, and concentrations of cell components.

When voltage is reported, it is important to specify the temperature, pressure, and cell concentrations at which the voltage applies.

A standard set of temperature, pressure, and concentrations has been adopted:

Standard State Conditions   –   Temperature = 25°C
                                 Gas pressures = 1.00 atm
                                 Solute concentrations = 1.00 M

$E°$ (*E*-zero) is the potential of a reaction at standard state conditions.

Table 1. Standard Reduction Potentials

	Reduction Electrode Reaction							$E°$
WEAK OXIDIZING AGENTS	$Li^+(aq)$		$+ e^-$	⟶		$Li(s)$	STRONG REDUCING AGENTS	–3.05
	$K^+(aq)$		$+ e^-$	⟶		$K(s)$		–2.93
	$Ba^{2+}(aq)$		$+ 2 e^-$	⟶		$Ba(s)$		–2.90
	$Ca^{2+}(aq)$		$+ 2 e^-$	⟶		$Ca(s)$		–2.87
	$Na^+(aq)$		$+ e^-$	⟶		$Na(s)$		–2.71
	$Mg^{2+}(aq)$		$+ 2 e^-$	⟶		$Mg(s)$		–2.37
	$Al^{3+}(aq)$		$+ 3 e^-$	⟶		$Al(s)$		–1.66
	$Mn^{2+}(aq)$		$+ 2 e^-$	⟶		$Mn(s)$		–1.18
	$2 H_2O(l)$		$+ 2 e^-$	⟶	$2 OH^-(aq) +$	$H_2(g)$		–0.83
	$Zn^{2+}(aq)$		$+ 2 e^-$	⟶		$Zn(s)$		–0.76
	$Cr^{3+}(aq)$		$+ 3 e^-$	⟶		$Cr(s)$		–0.74
	$Fe^{2+}(aq)$		$+ 2 e^-$	⟶		$Fe(s)$		–0.44
	$Cr^{3+}(aq)$		$+ e^-$	⟶		$Cr^{2+}(aq)$		–0.41
	$Cd^{2+}(aq)$		$+ 2 e^-$	⟶		$Cd(s)$		–0.40
	$PbSO_4(s)$		$+ 2 e^-$	⟶	$SO_4^{2-}(aq) +$	$Pb(s)$		–0.36
	$Tl^+(aq)$		$+ e^-$	⟶		$Tl(s)$		–0.34
	$Co^{2+}(aq)$		$+ 2 e^-$	⟶		$Co(s)$		–0.28
	$Ni^{2+}(aq)$		$+ 2 e^-$	⟶		$Ni(s)$		–0.25
	$AgI(s)$		$+ e^-$	⟶	$I^-(aq) +$	$Ag(s)$		–0.15
	$Sn^{2+}(aq)$		$+ 2 e^-$	⟶		$Sn(s)$		–0.14
	$Pb^{2+}(aq)$		$+ 2 e^-$	⟶		$Pb(s)$		–0.13
	$2 H^+(aq)$		$+ 2 e^-$	⟶		$H_2(g)$		0.00
	$AgBr(s)$		$+ e^-$	⟶	$Br^-(aq) +$	$Ag(s)$		+0.10
	$S(s)$	$+ 2 H^+(aq)$	$+ 2 e^-$	⟶		$H_2S(aq)$		+0.14
	$Sn^{4+}(aq)$		$+ 2 e^-$	⟶		$Sn^{2+}(aq)$		+0.15
	$Cu^{2+}(aq)$		$+ e^-$	⟶		$Cu^+(aq)$		+0.15
	$SO_4^{2-}(aq)$	$+ 4 H^+(aq)$	$+ 2 e^-$	⟶	$2 H_2O +$	$SO_2(g)$		+0.20
	$Cu^{2+}(aq)$		$+ 2 e^-$	⟶		$Cu(s)$		+0.34
	$Cu^+(aq)$		$+ e^-$	⟶		$Cu(s)$		+0.52
	$I_2(s)$		$+ 2 e^-$	⟶		$2 I^-(aq)$		+0.53
	$Fe^{3+}(aq)$		$+ e^-$	⟶		$Fe^{2+}(aq)$		+0.77
	$Hg_2^{2+}(aq)$		$+ 2 e^-$	⟶		$2 Hg(l)$		+0.79
	$Ag^+(aq)$		$+ e^-$	⟶		$Ag(s)$		+0.80
	$2 Hg^{2+}(aq)$		$+ 2 e^-$	⟶		$Hg_2^{2+}(aq)$		+0.92
	$NO_3^-(aq)$	$+ 4 H^+(aq)$	$+ 3 e^-$	⟶	$2 H_2O(l) +$	$NO(g)$		+0.96
STRONG OXIDIZING AGENTS	$AuCl_4^-(aq)$		$+ 3 e^-$	⟶	$4 Cl^-(aq) +$	$Au(s)$	WEAK REDUCING AGENTS	+1.00
	$Br_2(l)$		$+ 2 e^-$	⟶		$2 Br^-(aq)$		+1.07
	$O_2(g)$	$+ 4 H^+(aq)$	$+ 4 e^-$	⟶		$2 H_2O(l)$		+1.23
	$MnO_2(s)$	$+ 4 H^+(aq)$	$+ 2 e^-$	⟶	$2 H_2O(l) +$	$Mn^{2+}(aq)$		+1.23
	$Cr_2O_7^{2-}(aq)$	$+ 14 H^+(aq)$	$+ 6 e^-$	⟶	$7 H_2O(l) +$	$2 Cr^{3+}(aq)$		+1.33
	$Cl_2(g)$		$+ 2 e^-$	⟶		$2 Cl^-(aq)$		+1.36
	$ClO_3^-(aq)$	$+ 6 H^+(aq)$	$+ 5 e^-$	⟶	$3 H_2O(l) +$	$^1/_2 Cl_2(g)$		+1.47
	$Au^{3+}(aq)$		$+ 3 e^-$	⟶		$Au(s)$		+1.50
	$MnO_4^-(aq)$	$+ 8 H^+(aq)$	$+ 5 e^-$	⟶	$4 H_2O(l) +$	$Mn^{2+}(aq)$		+1.52
	$H_2O_2(aq)$	$+ 2 H^+(aq)$	$+ 2 e^-$	⟶		$2 H_2O(l)$		+1.77
	$Co^{3+}(aq)$		$+ e^-$	⟶		$Co^{2+}(aq)$		+1.82
	$F_2(g)$		$+ 2 e^-$	⟶		$2 F^-(aq)$		+2.87

⇑ oxidizing agents reducing agents ⇑

The cell potential results from the difference between the potentials of the cell's two electrodes. However, the potential of a single electrode cannot be measured. Only the potential difference between two electrodes (the cell potential) can be measured.

Yet, it is convenient to assign values to the individual electrode potentials. To set the value of the potential for a single electrode, one electrode is selected as a standard and its value defined.

The standard reference electrode is the hydrogen electrode. At this electrode, hydrogen gas is reduced to hydrogen ions in aqueous solution.

$$2 \, H^+(aq) + 2 \, e^- \longrightarrow H_2(g)$$

The standard potential of the hydrogen electrode has been set at zero volts.

Therefore, the potential of a cell constructed by combining the hydrogen electrode with a second electrode gives the potential of the second electrode.

Reversing the direction of an electrode reaction changes the sign of the electrode.
Reduction of silver ion
$$Ag^+ + e^- \longrightarrow Ag \qquad \text{Reduction potential is } +0.80 \text{ volt.}$$
Oxidation of silver
$$Ag \longrightarrow Ag^+ + e^- \qquad \text{Oxidation potential is } -0.80 \text{ volt.}$$

The differences in potential between the hydrogen electrode and other electrodes have been measured and some are listed in Table 1. These electrode potentials are often called **half-cell potentials**, and their reactions are called **half-reactions**.

The greater the half-cell potential, the greater the tendency of the half-reaction to occur.

$$Ni^{2+} + 2 \, e^- \longrightarrow Ni \qquad E° = -0.25 \text{ volt}$$
$$Al^{3+} + 3 \, e^- \longrightarrow Al \qquad E° = -1.66 \text{ volts}$$

−0.25 is greater than −1.66 therefore Ni^{2+} is more readily reduced than Al^{3+}.

$$Ni \longrightarrow Ni^{2+} + 2 \, e^- \qquad E° = +0.25 \text{ volt}$$
$$Al \longrightarrow Al^{3+} + 3 \, e^- \qquad E° = +1.66 \text{ volts}$$

+1.66 is greater than +0.25; therefore Al is more readily oxidized than Ni.

Exercise One

A.   Define a voltaic cell.

B.   Define anode and cathode.

C.   Define standard state conditions.

D.   Give the standard potential of each of the following half-reactions:

1. $Zn^{2+} + 2\,e^- \longrightarrow Zn$

2. $Cl_2 + 2\,e^- \longrightarrow 2\,Cl^-$

3. $Fe^{2+} \longrightarrow Fe^{3+} + e^-$

4. $Co^{2+} \longrightarrow Co^{3+} + e^-$

E.   Of the reactions in D above, which will occur more readily at standard state conditions,

1 or 2 ?

3 or 4 ?

RELATIVE STRENGTHS OF OXIDIZING AND REDUCING AGENTS

An **oxidizing agent** causes another substance to be oxidized.
   The oxidizing agent is reduced in the process.

$Ni^{2+} + 2\,e^- \longrightarrow Ni \qquad E° = -0.25 \text{ volt}$
$Al^{3+} + 3\,e^- \longrightarrow Al \qquad E° = -1.66 \text{ volt}$

   Ni^{2+} is more readily reduced than Al^{3+}, so Ni^{2+} is more likely to act as an oxidizing agent. Therefore Ni^{2+} is a stronger oxidizing agent than Al^{3+}.

A **reducing agent** causes another substance to be reduced.
   The reducing agent is oxidized in the process.

$Ni \longrightarrow Ni^{2+} + 2\,e^- \qquad E° = +0.25 \text{ volt}$
$Al \longrightarrow Al^{3+} + 3\,e^- \qquad E° = +1.66 \text{ volt}$

   Al is more readily oxidized than Ni, so Al is more likely to act as an reducing agent. Therefore Al is a stronger reducing agent than Ni.

Example 1

Arrange Fe^{2+}, I^-, Cr^{2+}, Sn^{2+} in order of decreasing strength as reducing agents.

A reducing agent is oxidized, so we compare oxidation potentials:

$$Fe^{2+} \longrightarrow Fe^{3+} + e^- \qquad E^\circ = -0.77 \text{ volt}$$

$$2\,I^- \longrightarrow I_2 + 2\,e^- \qquad E^\circ = -0.53 \text{ volt}$$

$$Cr^{2+} \longrightarrow Cr^{3+} + e^- \qquad E^\circ = +0.41 \text{ volt}$$

$$Sn^{2+} \longrightarrow Sn^{4+} + 2\,e^- \qquad E^\circ = -0.15 \text{ volt}$$

Cr^{2+} has the largest oxidation potential; therefore it is the strong reducing agent.

Then, in order of decreasing strength:

$$\text{strongest} \;=\; Cr^{2+} > Sn^{2+} > I^- > Fe^{2+} \;=\; \text{weakest}$$

Table 1 summarizes the relationships between the strengths of oxidizing and reducing agents. Oxidizing agents are in the column at the left, and reducing agents are on the right. The strong oxidizing agents are at the bottom of the column, but the strong reducing agents are at the top of their column. (In other tables, the half-reactions may be listed in different orders, so these relationships don't apply to all tables of electrode potentials.)

Exercise Two

A. Arrange the following ions in order of decreasing strength as oxidizing agents:

$$Au^{3+} \qquad Pb^{2+} \qquad Fe^{3+} \qquad NO_3^-$$

B. In each set below, select the strongest reducing agent:

1. Zn, Fe, Ni

2. H_2, Ag, Pb

3. Br^-, Cl^-, I^-

4. Cu^+, Br^-, Fe^{2+}

FINDING REACTION POTENTIALS

Example 2

Find the standard potential of the following reaction.

$$6 \, I^- + 2 \, NO_3^- + 8 \, H^+ \longrightarrow 3 \, I_2 + 2 \, NO + 4 \, H_2O$$

1. Identify the oxidation
    $$I^- \longrightarrow I_2$$

    Find the oxidation potential
    $$2 \, I^- \longrightarrow I_2 + 2 \, e^- \qquad E° = -0.53 \text{ V}$$

2. Identify the reduction
    $$NO_3^- \longrightarrow NO$$

    Find the reduction potential
    $$NO_3^- + 4 \, H^+ + 3 \, e^- \longrightarrow NO + 2 \, H_2O \qquad E° = +0.96 \text{ V}$$

3. To find the potential of the reaction, add the half-reaction potentials:
    $$-0.53 + 0.96 = 0.43 \text{ volt}$$

Exercise Three

Find the standard potentials for the following reactions:

A. $Cu^+ + Ag^+ \longrightarrow Cu^{2+} + Ag$

B. $3 \, H_2S + 2 \, NO_3^- + 2 \, H^+ \longrightarrow 3 \, S + 2 \, NO + 4 \, H_2O$

C. $Mn^{2+} + Br_2 + 2 \, H_2O \longrightarrow MnO_2 + 4 \, H^+ + 2 \, Br^-$

SPONTANEITY OF A REDOX REACTION

A **spontaneous reaction** is one that occurs in the direction its equation is written. Spontaneous reactions may be fast or slow.

The sign of a reaction potential indicates whether or not the reaction is spontaneous.

If the reaction potential is positive, the reaction is spontaneous.

If the reaction potential is negative, the reaction is non-spontaneous. In fact, the reverse reaction is spontaneous—i.e., the reaction will tend to proceed in the reverse of the way it is written.

In Exercise Three, reactions (a) and (b) are spontaneous, while reaction (c) is non-spontaneous.

Exercise Four

Determine whether each of the following reactions is spontaneous or non-spontaneous.

A. $H_2S + Sn^{2+} \longrightarrow 2 H^+ + S + Sn$

B. $2 I^- + Br_2 \longrightarrow 2 Br^- + I_2$

C. $SO_4^{2-} + 4 H^+ + Co \longrightarrow Co^{2+} + SO_2 + 2 H_2O$

A VOLTAIC CELL

Example 3

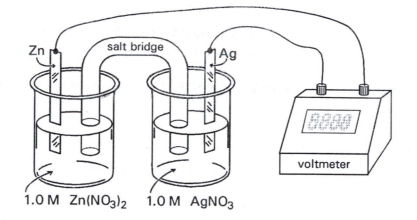

1.0 M Zn(NO₃)₂   1.0 M AgNO₃

This is a voltaic cell composed of a zinc electrode immersed in 1.0 M zinc nitrate and a silver electrode immersed in 1.0 M silver nitrate. A salt bridge connects the two solutions. What is the standard potential of this cell?

Reduction must occur at one electrode and oxidation at the other.
If oxidation occurs at the zinc electrode, reduction must occur at the silver electrode.
If oxidation occurs at the silver electrode, reduction must occur at the zinc electrode.

We do not need to know which of these is the case. We can guess, and if our guess turns out to be wrong, we can correct it.

If the oxidation is $Ag \longrightarrow Ag^+ + e^-$ $E° = -0.80$ V
then the reduction is $Zn^{2+} + 2\,e^- \longrightarrow Zn$ $E° = -0.76$ V

     This means $E°$ for the reaction is $-0.80 + (-0.76) = -1.56$ V.

The reaction in any voltaic cell is a spontaneous reaction. Therefore, its $E°$ must be greater than zero.

     Because $E°$ must be positive, the cell reactions occur in the opposite direction of the initial guess.

         $Ag^+ + e^- \longrightarrow Ag$ $E° = 0.80$ V
         $Zn \longrightarrow Zn^{2+} + 2\,e^-$ $E° = 0.76$ V

     The $E°$ for the cell is $+1.56$ V.

Now we know the oxidation and reduction half-reactions in this cell:

         Oxidation: $Zn \longrightarrow Zn^{2+} + 2\,e^-$
         Reduction: $Ag^+ + e^- \longrightarrow Ag$

We also know the overall cell reaction:

$$2 Ag^+ + Zn \longrightarrow 2 Ag + Zn^{2+}$$

Direction of electron flow in the external circuit:

$$Zn \longrightarrow Zn^{2+} + 2 e^-$$ electrons are released at the zinc electrode
$$Ag^+ + e^- \longrightarrow Ag$$ electrons are consumed at the silver electrode

So electrons travel away from the zinc electrode and toward the silver electrode.

Direction of flow of ions in the salt bridge:

The electrons carry negative charge to the silver half-cell.

The negative ions in the salt bridge flow away from the silver half-cell.

The positive ions in the salt bridge flow toward the silver half-cell.

Exercise Five

Consider a voltaic cell that includes a lead electrode dipped into a 1.0 M $Pb(NO_3)_2$ solution and a cadmium electrode dipped into a 1.0 M solution of $Cd(NO_3)_2$ with a salt bridge connecting the two solutions.

A.   Calculate the cell potential.

B.   Write the cell reaction.

C.   Write the oxidation and reduction half-reactions.

D.   Indicate the direction of electron flow in the external circuit.

E.   In what directions do the cations and anions move in the salt bridge?

NON-STANDARD CONDITIONS

Non-standard concentrations

When the concentrations of solutes are not 1.0 M, the cell potential can be affected. The effect of non-standard concentrations can be calculated by the **Nernst Equation**.

$$E_{cell} = E° - \frac{0.0257}{n} \ln Q$$

E_{cell} is the potential of the cell.

$E°$ is the potential of the cell at standard-state conditions.

n is the number of electrons transferred in the balanced cell reaction.

Q is the reaction quotient (described in Lesson 17).

Example 4

Calculate the potential of this cell.

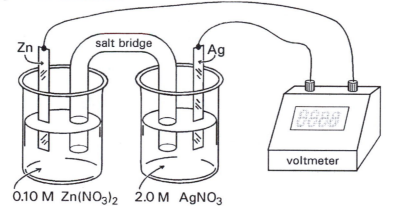

0.10 M $Zn(NO_3)_2$ 2.0 M $AgNO_3$

The concentrations are not 1.0 M, so the Nernst equation must be used to calculate the cell potential.

$$E_{cell} = E° - \frac{0.0257}{n} \ln Q$$

To calculate E_{cell}, we need $E°$, n, and the value of Q.

From Example 3, $E° = 1.56$ volts.

To find n, we must examine the balanced cell reaction and its half reactions.

At standard conditions, the cell reaction is
$$Zn + 2 Ag^+ \longrightarrow Zn^{2+} + 2 Ag$$

The half-reactions are
$$Zn \longrightarrow Zn^{2+} + 2 e^-$$
$$2 Ag^+ + 2 e^- \longrightarrow 2 Ag$$

Two electrons are transferred in the *balanced* equation; therefore $n = 2$.

Q has the form of the equilibrium constant of the cell reaction.
$$Q = \frac{[Zn^{2+}]}{[Ag^+]^2} = \frac{(0.10)}{(2.0)^2} = 0.025$$

From the Nernst equation:
$$E_{cell} = 1.56 - \frac{0.0257}{2} \ln 0.025$$
$$E_{cell} = 1.56 - (0.0129)(-3.69) = 1.61 \text{ volts}$$

These non-standard concentrations had only a small effect on the cell potential.

Exercise Six

Calculate the potential of a voltaic cell that uses a lead electrode dipped into a 2.0 M solution of $Pb(NO_3)_2$ and a silver electrode dipped into a 0.0010 M solution of $AgNO_3$.

FINDING CONCENTRATIONS FROM CELL POTENTIALS

Example 5

The measured cell potential is 0.900 volt. What is $[Cu^{2+}]$ in the copper half-cell?

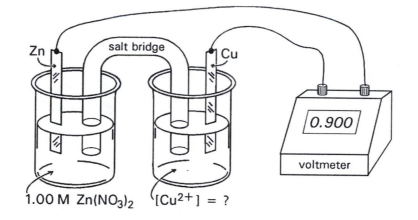

1.00 M $Zn(NO_3)_2$ $[Cu^{2+}] = ?$

The half-reactions are

$$Zn \longrightarrow Zn^{2+} + 2\,e^- \qquad E° = 0.76 \text{ volt}$$
$$Cu^{2+} + 2\,e^- \longrightarrow Cu \qquad E° = 0.34 \text{ volt}$$

The overall cell reaction is

$$Zn + Cu^{2+} \longrightarrow Zn^{2+} + Cu \qquad E° = 1.10 \text{ volts}$$

For this non-standard cell

$$n = 2$$
$$E_{cell} = 0.90 \text{ volt}$$
$$Q = \frac{[Zn^{2+}]}{[Cu^{2+}]} = \frac{1.00}{[Cu^{2+}]}$$

From the Nernst equation: $E_{cell} = E° - \dfrac{0.0257}{n} \ln Q$

$$0.900 = 1.10 - \frac{0.0257}{2} \ln \frac{1.00}{[Cu^{2+}]}$$

$$-0.200 = -0.0129 \ln \frac{1.00}{[Cu^{2+}]}$$

$$15.5 = \ln \frac{1.00}{[Cu^{2+}]}$$

$$\frac{1.00}{[Cu^{2+}]} = e^{15.5} = 5.39 \times 10^6$$

$$[Cu^{2+}] = 1.8 \times 10^{-7} \text{ M}$$

Exercise Seven

Calculate the concentration of Pb^{2+} ion in a half-cell that gives a potential of 1.00 volt when connected to a 0.10 M silver ion half-cell. (Use the results from Exercise Six.)

DETERMINING THE VALUE OF EQUILIBRIUM CONSTANTS

The reaction quotient, Q, has the *form* of the equilibrium constant expression.

When E_{cell} is greater than zero, the reaction is spontaneous in the forward direction.

When E_{cell} is less than zero, the reaction is spontaneous in the reverse direction.

When E_{cell} is equal to zero, the reaction is spontaneous in neither direction — the reaction is at equilibrium.

When E_{cell} is zero, Q has the *value* of the equilibrium constant.

Example 6

Calculate the value of the equilibrium constant for the following reaction.

$$2\, Fe^{3+} + Sn^{2+} \rightleftharpoons 2\, Fe^{2+} + Sn^{4+}$$

The half reactions are

$$Fe^{3+} + e^- \longrightarrow Fe^{2+} \qquad\qquad E° = +0.77 \text{ volt}$$
$$Sn^{2+} \longrightarrow Sn^{4+} + 2\, e^- \qquad\qquad E° = -0.15 \text{ volt}$$

The overall cell reaction and standard potential are

$$2\, Fe^{3+} + Sn^{2+} \rightleftharpoons 2\, Fe^{2+} + Sn^{4+} \qquad E° = +0.62 \text{ volt}$$

For the cell reaction, two moles of electrons are transferred, so

$$n = 2.$$

When $E_{cell} = 0$, then $Q = K$.

From the Nernst equation

$$E_{cell} = E° - \frac{0.0257}{n} \ln Q$$

$$0 = +0.62 - \frac{0.0257}{2} \ln K$$

$$-0.62 = -0.0129 \ln K$$

$$\ln K = 48.06$$

$$K = e^{48.06} = 7.4 \times 10^{20}$$

Exercise Eight

Calculate the value of the equilibrium constant for the following reaction.

$$2\,Fe^{3+}(aq) + 2\,I^-(aq) \rightleftharpoons 2\,Fe^{2+}(aq) + I_2(aq)$$

PROBLEMS

1. Consider the following list:

$$Mg \quad Mg^{2+} \quad I_2 \quad I^- \quad Co^{3+} \quad Co^{2+} \quad Cl_2 \quad Cl^-$$

   (a) Which is the strongest reducing agent?
   (b) Which is the weakest oxidizing agent?
   (c) Which is the strongest oxidizing agent?
   (d) Which is the weakest reducing agent?

2. For the redox reaction

$$H_2O_2 + 2\,I^- + 2\,H^+ \longrightarrow 2\,H_2O + I_2$$

   (a) calculate $E°$.
   (b) calculate the value of the equilibrium constant at 25°C.

3. Calculate the K_{sp} of $PbSO_4$ using the table of standard reduction potentials.

4. What is the voltage at 25°C for the voltaic cell of Ni and Ag electrodes if the Ni electrode is dipped into a solution of 0.050 M $NiCl_2$ and the Ag electrode is dipped into a solution of 2.0 M $AgNO_3$.

5. A zinc-copper voltaic cell has a measured E_{cell} of 0.71 volt when $[Zn^{2+}]$ is 1.0 M.

   (a) Write the overall cell reaction.
   (b) What is $E°$?
   (c) What is $[Cu^{2+}]$ in the copper half-cell?

6. The two half-cell reactions in the lead storage battery are:

$$PbO_2(s) + 3\,H^+ + HSO_4^- + 2\,e^- \longrightarrow PbSO_4(s) + 2\,H_2O \qquad E° = 1.64\ V$$

$$PbSO_4(s) + H^+ + 2\,e^- \longrightarrow Pb(s) + HSO_4^- \qquad E° = -0.36\ V$$

   (a) Write the balanced equation for the overall cell reaction during use of the battery.
   (b) Calculate the cell potential.
   (c) How many such cells does a 12-volt car battery contain?
   (d) Which reaction occurs at the anode and which at the cathode?
   (e) What is the equilibrium constant of the cell reaction that occurs during use of the battery?

7. Calculate the value of the equilibrium constant that corresponds to the reaction produced when the following two half-reactions are combined.

$$AgBr(s) + e^- \longrightarrow Ag(s) + Br^-(aq)$$
$$Ag(s) \longrightarrow Ag^+ + e^-$$

   What is the name generally given to this sort of equilibrium constant?

ELECTROLYTIC CELLS

This lesson deals with:

1. Writing anode and cathode half-reactions and the overall cell reaction for an electrolytic cell containing either a molten salt or the aqueous solution of a salt.
2. Calculating the moles of electrons needed to produce a given amount of a substance in an electrolytic cell.
3. Converting the various units used in measuring electricity (faraday, coulomb, ampere, and mole).
4. Using Faraday's Laws to calculate the amount of substance produced by a given current and the change in mass of the electrodes during an electrolytic process.

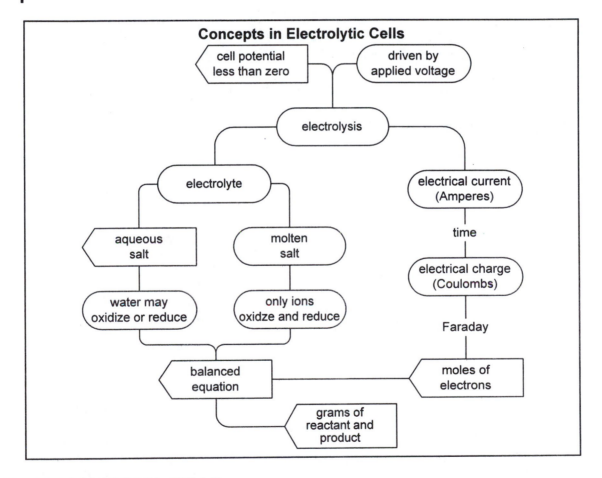

ELECTROCHEMICAL CELLS

An **electrochemical cell** is a device in which an oxidation-reduction reaction takes place, such that the oxidation process occurs in a different location than the reduction process. The oxidation takes place at the surface of an electrically conductive material, called the **anode**. The reduction takes place at another electrically conductive surface called the **cathode**.

Together, the anode and cathode are called **electrodes**. The electrons released at the anode travel through an **external circuit** to the cathode. The cell itself contains an electrically conductive medium called the **electrolyte**.

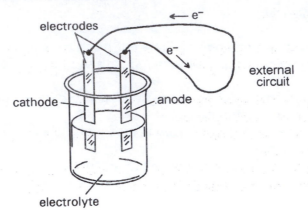

There are two kinds of electrochemical cells:
1. A **voltaic cell** uses a spontaneous oxidation-reduction reaction to produce electrical energy. Automobile batteries and AA alkaline cells are voltaic cells.
2. An **electrolytic cell** uses electrical energy to produce a non-spontaneous oxidation-reduction reaction. The process is called **electrolysis**.

ELECTROLYSIS OF A MOLTEN SALT

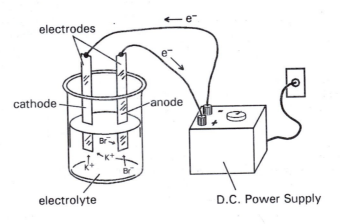

Electrolysis of molten KBr

The power supply pushes electrons into one of the electrodes, the cathode. This gives the cathode a negative charge, attracting positive ions (cations) in the electrolyte. In the diagram, the electrolyte is molten potassium bromide, and potassium ions move toward the cathode. Potassium ions combine with electrons at the cathode and form potassium metal.

$$\text{Cathode reaction: } K^+(l) + e^- \longrightarrow K(l)$$

The power supply pulls electrons from the other electrode, the anode. This gives the anode a positive charge, attracting negative ions (anions) in the electrolyte. In molten KBr, bromide ions move toward the anode. Bromide ions release electrons at the anode and form bromine gas.

Anode reaction: $2\ Br^-(l) \longrightarrow Br_2(g) + 2\ e^-$

The overall cell reaction is the sum of the electrode reactions. The number of electrons involved at the cathode must be the same as that at the anode. Therefore, the cathode reaction is multiplied by 2.

Cathode:	$2\ K^+(l)\ +\ 2\ e^-\ \longrightarrow$	$2\ K(l)$
Anode:	$2\ Br^-(l)\ \longrightarrow$	$Br_2(g)\ +\ 2e^-$
Overall cell reaction:	$2\ K^+(l)\ +2\ Br^-(l)\ \longrightarrow$	$2\ K(l)\ +\ Br_2(g)$

In the electrolysis of the molten salt of a metal,
  the *cation* is reduced to the metal at the *cathode*, and
  the *anion* is oxidized at the *anode*.

Exercise One

Write the cathode and anode half-reactions and the overall cell reaction that occur in the electrolysis of each of the following molten salts:

A. molten NaF

B. molten $CaCl_2$

C. molten Li_2O

D. molten Al_2S_3

ELECTROLYSIS OF AQUEOUS SOLUTIONS

In the electrolysis of aqueous KBr,
> bromine is produced at the anode, and
> hydrogen gas is produced at the cathode.

This is different than the electrolysis of molten KBr, in which potassium metal is formed at the cathode.

In the electrolysis of aqueous solutions:

> If the cations are more difficult to reduce than water, then water will be reduced.
> Determine if the cation is more difficult to reduce by comparing its standard reduction potential to that of water.

$$2\ H_2O(l)\ +\ 2\ e^-\ \longrightarrow\ H_2(g)\ +\ 2\ OH^-\qquad E° = -0.83\ \text{volt}$$

$$K^+(aq)\ +\ e^-\ \longrightarrow\ K(s)\qquad E° = -2.93\ \text{volt}$$

> The reduction potentials indicate that K^+ is more difficult to reduce than water, so water is reduced during electrolysis of aqueous KBr.

> Generally, cations that cannot be reduced in aqueous solution include the cations of A-group metals, such as K^+ and Ca^{2+}.

> If the anions are more difficult to oxidize than water, then water will be oxidized.
> Determine if the anion is more difficult to oxidize by comparing its standard oxidation potential to that of water.

$$2\ H_2O(l)\ \longrightarrow\ O_2(g) + 4\ H^+ + 4\ e^-\qquad E° = -1.23\ \text{volts}$$

$$2\ Br^-(aq)\ \longrightarrow\ Br_2(l)\ +\ 2\ e^-\qquad E° = -1.07\ \text{volts}$$

> The oxidation potentials indicate that H_2O is more difficult to oxidize than Br^-, so the bromide ions are oxidized in the electrolysis of aqueous KBr.

> An exception to this principle: the chloride ion.
> The aqueous chloride ion is easier to oxidize than water, although its oxidation potential would indicate otherwise. This happens because standard electrode potentials are measured under equilibrium conditions, which are not the situation during electrolysis. The difference between the electrode potential at equilibrium and during electrolysis is called **overpotential**. Because the oxidation potential of chloride ions is close to that of water, the overpotential is significant. The overpotential of water makes it more difficult to oxidize than chloride ions. Therefore, **chloride ions are oxidized in the electrolysis of aqueous chloride solutions**.

> Here's a general rule for oxygen-containing anions.
> In some of these anions, the central atom has its highest possible oxidation number. For example, in SO_4^{2-} the oxidation number of S is +6, and in NO_3^- the oxidation number of N is +5. Such anions are not oxidized in the electrolysis of solutions of their salts. Therefore, the electrolysis of aqueous copper sulfate produces oxygen gas at the anode.

Example I

What are the electrode reactions and the overall reaction in the electrolysis of an aqueous copper sulfate solution?

Identify the ions in copper sulfate.
$CuSO_4$ contains Cu^{2+} and SO_4^{2-}

Anions migrate to the anode where oxidation takes place.
SO_4^{2-} anions are not oxidized in aqueous solution. Therefore, water is oxidized.

Anode reaction: $2 H_2O(l) \longrightarrow O_2(g) + 4 H^+ + 4 e^-$

Cations migrate to the cathode, where reduction takes place.
Cu^{2+} can be reduced in aqueous solution:

Cathode reaction: $Cu^{2+} + 2 e^- \longrightarrow Cu(s)$

The overall cell reaction is the sum of the two electrode reactions.
Both electrode reactions must involve the same number of electrons.
The cathode reaction should be multiplied by 2.

$$2 H_2O \longrightarrow O_2 + 4 H^+ + 4 e^-$$
$$2 Cu^{2+} + 4 e^- \longrightarrow 2 Cu$$

Overall cell reaction: $2 Cu^{2+} + 2 H_2O \longrightarrow 2 Cu + O_2 + 4 H^+$

Exercise Two

Write the anode and cathode reactions and the overall cell reaction for the electrolysis of each of the following salts in aqueous solution.

A.   aqueous $CuBr_2$

B.   aqueous $AgNO_3$

C.   aqueous $CaCl_2$

D.   aqueous $Mg(NO_3)_2$

STOICHIOMETRY OF ELECTRODE REACTIONS

Example 2

If we want 10 grams of Ca from molten $CaCl_2$, how many moles of electrons must be used?

$$Ca^{2+} + 2\,e^- \longrightarrow Ca$$

$$(10 \text{ grams Ca}) \left(\frac{1 \text{ mol Ca}}{40.0 \text{ g Ca}} \right) = 0.25 \text{ mol Ca}$$

$$(0.25 \text{ mol Ca}) \left(\frac{2 \text{ mol e}^-}{1 \text{ mol Ca}} \right) = 0.50 \text{ mol e}^-$$

Exercise Three

A.   How many moles of electrons are needed to produce 38 grams of F_2 by electrolysis of *molten* NaF?

B.   How many grams of oxygen can be produced with two moles of electrons in the electrolysis of *aqueous* NaF?

UNITS FOR MEASURING ELECTRICITY

Units of electrical charge

Coulomb (C): one mole of electrons has a charge of 96500 coulombs (9.65×10^4 coulombs).

$$96500 \text{ coulombs } = 1 \text{ mole e}^-$$

Faraday (F): the charge on one mole of electrons.

$$1 \text{ F } = 1 \text{ mole e}^-$$
$$1 \text{ F } = 96500 \text{ coulombs}$$

Unit of electric current

Ampere (A): a current equal to 1 coulomb per second

$$1 \text{ ampere} = 1 \text{ coulomb/second}$$
$$(\text{amperes}) \times (\text{seconds}) = \text{coulombs}$$

Important Electrochemical Unit Relationships
1 mole electrons = 1 Faraday
1 Faraday = 96500 coulombs
1 ampere = 1 coulomb/second

Example 3

A. Convert 3120 coulombs to moles of electrons.

Use these relationships:

$$96500 \text{ C } = 1 \text{ F} \qquad \text{and} \qquad 1 \text{ F } = 1 \text{ mole e}^-$$

$$(3120 \text{ C})\left(\frac{1 \text{ F}}{96500 \text{ C}}\right)\left(\frac{1 \text{ mole e}^-}{1 \text{ F}}\right) = 0.0323 \text{ mole e}^-$$

B. How long must a current of 5.0 amperes flow to produce 10.0 moles of electrons?

$$(10.0 \text{ mol e}^-)\left(\frac{1 \text{ F}}{1 \text{ mol e}^-}\right) = 10.0 \text{ F}$$

$$(10.0 \text{ F})\left(\frac{9.65 \times 10^4 \text{ C}}{1 \text{ F}}\right) = 9.65 \times 10^5 \text{ C}$$

$$(9.65 \times 10^5 \text{ C})\left(\frac{1 \text{ second}}{5.0 \text{ C}}\right) = 1.9 \times 10^5 \text{ seconds}$$

Exercise Four

A. How many coulombs would be equivalent to 4.2 moles of electrons?

B. How many moles of electrons would be transferred by a current of 8.0 amperes flowing for 6.0 hours?

C. What current is needed to produce 0.30 mole of electrons in twenty seconds?

D. How long would it take to produce 5000 coulombs from a current of 0.10 amperes?

Example 4

What current is necessary to produce 1.0 gram of nickel per hour from an aqueous solution of $NiCl_2$?

$$Ni^{2+} + 2\,e^- \longrightarrow Ni$$

Current is expressed as amperes, which are coulombs per second. We must calculate both coulombs and seconds.

1. Find coulombs.

$$1.0 \text{ g Ni} \left(\frac{1 \text{ mol Ni}}{58.71 \text{ g Ni}} \right) \left(\frac{2 \text{ mol e}^-}{1 \text{ mol Ni}} \right) = 0.034 \text{ mol e}^-$$

$$0.034 \text{ mol } e^- \left(\frac{1 \text{ F}}{1 \text{ mol } e^-}\right)\left(\frac{96500 \text{ C}}{1 \text{ F}}\right) = 3.3 \times 10^3 \text{ C}$$

2. Find seconds.

$$1.0 \text{ hour} \left(\frac{60 \text{ min}}{1 \text{ hour}}\right)\left(\frac{60 \text{ sec}}{1 \text{ min}}\right) = 3600 \text{ sec}$$

3. Find current.

$$\frac{3.3 \times 10^3 \text{ C}}{3600 \text{ sec}} = 0.92 \text{ C/sec} = 0.92 \text{ ampere}$$

Example 5

In the electrolysis of aqueous NaCl, how many grams of NaOH can be produced by 15400 coulombs?

$$2 \text{ H}_2\text{O} + 2 e^- \longrightarrow \text{H}_2 + 2 \text{ OH}^-$$

$$15400 \text{ C} \left(\frac{1 \text{ F}}{96500 \text{ C}}\right)\left(\frac{1 \text{ mol } e^-}{1 \text{ F}}\right) = 0.160 \text{ mol } e^-$$

$$0.160 \text{ mol } e^- \left(\frac{2 \text{ mol OH}^-}{2 \text{ mol } e^-}\right)\left(\frac{1 \text{ mol NaOH}}{1 \text{ mol OH}^-}\right) = 0.160 \text{ mol NaOH}$$

$$0.160 \text{ mol NaOH} \left(\frac{40.0 \text{ g NaOH}}{1 \text{ mol NaOH}}\right) = 6.40 \text{ g NaOH}$$

Exercise Five

A. How many grams of silver can be produced by passing a current of 5.0 amperes through an aqueous solution of $AgNO_3$ for 2.0 hours?

B.   How long would it take to produce 80.0 grams of Br_2 from aqueous KBr using a current of 5.0 amperes?

Example 6

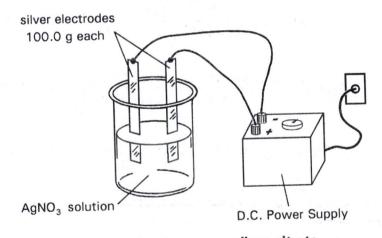

silver electrodes
100.0 g each

$AgNO_3$ solution

D.C. Power Supply

Electrolysis of aqueous silver nitrate

An electrolytic cell has two 100.0-gram silver electrodes dipped into a silver nitrate solution. If an 0.80-ampere current flows through this cell for 1.2 hours, how much will each electrode weigh at the end of the process?

$$\text{cathode: } Ag^+ + e^- \longrightarrow Ag$$

$$\text{anode: } Ag \longrightarrow Ag^+ + e^-$$

0.80 ampere = 0.80 coulomb/second

$$1.2 \text{ hours} \left(\frac{3600 \text{ sec}}{1 \text{ hour}} \right) \left(\frac{0.80 \text{ C}}{1 \text{ sec}} \right) = 3.46 \times 10^3 \text{ C}$$

$$3.46 \times 10^3 \text{ C} \left(\frac{1 \text{ F}}{96500 \text{ C}} \right) \left(\frac{1 \text{ mol e}^-}{1 \text{ F}} \right) = 0.036 \text{ mol e}^-$$

Ag is removed from the anode, so the anode loses mass.

weight of anode = 100.0 grams − 3.9 grams = 96.1 grams

Ag is formed on the cathode, so the cathode gains mass.

weight of cathode = 100.0 grams + 3.9 grams = 103.9 grams

Exercise Six

Each one of two silver electrodes weighs 20.0 grams. Find the final weight of each after they are used to pass a current of 0.460 ampere through a solution of silver nitrate for 6.00 hours.

PROBLEMS

1. An early electricity meter consisted of a cell in which a solution of $AgNO_3$ was electrolyzed with silver electrodes. The monthly bill was calculated on the basis of the gain in weight of the cathode. In 30 days, a given cathode gained 5796 grams.
    (a) What reaction occurs at the cathode?
    (b) Calculate the average current during the 30 days.

2. When a solution of a certain uranium salt is electrolyzed, a current of 5.0 amperes deposits 0.0233 mole of uranium in 30.0 minutes. Calculate the charge on the uranium cation.

3. The cathode reaction in a typical flashlight battery is

    $$2\ MnO_2\ +\ Zn^{2+}\ +\ 2\ e^-\ \longrightarrow\ \ ZnMn_2O_4$$

    If the cathode in a typical flashlight battery contains 4.35 g MnO_2, how many hours can such a battery deliver a current of 2 milliampere (2×10^{-3} ampere) before it exhausts all the MnO_2?

4. If a current of 100.0 ampere is used in the electrolysis of molten Al_2O_3, how many hours does it take to produce 1.00 kilogram of aluminum?

5. What current is required to deposit lead metal at a rate of 2.00 kg per hour from a solution containing Pb^{2+} ions?

6. How many kilograms of copper will dissolve from a copper anode over a period of 8.00 hours using a current of 3.00×10^3 ampere?

7. Chromium plating of the kind found on automobiles from the 1950s and 1960s is performed using a solution in which chromium has an oxidation number of +6. The plating process is only 15% efficient – only 15% of the current produces chromium metal; the remainder causes other processes. How many kilograms of chromium will be plated using such a solution using 1.80×10^4 ampere over a period of 8.00 hours?

8. Sodium metal can be prepared by electrolysis of molten sodium chloride. What current must be used to produce sodium at a rate of 9.0 kg per hour?

NUCLEAR REACTIONS AND RADIOACTIVITY

This lesson deals with:

1. Writing balanced equations for nuclear reactions.
2. Calculating mass decrement and binding energy.
3. Calculating the amount of energy involved in a nuclear reaction.
4. Distinguishing between fission and fusion processes.
5. Interpreting the rate of a nuclear decay reaction using the first order rate equation.
6. Calculating the age of organic material by the carbon-14 method.

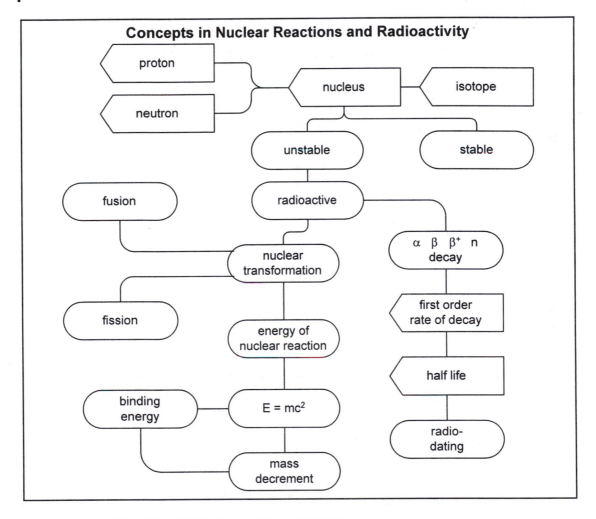

Concepts in Nuclear Reactions and Radioactivity

ATOMIC NUCLEI AND NUCLEAR REACTIONS

Nuclear reactions are changes that take place within atomic nuclei.

Atomic nuclei are composed of protons and neutrons.
  atomic number = number of protons in nucleus
  mass number = number of protons + number of neutrons = number of nucleons

Mass number and atomic number are indicated with the symbol of the element.

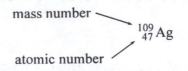

mass number

$^{109}_{47}Ag$

atomic number

Atoms having the same atomic number but different mass numbers are **isotopes** of an element.

Unstable nuclei decompose by changing their composition and by simultaneously emitting radiation.

Types of Radiation

1. **alpha rays**:   made of alpha particles
   Each alpha particle contains 2 neutrons and 2 protons.

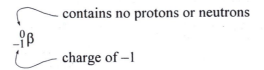

— contains 2 protons and 2 neutrons

$^{4}_{2}\alpha$

— contains 2 protons (charge of +2)

$^{4}_{2}He^{2+}$ also represents an alpha particle

2. **beta rays**: made of beta particles
   A beta particle is an electron.

— contains no protons or neutrons

$^{0}_{-1}\beta$

— charge of –1

$^{0}_{-1}e$ also represents a beta particle

3. **gamma rays**:   high energy electromagnetic radiation

   $^{0}_{0}\gamma$ no charge and no mass

4. **positrons**: positively charged beta particles

   $^{0}_{+1}\beta$ or $^{0}_{+1}e$

5. **neutrons**: mass similar to proton's, but has no charge

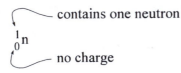

— contains one neutron

$^{1}_{0}n$

— no charge

EQUATIONS FOR NUCLEAR REACTIONS

Example 1

A. Uranium-238 decomposes by alpha emission to form thorium-234.

$$^{238}_{92}U \longrightarrow ^{4}_{2}\alpha + ^{234}_{90}Th$$

In nuclear equations, the sum of all atomic numbers on one side of the equation must equal the sum of all atomic numbers on the other side of the equation.

$$92 = 2 + 90$$

Similarly, the sums of the mass numbers on each side of the equation must be the same.

$$238 = 4 + 234$$

B. Strontium-90 decomposes by beta decay.

$$^{90}_{38}Sr \longrightarrow ^{0}_{-1}\beta + ?$$

$$
\begin{array}{lllll}
\text{mass no:} & 90 & = & 0 & + 90 \\
\text{atomic no:} & 38 & = & -1 & + 39
\end{array}
\left.\right\}\ ^{90}_{39}Y
$$

$$^{90}_{38}Sr \longrightarrow ^{0}_{-1}\beta + ^{90}_{39}Y$$

Exercise One

A. Write the symbol for each of the following.

1. a positron _____   2. a neutron _____

3. an alpha particle _____   4. a beta particle _____

5. a gamma ray _____

B. Complete the following equations.

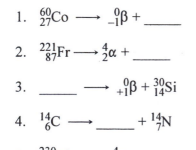

1. $^{60}_{27}Co \longrightarrow ^{0}_{-1}\beta +$ _____

2. $^{221}_{87}Fr \longrightarrow ^{4}_{2}\alpha +$ _____

3. _____ $\longrightarrow ^{0}_{+1}\beta + ^{30}_{14}Si$

4. $^{14}_{6}C \longrightarrow$ _____ $+ ^{14}_{7}N$

5. $^{230}_{90}Th \longrightarrow ^{4}_{2}He +$ _____

6. $^{11}_{6}C \longrightarrow$ _____ $+ ^{11}_{5}B$

C. Write a balanced nuclear equation for each of the following processes.

1. Tin-121 decaying by beta emission.

2. Nitrogen-13 undergoing positron decay.

3. Lead-210 decaying to mercury-206.

TYPES OF RADIOACTIVITY

Natural radioactivity: decomposition of nuclei that occur in nature.

Induced radioactivity: decomposition of nuclei that have been prepared in the laboratory.

A stable nucleus may be converted to an unstable nucleus by bombardment with high-energy particles or radiation.

Example 2

The stable isotope aluminum-27 absorbs a neutron to form $^{28}_{13}Al$, which is unstable and decomposes by beta emission to give a stable isotope of silicon.

1. $^{27}_{13}Al + ^{1}_{0}n \longrightarrow ^{28}_{13}Al$
2. $^{28}_{13}Al \longrightarrow ^{0}_{-1}\beta + ^{28}_{14}Si$

New elements have been made by bombarding nuclei of heavy elements with nuclei of light elements. The **transuranium elements**, elements with atomic number greater than 92, are made this way.

Example 3

When uranium-238 is bombarded by a carbon-12 nucleus, the transuranium isotope $^{246}_{98}Cf$ is formed, along with several neutrons.

$$^{238}_{92}U + ^{12}_{6}C \longrightarrow ^{246}_{98}Cf + 4\,^{1}_{0}n$$

atomic no: $92 + 6 = 98 + 4(0)$
mass no: $238 + 12 = 246 + 4(1)$

Exercise Two

A. Complete the following.

1. $_{26}^{54}\text{Fe} + _2^4\alpha \longrightarrow _0^1\text{n} + \underline{\hspace{1.5cm}}$

2. $\underline{\hspace{1.5cm}} + \gamma \longrightarrow _0^1\text{n} + _4^8\text{Be}$

3. $_{92}^{238}\text{U} + _0^1\text{n} \longrightarrow _{92}^{239}\text{U} \longrightarrow _{-1}^0\beta + \underline{\hspace{1cm}}$

B. Write equations representing the following nuclear processes.

1. Bombarding $_{99}^{253}\text{Es}$ with an alpha particle yields one neutron plus another transuranium isotope.

2. Carbon-14 is generated in the bombardment of nitrogen-14 by a neutron.

MASS DECREMENT AND BINDING ENERGY

The mass of a nucleus is less than the sum of the masses of its constituent protons and neutrons.

Mass decrement = (sum of the masses of constituent protons and neutrons)

$-$ (mass of nucleus)

Example 4

Use the data given in Table 1 to calculate the mass decrement of the $_2^4\text{He}$ nucleus.

$_2^4\text{He}$ contains 2 protons and 2 neutrons

mass of 2 protons ($_1^1\text{H}$) = 2(1.00728 g/mol) =	2.01456 g/mol
mass of 2 neutrons ($_0^1\text{n}$) = 2(1.00867 g/mol) =	2.01734 g/mol
	4.03190 g/mol

mass of $_2^4\text{He}$ = 4.00150 g/mol

mass decrement   = 4.03190 g/mol – 4.00150 g/mol
                  = 0.03040 g/mol

THE EINSTEIN EQUATION

In nuclear transformations, mass can be transformed into energy, and energy can be transformed into mass. The Einstein equation indicates the relationship between mass and energy.

$$E = mc^2$$

where   m = mass
        E = energy
        c = speed of light

If the speed of light is expressed in meters per second and the mass change in kilograms, then the energy calculated using the Einstein equation has units of joules.

$$E \text{ (in joules)} = m \text{ (in kilograms)} \times (3.0 \times 10^8 \text{ m/s})^2$$

$$\text{because } 1 \text{ joule} = 1 \text{ kg m}^2/\text{s}^2$$

The mass decrement of a nucleus corresponds to the amount of energy that would be required to break the nucleus into its constituent protons and neutrons. This amount of energy is called the **binding energy** of the nucleus. The Einstein equation relates the binding energy to the mass decrement of a nucleus.

In a process in which there is a change in mass, the Einstein equation can also be represented as

$$\Delta E = \Delta mc^2$$

where Δm = change in mass during process
              = (mass of products) − (mass of reactants)
        ΔE = energy absorbed in process
        c = speed of light

Calculating Binding Energy

Example 5

Calculate the total binding energy in joules for one $^4_2$He nucleus.

From Example 4, the mass decrement of $^4_2$He is 0.03040 g/mol. To get energy in joules, we must convert this mass to kilograms.

$$\Delta m = 0.03040 \text{ g/mol} \left(\frac{1 \text{ kg}}{1000 \text{ g}} \right) = 3.040 \times 10^{-5} \text{ kg/mol}$$

Then,

$$\Delta E = (3.040 \times 10^{-5} \text{ kg/mol})(3.00 \times 10^8 \text{ m/s})^2$$

$$= 2.74 \times 10^{12} \text{ J/mol of } ^4_2\text{He}$$

Because 1 mole of $^4_2$He contains 6.022×10^{23} nuclei,

$$\Delta E = 2.74 \times 10^{12} \frac{\text{J}}{\text{mol}} \left(\frac{1 \text{ mol}}{6.022 \times 10^{23} \text{ nuclei}} \right) = 4.55 \times 10^{-12} \text{ J/nucleus}$$

Table 1.   Nuclear Molar Masses in grams per mole*

	At.No.	Mass No.	Mass		At.No.	Mass No.	Mass
n	0	1	1.00867	Br	35	79	78.8992
H	1	1	1.00728		35	81	80.8971
	1	2	2.01355		35	87	86.9028
	1	3	3.01550	Rb	37	89	88.8909
He	2	3	3.01493	Sr	38	90	89.8864
	2	4	4.00150	Mo	42	99	98.8849
Li	3	6	6.01348	Ru	44	106	105.8829
	3	7	7.01436	Ag	47	109	108.8789
Be	4	9	9.00999	Cd	48	109	108.8786
	4	10	10.01134		48	115	114.8793
B	5	10	10.01019	Sn	50	120	119.8747
	5	11	11.00656	Ce	58	144	143.8816
C	6	11	11.00814		58	146	145.8865
	6	12	11.99671	Pr	59	144	143.8807
	6	13	13.00006	Sm	62	152	151.8853
	6	14	13.99995	Eu	63	157	156.8914
O	8	16	15.99052	Er	68	168	167.8941
	8	17	16.99474	Hf	72	179	178.9048
	8	18	17.99477	W	74	186	185.9107
F	9	18	17.99601	Os	76	192	191.9187
	9	19	18.99346	Au	79	196	195.9231
Na	11	23	22.98373	Hg	80	196	195.9219
Mg	12	24	23.97845	Pb	82	206	205.9295
	12	25	24.97925		82	207	206.9309
	12	26	25.97600		82	208	207.9316
Al	13	26	25.97977	Po	84	210	209.9368
	13	27	26.97439		84	218	217.9628
	13	28	27.97477	Rn	86	222	221.9703
Si	14	28	27.96924	Ra	88	226	225.9771
S	16	32	31.96329	Th	90	230	229.9837
Cl	17	35	34.95952	Pa	91	234	233.9934
	17	37	36.95657	U	92	233	232.9890
Ar	18	40	39.95250		92	235	234.9934
K	19	39	38.95328		92	238	238.0003
	19	40	39.95358		92	239	239.0038
Ca	20	40	39.95162	Np	93	237	236.9971
Ti	22	48	47.93588	Pu	94	239	239.0006
Cr	24	52	51.92734		94	241	241.0051
Fe	26	56	55.92066	Am	95	241	241.0045
Co	27	59	58.91837	Cm	96	242	242.0061
Ni	28	59	58.97897	Bk	97	245	245.0129
Zn	30	64	63.91268	Cf	98	248	248.0186
	30	72	71.91128	Es	99	251	251.0255
Ge	32	76	75.90380	Fm	100	252	252.0278
As	33	79	78.90288		100	254	254.0331

*These are nuclear masses. The masses of the corresponding atoms can be calculated by adding the mass of the extranuclear electrons (mass of electron = 0.000549 g/mol). For example, the molar mass of $_2^4He$ is 4.00150 g/mol + 2(0.000549) g/mol, which is 4.00260 g/mol. Similarly, the molar mass of $_6^{12}C$ is 11.99671 g/mol + 6(0.000549) g/mol, or 12.00000 g/mol.

Exercise Three

Using the information in Table 1, calculate the mass decrement in grams per mole and binding energy in joules per nucleus for each of the following nuclei.

A. $^{12}_{6}C$

B. $^{238}_{92}U$

ENERGY OF NUCLEAR REACTIONS

In nuclear decay reactions, less stable nuclei decompose to more stable nuclei.

The difference between the total masses of the original nuclei and the total masses of the product nuclei is converted to energy.

Calculating the Amount of Energy Given off in a Nuclear Reaction

Example 6

A. Calculate ΔE in joules when the nuclei in one mole of $^{226}_{88}Ra$ decay.

$$^{226}_{88}Ra \longrightarrow {}^{4}_{2}He + {}^{222}_{86}Rn$$

$$\Delta m = \text{(mass of one mole } ^{222}_{86}\text{Rn} + \text{mass of one mole } ^{4}_{2}\text{He})$$
$$- \text{(mass of one mole } ^{226}_{88}\text{Ra)}$$

$$\Delta m = (221.9703 \text{ g/mol} + 4.0015 \text{ g/mol}) - (225.9771 \text{ g/mol})$$
$$= -0.0053 \text{ g/mol}$$

$$\Delta m = -0.0053 \text{ g/mol} \left(\frac{1 \text{ kg}}{1000 \text{ g}} \right) = -5.3 \times 10^{-6} \text{ kg/mol}$$

$$\Delta E = (-5.3 \times 10^{-6} \text{ kg/mol})(3.00 \times 10^{8} \text{ m/s})^{2} = -4.8 \times 10^{11} \text{ J/mol}$$

The negative value indicates that energy is released in the process.

B. Calculate ΔE in joules when one nucleus of $^{226}_{88}\text{Ra}$ decomposes by alpha decay.

From part A:

$$\Delta E = -4.8 \times 10^{11} \text{ J/mol}$$

One mole of $^{226}_{88}\text{Ra}$ contains 6.022×10^{23} nuclei, so for one nucleus,

$$\Delta E = -4.8 \times 10^{11} \text{ J/mol} \left(\frac{1 \text{ mol}}{6.022 \times 10^{23} \text{ nuclei}} \right) = -8.0 \times 10^{-13} \text{ J/nucleus}$$

C. Calculate the amount of energy in kJ evolved when the nuclei in 1.0 gram of $^{226}_{88}\text{Ra}$ decay.

From part A, the energy evolved when one mole of $^{226}_{88}\text{Ra}$ decays is 4.8×10^{11} J/mol.

The number of moles of $^{226}_{88}\text{Ra}$ in 1.0 gram is determined from its molar mass, namely, 226 grams/mole.

$$1.0 \text{ g} \left(\frac{1 \text{ mol}}{226 \text{ g}} \right) = 4.4 \times 10^{-3} \text{ mol}$$

Then, for 1.0 gram of $^{226}_{88}\text{Ra}$,

$$\Delta E = 4.4 \times 10^{-3} \text{ mol} \left(\frac{-4.8 \times 10^{11} \text{ J}}{1 \text{ mol}} \right) \left(\frac{1 \text{ kJ}}{1000 \text{ J}} \right) = -2.1 \times 10^{6} \text{ kJ}$$

Note:

When one gram of CH_4 is burned, 2.9 kJ are produced.

When one gram of $^{226}_{88}\text{Ra}$ decays, 2.1×10^{6} kJ are produced.

This nuclear reaction releases 700,000 times as much energy per gram as a very exothermic chemical reaction.

Exercise Four

A. Calculate ΔE in joules for the reaction of one nucleus of polonium-210:

$$^{210}_{84}\text{Po} + 2\,^{1}_{0}\text{n} \longrightarrow\ ^{208}_{82}\text{Pb} + ^{4}_{2}\text{He}$$

B. Calculate ΔE in kJ when the nuclei in 1.0 gram of polonium-210 react by the equation above.

C. How many joules are released when one atom of americium-241, the isotope used in ionization-type smoke detectors, undergoes alpha emission?

BINDING ENERGY PER NUCLEON

Nucleon is the general name for a neutron or proton

$$\text{binding energy per nucleon} = \frac{\text{binding energy of nucleus}}{\text{mass number of nucleus}}$$

Example 7

Calculate the binding energy per nucleon for the $^{4}_{2}\text{He}$ nucleus.

binding energy for $^{4}_{2}\text{He} = 4.55 \times 10^{-12}$ J/nucleus (from Example 5)

The $^{4}_{2}\text{He}$ nucleus contains 2 protons and 2 neutrons, a total of 4 nucleons.

$$\text{binding energy per nucleon} = 4.55 \times 10^{-12} \ \frac{J}{\text{nucleus}} \left(\frac{1 \ ^4\text{He nucleus}}{4 \ \text{nucleons}} \right)$$

$$= 1.14 \times 10^{-12} \ \text{J/nucleon}$$

The binding energy per nucleon varies from one nucleus to another. Figure 1 shows a plot of binding energy per nucleon versus the mass number of the nucleus for the known nuclei.

Light nuclei and heavy nuclei have less binding energy per nucleon than nuclei in the middle of the range.

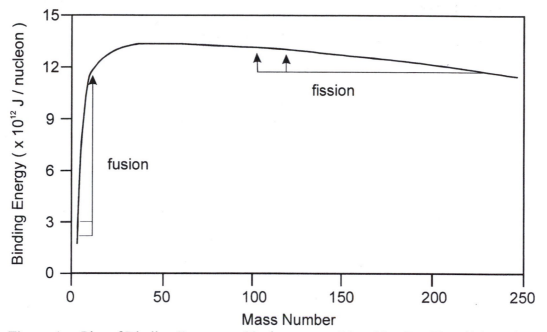

Figure 1.   Plot of Binding Energy per Nucleon versus Mass Number. Very light and very heavy nuclei are unstable relative to those of moderate mass.

FISSION AND FUSION

In nuclear reactions, less stable nuclei form more stable nuclei.

Combining two small nuclei into a larger nucleus forms a more stable nucleus and releases energy (fusion in Figure 1).

Breaking a very large nucleus into two smaller ones forms more stable nuclei and releases energy (fission in Figure 1).

Fusion: combining two small nuclei into a larger one.
Fission: breaking a large nucleus into smaller ones.

Fission is triggered by neutron bombardment of certain isotopes, e.g., $^{235}_{92}\text{U}$ or $^{239}_{94}\text{Pu}$.

$$^{235}_{92}\text{U} + {}^{1}_{0}\text{n} \longrightarrow {}^{90}_{37}\text{Rb} + {}^{144}_{55}\text{Cs} + 2 \ {}^{1}_{0}\text{n}$$

$$^{235}_{92}\text{U} + {}^{1}_{0}\text{n} \longrightarrow {}^{87}_{35}\text{Br} + {}^{146}_{57}\text{La} + 3 \ {}^{1}_{0}\text{n}$$

A chain reaction is a series of reactions in which the product of an earlier reaction is involved in a subsequent reaction.

In fusion, two smaller nuclei combine into a larger nucleus. For example,

$$\text{}^{2}_{1}\text{H} + \text{}^{3}_{1}\text{H} \longrightarrow \text{}^{4}_{2}\text{He} + \text{}^{1}_{0}\text{n}$$

Fusion is the source of the energy of the sun. Fusion requires extremely high temperatures and pressures to force two positively-charged nuclei together.

Exercise Five

A. Calculate the binding energy per nucleon for the $^{56}_{26}\text{Fe}$ nucleus.

B. Calculate how much energy in joules is produced when one nucleus of $^{239}_{94}\text{Pu}$ undergoes fission according to

$$\text{}^{239}_{94}\text{Pu} + \text{}^{1}_{0}\text{n} \longrightarrow \text{}^{87}_{35}\text{Br} + \text{}^{144}_{59}\text{Pr} + 9\,\text{}^{1}_{0}\text{n}$$

C. Calculate how much energy in kJ is produced when one mole of $^{2}_{1}\text{H}$ fuses with one mole of $^{3}_{1}\text{H}$:

$$\text{}^{2}_{1}\text{H} + \text{}^{3}_{1}\text{H} \longrightarrow \text{}^{4}_{2}\text{He} + \text{}^{1}_{0}\text{n}$$

RATES OF RADIOACTIVE DECAY

Radioactive decay reactions follow first order rate law. The integrated form of the first order rate law is represented by the following equation.

$$\ln \frac{X_t}{X_0} = -kt$$

where X_t = amount of radioactive isotope present at a later time, t
X_0 = initial amount of radioactive isotope
k = rate constant of nuclear decay

The ratio X_t / X_0 is the fraction remaining after time t.

The amounts of radioactive isotope used in this equation, X_t and X_0, can be expressed in a variety of units, such as numbers of atoms, moles of atoms, mass, or radioactivity (counts). However, the units of both X_t and X_0 must be the same.

The **half-life** ($t_{1/2}$) of a radioactive decay of an isotope is the time required for half of a sample of the isotope to decay. As for any first-order process, the half-life is related to the rate constant by the equation

$$t_{1/2} = \frac{0.693}{k}$$

Example 8

What percent of a sample of $^{226}_{88}Ra$ remains after 100 years? The half-life of $^{226}_{88}Ra$ is 1620 years.

$$\ln \frac{X_t}{X_0} = -kt$$

The ratio X_t / X_0 is the fraction remaining after time t. Multiplying this by 100% will give the percent remaining.

$$k = \frac{0.693}{t_{1/2}} = \frac{0.693}{1620 \text{ yr}} = 4.28 \times 10^{-4} \text{ yr}^{-1}$$

$$\ln \frac{X_t}{X_0} = -(4.28 \times 10^{-4} \text{ yr}^{-1})(100 \text{ yr})$$

$$\ln \frac{X_t}{X_0} = -4.28 \times 10^{-2}$$

$$\frac{X_t}{X_0} = e^{-0.0428} = 0.958$$

$$0.958 \times 100\% = 95.8\%$$

Exercise Six

A. A sample contains 4.6 mg of $^{131}_{53}I$. How many mg will remain after 3.0 days? The half-life of $^{131}_{53}I$ is 8.0 days.

B. The majority of naturally occurring rhenium is $^{187}_{75}Re$, which is radioactive and has a half-life of 7×10^{10} years. In how many years will 5% of the earth's $^{187}_{75}Re$ decompose?

CARBON-14 DATING

$^{14}_{6}C$ may be used to determine the age of old organic materials.

Carbon-14 is produced in the upper atmosphere from the bombardment of nitrogen-14 by neutrons from cosmic radiation.

$$^{14}_{7}N + ^{1}_{0}n \longrightarrow ^{14}_{6}C + ^{1}_{1}H$$

Carbon-14 is radioactive and undergoes beta decay.

$$^{14}_{6}C \longrightarrow ^{0}_{-1}\beta + ^{14}_{7}N \qquad\qquad t_{1/2} = 5720 \text{ years}$$

The rate of production of $^{14}_{6}C$ is nearly the same as its decay rate. The $^{14}_{6}C/^{12}_{6}C$ ratio is constant in the CO_2 in the air. A living plant has the same $^{14}_{6}C/^{12}_{6}C$ ratio as the CO_2 in the air. After a plant dies, $^{14}_{6}C/^{12}_{6}C$ ratio diminishes continuously.

Example 9

A piece of paper from the Dead Sea scrolls was found to have a $^{14}_{6}C/^{12}_{6}C$ ratio 79.5% of that in a plant living today. Estimate the age of the paper.

Radioactive decay is a first order process following the first-order rate equation.

$$\ln \frac{X_0}{X_t} = kt$$

The value of the rate constant k can be found from the half life of a first order process.

$$k = \frac{0.693}{t_{1/2}} = \frac{0.693}{5720 \text{ yr}} = 1.21 \times 10^{-4} \text{ yr}^{-1}$$

t = age of paper

X_t = 79.5% ^{14}C remaining

X_0 = 100% of ^{14}C initially

$$\ln \frac{100\%}{79.5\%} = (1.21 \times 10^{-4} \text{ yr}^{-1})(t)$$

$$0.223 = (1.21 \times 10^{-4} \text{ yr}^{-1})(t)$$

$$t = \frac{0.223}{1.21 \times 10^{-4} \text{ yr}^{-1}} = 1890 \text{ yr}$$

Limitations of Carbon-14 Dating:
1. Very old materials cannot be dated because only immeasurable amounts of $^{14}_{6}C$ remain.
2. Very new materials cannot be dated because no appreciable change in $^{14}_{6}C$ content has yet occurred.

Exercise Seven

The charcoal from ashes found in a cave gave a ^{14}C activity of 8.6 counts per gram per minute. Calculate the age of the charcoal (wood from a growing tree gives a comparable count of 15.3). For ^{14}C, $t_{1/2}$ = 5720 years.

PROBLEMS

1. Complete the following equations:

    (a) $^{23}_{11}Na + ^{2}_{1}H \longrightarrow ^{24}_{11}Na + \underline{\hspace{1cm}}$

    (b) $^{44}_{20}Ca + \underline{\hspace{1cm}} \longrightarrow ^{44}_{21}Sc + ^{1}_{0}n$

    (c) $^{250}_{98}Cf + ^{11}_{5}B \longrightarrow \underline{\hspace{1cm}} + 4\,^{1}_{0}n$

    (d) $^{7}_{3}Li + \underline{\hspace{1cm}} \longrightarrow 2\ He$

2. Write equations for

    (a) the production of $^{17}_{8}O$ from $^{14}_{7}N$ by alpha bombardment.

    (b) the production of $^{254}_{102}No$ from $^{246}_{96}Cm$ by bombardment with $^{12}_{6}C$.

    (c) the formation of $^{13}_{7}N$ from $^{10}_{5}B$ by alpha bombardment and the subsequent decay of $^{13}_{7}N$ by positron emission. (2 equations)

3. Calculate the amount of energy evolved in kJ when 1.00 gram of $^{2}_{1}H$ undergoes fusion according to the equation

    $$^{2}_{1}H + ^{3}_{1}H \longrightarrow ^{4}_{2}He + ^{1}_{0}n$$

4. In a certain activity meter, a pure sample of $^{90}_{38}Sr$ has an activity (rate of decay) of 1000.0 disintegrations per minute. If the activity of this sample after 2.00 years is 953.2 disintegrations per minute, what is the half-life of $^{90}_{38}Sr$?

5. A sample of a wooden artifact from an Egyptian tomb has a $^{14}C/^{12}C$ ratio 54.2 percent that of freshly cut wood. In approximately what year was the old wood cut? For ^{14}C, $t_{1/2} = 5720$ years.

6. A typical ionization type smoke detector contains 0.20 mg of americium-241, which decays by alpha emission.

    $$^{241}Am \longrightarrow ^{4}He + ^{237}Np$$

    (a) How many joules of energy are released when one atom of ^{241}Am decays?
    (b) How many joules are released when all of the atoms of ^{241}Am in the smoke detector decay?

STRUCTURES AND FORMULAS OF ORGANIC COMPOUNDS

This lesson deals with:

1. Interpreting condensed structural formulas in terms of Lewis structures.
2. Interpreting bond-line formulas in terms of Lewis structures and condensed structural formulas.
3. Classifying organic compounds by functional group.

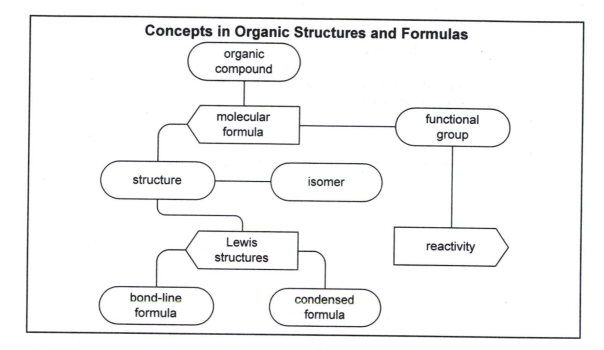

REPRESENTING THE STRUCTURE OF ORGANIC COMPOUNDS

Organic chemistry is the study of the compounds of carbon.
    Millions of carbon compounds are known.
    The structure of the molecules determine the identity of the compound.

Structures are represented by several methods.

1. Lewis structures:

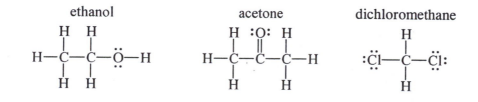

2 Abbreviated Lewis structures (omit lone pairs):

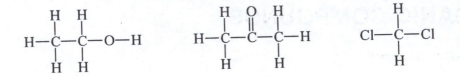

3. Condensed structural formulas:

$$CH_3CH_2OH \qquad CH_3\overset{\overset{\displaystyle O}{\|}}{C}CH_3 \qquad CH_2Cl_2$$

Conventions for condensed formulas:

1. Atoms bonded to one atom are written immediately after that atom, with a subscript indicating how many are bonded.

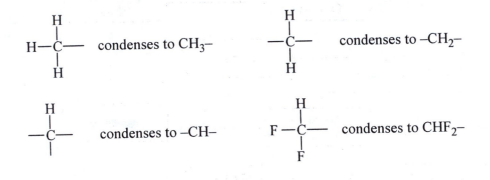

2. Bond lines between atoms written horizontally are omitted.

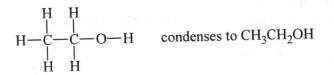

condenses to CH_3CH_2OH

3. Vertical bond lines are used when an atom is written above or below the atom to which it is bonded.

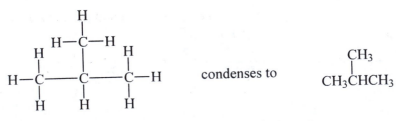

condenses to

$$\overset{\overset{\displaystyle CH_3}{|}}{CH_3}CHCH_3$$

4. Multiple bonds are shown explicitly.

$$H-\overset{\overset{\displaystyle H}{|}}{C}=\overset{\overset{\displaystyle H}{|}}{C}-\underset{\underset{\displaystyle H}{|}}{\overset{\overset{\displaystyle H}{|}}{C}}-H \qquad \text{condenses to} \qquad CH_2{=}CHCH_3$$

Example 1

Write a condensed formula for the molecule whose abbreviated Lewis structure is below.

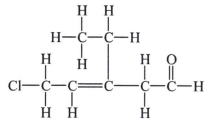

Begin at the leftmost carbon atom. The double bonds should be explicitly indicated.

$$\overset{\overset{\displaystyle CH_3CH_2}{|}}{\underset{}{CH_2ClCH{=}C}}CH_2\overset{\overset{\displaystyle O}{\|}}{C}H$$

Formulas are frequently written with both condensed and expanded sections.

$$Cl{-}CH_2{-}\underset{\underset{\displaystyle H}{|}}{C}{=}\overset{\overset{\displaystyle CH_2CH_3}{|}}{C}{-}CH_2{-}\underset{\underset{\displaystyle O}{\|}}{C}{-}H$$

This emphasizes the portions of the molecule that are written in expanded format.

Exercise One

A. Write a condensed formula for each of the structures below.

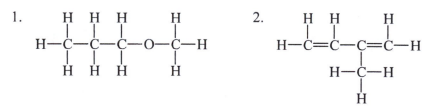

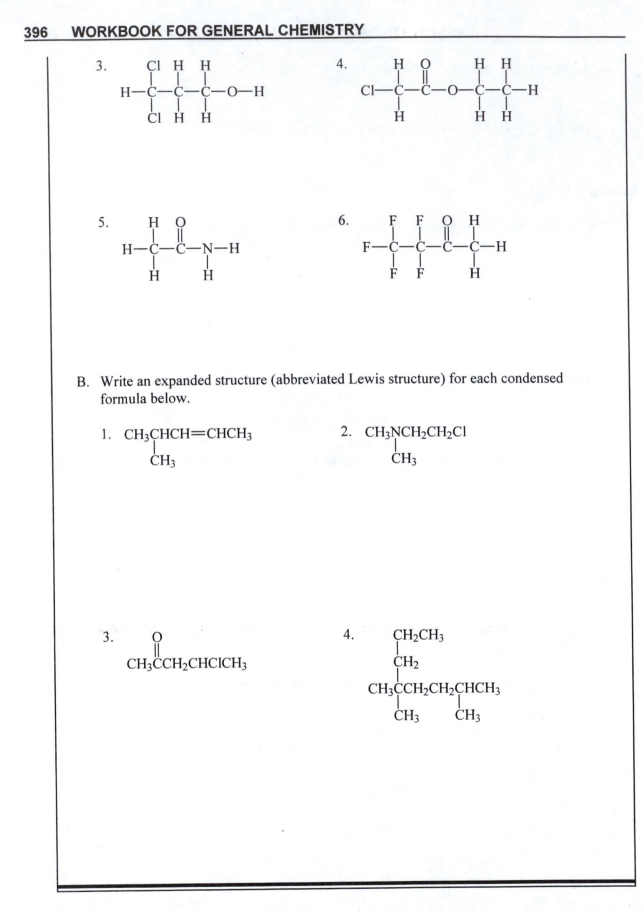

3.

Cl H H
| | |
H—C—C—C—O—H
| | |
Cl H H

4.

H O   H H
| ‖ | |
Cl—C—C—O—C—C—H
| | |
H   H H

5.

H O
| ‖
H—C—C—N—H
| |
H   H

6.

F F O H
| | ‖ |
F—C—C—C—C—H
| | |
F F   H

B. Write an expanded structure (abbreviated Lewis structure) for each condensed formula below.

1. CH₃CHCH=CHCH₃
        |
        CH₃

2. CH₃NCH₂CH₂Cl
        |
        CH₃

3.
         O
         ‖
    CH₃CCH₂CHClCH₃

4.
         CH₂CH₃
         |
         CH₂
    CH₃CCH₂CH₂CHCH₃
         | |
         CH₃   CH₃

BOND-LINE FORMULAS

In bond-line formulas, bonds between carbon atoms represent not only the bonds but the bonded carbon atoms as well. The symbols for the carbon atoms and the hydrogen atoms bonded to the carbon atoms are not written.

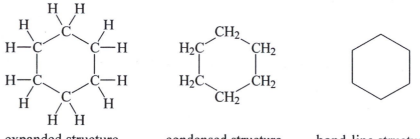

expanded structure   condensed structure   bond-line structure

Example 2

Write the bond-line structure for the molecule whose Lewis structure is below.

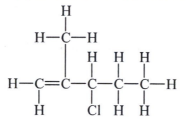

First, draw the carbon-carbon bonds. They must be drawn at angles to each other, to make clear where the carbon atoms are located.

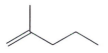

The double bond is shown explicitly.

At this stage, the structure implies a carbon atom at the end of every line segment. Each carbon atom has as many hydrogen atoms as necessary to provide 4 bonds to each carbon atom. Where hydrogen atoms are replaced by other atoms, these atoms must be shown. Therefore, the Cl atom must be added.

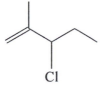

B. Rewrite the following bond-line formula as an expanded structure.

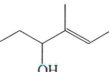

First, position the carbon atoms for the expanded structure. There is a C at the end of each line segment where no other atom is indicated.

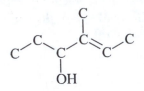

Then, add hydrogen atoms to each carbon atom, so that each carbon atom has four bonds. A double bond counts as two bonds.

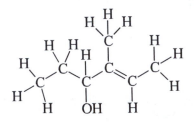

Exercise Two

A. Write bond-line formulas for each of the following structures.

1.

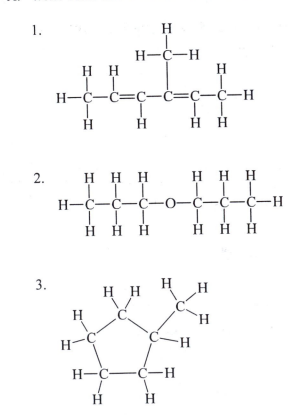

2.

3.

4.   Cl
              |
   $CH_3CH_2CHCH_2CHCH_2CH_3$
                     |
                   CH_3

B.   Write a condensed formula for each of the bond-line structures below.

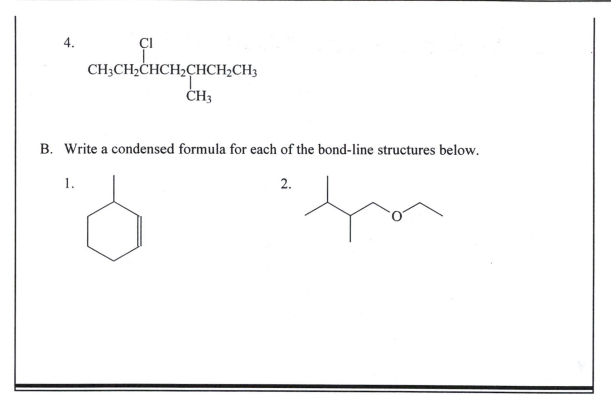

1.   2.

CLASSIFICATION OF ORGANIC COMPOUNDS

Organic compounds are classified into types based on the functional groups they contain.

A **functional group** is a small group of atoms that frequently occurs in organic molecules. (A functional group is any part of an organic molecule that is not made up of carbon and hydrogen atoms singly bonded together.)

Table 1 contains a list of common functional groups.

Example 3

Identify the functional groups in the following structural formulas.

A.   O
             ||
   $CH_3CHCCH_2CH_3$
         |
         CH_3

       The structure may be easier to analyze if it is redrawn in expanded form without the hydrogen atoms:

                              O
                             ||
           C—C—C—C—C
                 |
                 C

       The molecule contains a ketone functional group.

Table 1. Classes of Organic Compounds and their Functional Groups

Class	Functional Group	Example	Name of Example
alkane	none	$CH_3CH_2CH_3$	propane
alkene	$\diagdown C{=}C \diagup$	$CH_2{=}CHCH_3$	propene
alkyne	$-C{\equiv}C-$	$HC{\equiv}CCH_3$	propyne
aromatic			benzene
alkyl halide	$-\overset{\|}{\underset{\|}{C}}-X$ where X=F,Cl,Br, or I	$CH_3CH_2CH_2Br$	1-bromopropane
amine	$-\overset{\|}{\underset{\|}{C}}-\overset{\|}{N}-$	$CH_3CH_2CH_2NH_2$	propylamine
alcohol	$-\overset{\|}{\underset{\|}{C}}-OH$	$CH_3CH_2CH_2OH$	1-propanol
ether	$-\overset{\|}{\underset{\|}{C}}-O-\overset{\|}{\underset{\|}{C}}-$	$CH_3OCH_2CH_3$	methyl ethyl ether
aldehyde	$-\overset{O}{\overset{\|\|}{C}}-H$	$CH_3CH_2\overset{O}{\overset{\|\|}{C}}H$	propanal
ketone	$-\overset{\|}{\underset{\|}{C}}-\overset{O}{\overset{\|\|}{C}}-\overset{\|}{\underset{\|}{C}}-$	$CH_3\overset{O}{\overset{\|\|}{C}}CH_3$	propanone
carboxylic acid	$-\overset{O}{\overset{\|\|}{C}}-O-H$	$CH_3CH_2\overset{O}{\overset{\|\|}{C}}OH$	propanoic acid
ester	$-\overset{O}{\overset{\|\|}{C}}-O-\overset{\|}{\underset{\|}{C}}-$	$CH_3CH_2\overset{O}{\overset{\|\|}{C}}OCH_3$	methyl propanoate
amide	$-\overset{O}{\overset{\|\|}{C}}-\overset{\|}{N}-$	$CH_3CH_2\overset{O}{\overset{\|\|}{C}}NH_2$	propanoyl amide

B. $CH_3CH_2NHCH_3$

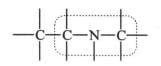   an amine functional group

C.

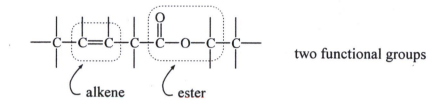

Example 4

Draw a structural formula for an ester with a molecular formula of $C_3H_6O_2$.

1. First, draw the functional group of an ester.

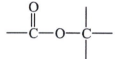

The functional group uses two C and two O. This leaves one C and six H.

2. Add the remaining C. It can be attached either at the left or at the right.

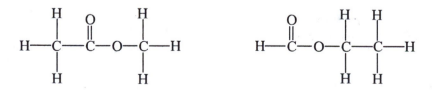

3. Then add the hydrogen atoms so each carbon atom has 4 bonds.

This shows that there are two possible esters with the molecular formula $C_3H_6O_2$. Two compounds with the same molecular formula are called **isomers**.

The isomer on the left is methyl acetate, and the isomer on the right is ethyl formate.

Exercise Three

A. Identify the functional groups in the following molecules.

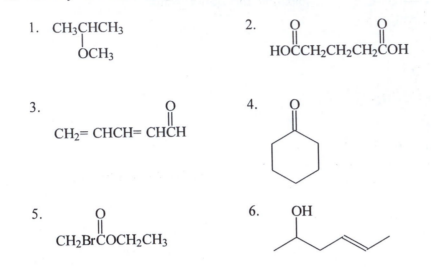

1. CH_3CHCH_3
    $\quad\quad |$
    $\quad\quad OCH_3$

2. $\underset{\displaystyle \overset{O}{\|}}{HOCCH_2CH_2CH_2}\underset{\displaystyle \overset{O}{\|}}{COH}$

3. $CH_2= CHCH= CH\overset{\displaystyle \overset{O}{\|}}{C}H$

4.

5. $CH_2Br\overset{\displaystyle \overset{O}{\|}}{C}OCH_2CH_3$

6. OH

B. 1. Draw the structural formula for an aldehyde that has a molecular formula of C_4H_8O.

2. Draw the structural formula for a ketone that has a molecular formula of C_4H_8O.

3. Draw the structural formula of another isomer with formula C_4H_8O that is neither an aldehyde nor a ketone.

AROMATIC COMPOUNDS

The class of carbon compounds called aromatic compounds was originally given its name because many of them do in fact have odors, usually not unpleasant. However, today, aromatic compounds are classified by a common structural feature (functional group) in their molecules. This feature is a ring of six carbon atoms, each of which is bonded to three other atoms, two from the ring and one other. In the simplest aromatic compound, each of the carbon atoms is bonded to a hydrogen atom. Its *Lewis structure* is

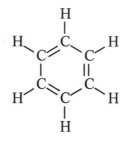

This compound is called benzene. This formula shows a ring with three double bonds alternating around the ring. However, in fact there are no double bonds in benzene. The bonding is better represented by a pair of resonance structures.

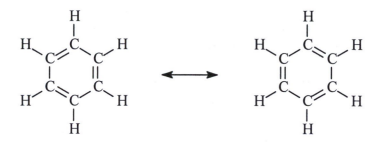

These structures show that the bonds between the carbon atoms are intermediate between single and double bonds. This situation, and the two resonance structures, are frequently represented by a circle drawn inside the ring of carbon atoms. The *bond-line formula* of benzene is often drawn as

Many compounds can be considered as derivative of benzene produced by replacing one or more of the hydrogen atoms with other atoms or groups.

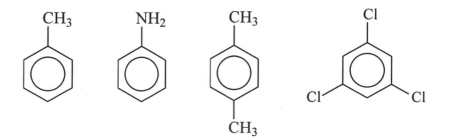

The derivatives of benzene are said to contain a benzene ring, the ring of six carbon atoms. In the *condensed formulas* for compounds containing benzene rings, the ring is often represented as $C_6H_{6-n}X_n$, where n is the number of substituted hydrogens. The condensed structural formulas for the compounds above are

$C_6H_5CH_3$ $\qquad$ $C_6H_5NH_2$ $\qquad$ $C_6H_4(CH_3)_2$ $\qquad$ $C_6H_3Cl_3$

Example 5

A. Draw a bond-line formula for a molecule whose condensed formula is $C_6H_3(NH_2)_3$.

We recognize from the C_6H_3 portion of the formula, that the molecule contains a benzene ring with three substituted hydrogen atoms. The $(NH_2)_3$ portion indicates the three substituted groups are each NH_2 groups. A bond-line formula would be

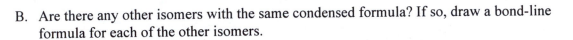

B. Are there any other isomers with the same condensed formula? If so, draw a bond-line formula for each of the other isomers.

The condensed formula does not indicate the relative locations of the three replaced hydrogens, so yes, there are other isomers. They are

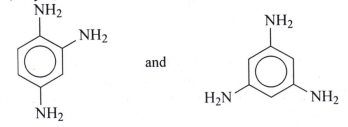

and

Exercise Four

A. Draw a bond-line formula for each of the following.

1. C_6H_5OH $\qquad\qquad\qquad\qquad\qquad$ 2. $CH_3CH_2C_6H_5$

B. Write a condensed formula for each of the following.

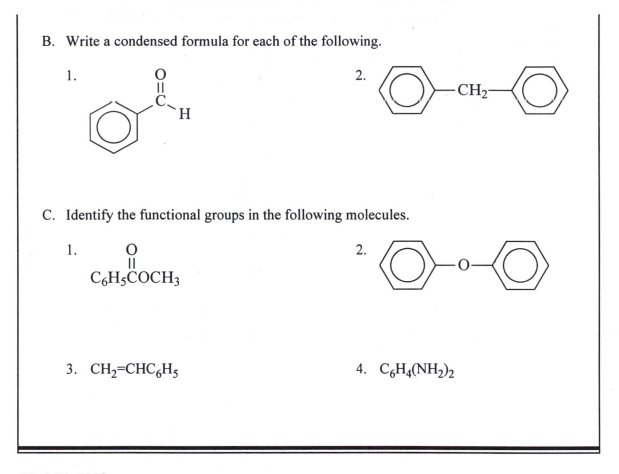

1.

2.

C. Identify the functional groups in the following molecules.

1.

$$C_6H_5\overset{\overset{\displaystyle O}{\|}}{C}OCH_3$$

2.

3. $CH_2=CHC_6H_5$

4. $C_6H_4(NH_2)_2$

PROBLEMS

1. Identify the functional groups in the following molecules.

(a) $CH_3CH_2NHCH_3$

(b) $ClCH_2CH_2OCH_2CH_2Cl$

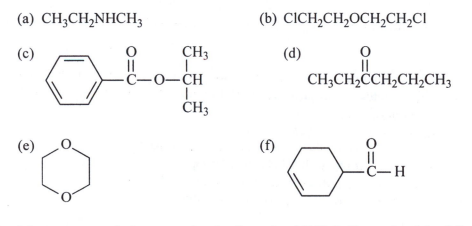

(c)

(d) $CH_3CH_2\overset{\overset{\displaystyle O}{\|}}{C}CH_2CH_2CH_3$

(e)

(f)

2. Many compounds have a molecular formula of C_3H_6O. For each of the following functional groups, draw a structural formula for a molecule that contains that functional group and has a molecular formula of C_3H_6O.
    (a) a ketone
    (b) an alcohol
    (c) an ether

3. What are the functional groups and what is the molar mass of each of the following compounds?

    (a)

    (b)

    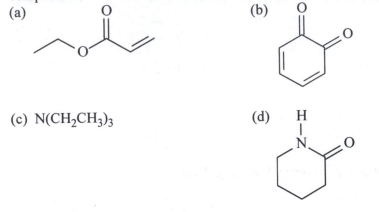

    (c) N(CH$_2$CH$_3$)$_3$

    (d)

4. Draw a condensed structural formula for each of the following.

    (a)

    (b)

    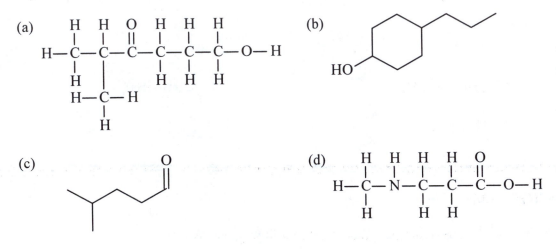

    (c)

    (d)

5. Identify the functional groups in the molecules whose structures are shown in question 4.

6. There are many compounds that have a molecular formula of C$_5$H$_{10}$O. These compounds are isomers of each other. Draw a structural formula for one of these isomers that fits each of the following descriptions.
    (a) This isomer contains a ketone functional group.
    (b) This isomer contains a double bond between carbon atoms.
    (c) This isomer contains an alcohol functional group.
    (d) This isomer contains an ether functional group.
    (e) This isomer contains **no** double bonds of any kind.

NAMING SIMPLE ORGANIC COMPOUNDS

This lesson deals with:

1. Drawing a structural formula of an alkane given its name.
2. Finding the longest chain of carbon atoms and the branch positions in the structural formula of an alkane.
3. Naming an alkane from its structural formula.
4. Naming an alkene or alkyne from its structural formula.
5. Naming an alkyl halide from its structural formula
6. Naming alcohols, aldehydes, ketones, carboxylic acids, ethers, amines, and esters using systematic nomenclature.
7. Naming aromatic compounds containing benzene rings.
8. Naming various organic compounds using common nomenclature.

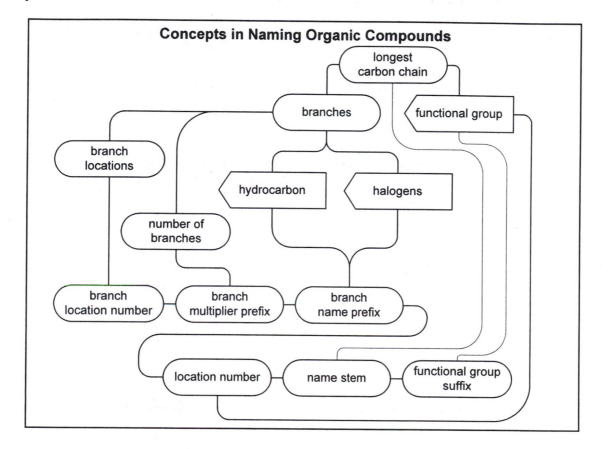

SYSTEMATIC METHOD OF NOMENCLATURE

Organic compounds are divided into classes and named as members of a class.

The fundamental class is alkanes. **Alkanes** are hydrocarbons (compounds containing only carbon and hydrogen) in which all bonds are single bonds. In alkanes, all carbon atoms are bonded to four other atoms. Because four is the maximum number of atoms to which carbon atoms can be bonded, alkanes are called **saturated hydrocarbons**.

Alkanes

In naming alkanes, the molecule is treated as having a main chain of carbon atoms with branches attached to it.

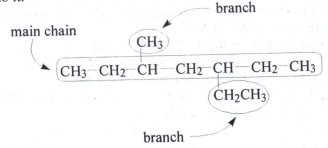

The main chain is the longest chain of carbon atoms in the molecule.

The name of an alkane indicates the number of carbon atoms in this longest chain, and the numbers of carbon atoms in each of the branches.

Table 1 lists the names of main chains and branches for chains of 1 through 10 carbon atoms.

Table 1. Names of Alkanes and Branches		
Carbon atoms in chain	Name of Alkane	Name of Branch
1	methane	methyl
2	ethane	ethyl
3	propane	propyl
4	butane	butyl
5	pentane	pentyl
6	hexane	hexyl
7	heptane	heptyl
8	octane	octyl
9	nonane	nonyl
10	decane	decyl

The name of an alkane also indicates the positions of the branches along the main chain. These positions are indicated with numerals placed before the name of the branch.

Example 1

Draw a condensed structural formula for 4-ethyl-2,5-dimethylheptane.

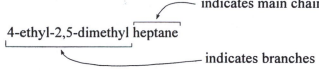

indicates main chain

4-ethyl-2,5-dimethyl heptane

indicates branches

1. Draw the longest chain.
    The end of the name indicates the length of the main chain.
    Heptane has 7 carbon atoms.

$$C—C—C—C—C—C—C$$

2. Attach the branches.

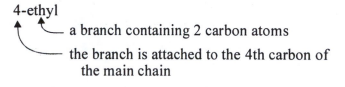

4-ethyl

a branch containing 2 carbon atoms

the branch is attached to the 4th carbon of
the main chain

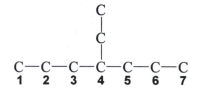

$$
\begin{array}{ccccccc}
 & & & C & & & \\
 & & & | & & & \\
 & & & C & & & \\
 & & & | & & & \\
C & C & C & C & C & C & C \\
1 & 2 & 3 & 4 & 5 & 6 & 7
\end{array}
$$

Multiple branches are indicated by prefixes to the branch name. The prefixes are listed in
Table 2.

Table 2. Prefixes for Multiple Branches

Number of Branches	Prefix
1	(none)
2	di
3	tri
4	tetra
5	penta
6	hexa

2,5 - di methyl

branch containing 1 carbon atom

two such branches

attached to the 2nd and 5th atoms of the main chain

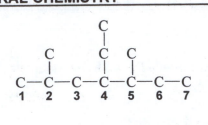

3. Add the hydrogen atoms.

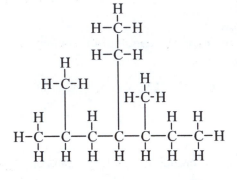

4. Condense the formula.

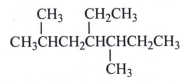

Exercise One

Give a structural formula for each of the compounds named below.

A. 3-ethylhexane

B. 2,2,4-trimethylpentane

C. 2,3-dimethyl-4-propyloctane

D. 2,2,3,3,4-pentamethylpentane

FINDING THE LONGEST CHAIN

The name of an alkane is determined by the longest chain of carbon atoms in its molecules. Therefore, the first step in naming an alkane is to find this longest chain. Generally, the easiest way to do this is to find the various chains and compare their lengths.

Example 2

Find the longest chain and branch-position numbers for the following structure.

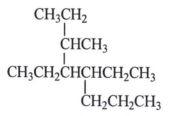

1. Locate the chain ends. These are the methyl groups, where carbon atoms are bonded to only one other carbon atom.

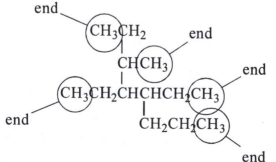

2. Find the chains that connect each pair of end-carbons. In this molecule there are ten such chains. The five of the longest ones are outlined below. For practice, you should find the others.

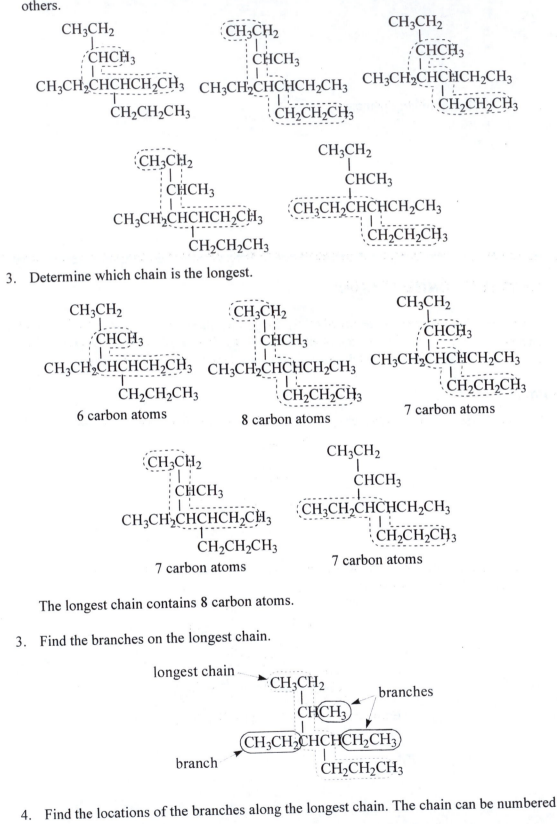

3. Determine which chain is the longest.

6 carbon atoms

8 carbon atoms

7 carbon atoms

7 carbon atoms

7 carbon atoms

The longest chain contains 8 carbon atoms.

3. Find the branches on the longest chain.

longest chain $\longrightarrow$ CH$_3$CH$_2$

branches

CH CH$_3$

CH$_3$CH$_2$CHCHCH$_2$CH$_3$

branch

CH$_2$CH$_2$CH$_3$

4. Find the locations of the branches along the longest chain. The chain can be numbered

from either end. Which end is numbered 1 is determined by the resulting branch locations. The chain is numbered to give the lowest sum of branch locations.

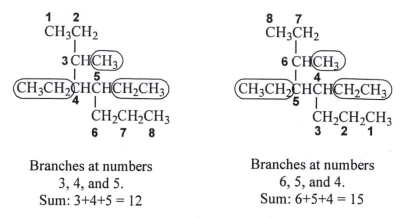

Branches at numbers
3, 4, and 5.
Sum: 3+4+5 = 12

Branches at numbers
6, 5, and 4.
Sum: 6+5+4 = 15

The numbering on the left gives the lower sum, so it is used.

Exercise Two

Find the longest chain and the branch position numbers for each structural formula below.

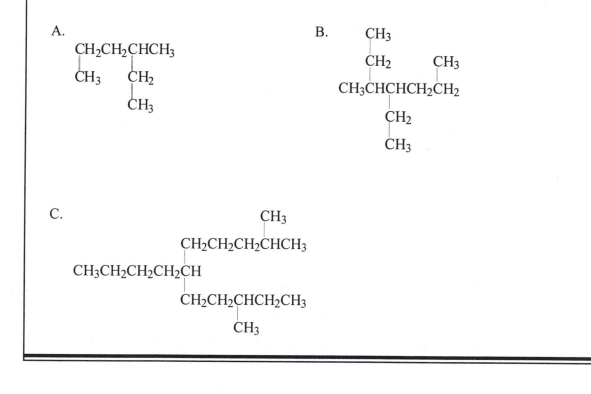

NAMING AN ALKANE

Example 3

Name the following compounds.

A.
$$CH_3$$
$$CH_2$$
$$CHCH_2CH_3$$
$$CH_3CH_2—C—CH_3$$
$$CH_3CH_2CH_2CHCH_2CH_2CH_3$$

1. Find the longest chain and the branch positions.

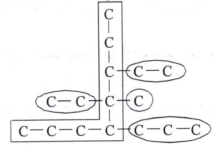

The longest chain is 8 carbons long: octane.

Starting from the top, the branch positions are 3, 4, 4, and 5. This adds to 16.
Starting from the left, the branch positions are 4, 5, 5, and 6. This adds to 20.
    Use the former, because the sum is smaller.

2. Identify the branches.

    —C—C—C   propyl

    —C   methyl

    —C—C   ethyl

3. Construct the name.

    5-propyl
    4-methyl
    3-ethyl and 4-ethyl combine to 3,4-diethyl.

Arrange the names of branches in alphabetical order, ethyl before methyl before propyl. (The prefixes, di, tri, tetra, etc., do not affect the order.)

The name is

3,4-diethyl-4-methyl-5-propyloctane

B.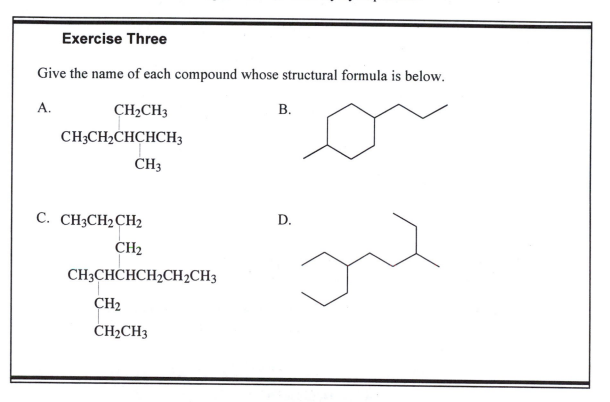

1. A ring of carbon atoms is named by the number of carbon atoms in the ring.
    In this structure the ring contains five atoms. Therefore, it is pentane.

    To indicate a ring, the prefix *cyclo* is used.
    This is cyclopentane.

    There are two branches from the ring. Positions on the ring are numbered starting
    with one branch being 1, and the others numbered from this, using the smaller set. In
    one direction around the ring, the numbers are 1 and 3, and in the other direction,
    they are 1 and 4. Threfore, the position numbers are 1 and 3.

2. The two branches are methyl groups.
    Because there are two, it is dimethyl-.

3. The name of the compound is 1,3-dimethylcyclopentane.

Exercise Three

Give the name of each compound whose structural formula is below.

A.　　　　　CH_2CH_3
　　　　　　　|
　　$CH_3CH_2CHCHCH_3$
　　　　　　　　|
　　　　　　　CH_3

B.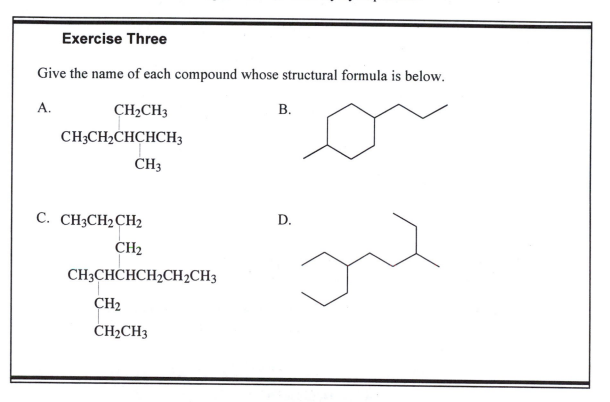

C.　$CH_3CH_2CH_2$
　　　　　　|
　　　　　CH_2
　　　　　　|
　　$CH_3CHCHCH_2CH_2CH_3$
　　　　|
　　　CH_2
　　　　|
　　　CH_2CH_3

D.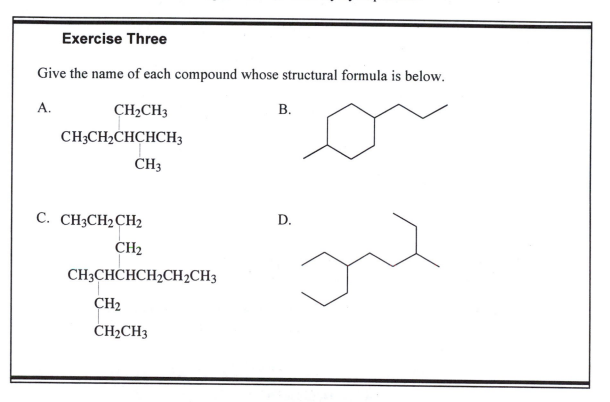

NAMING UNSATURATED HYDROCARBONS

Unsaturated hydrocarbons are those in which some carbon atoms are bonded to fewer than
four other atoms. This occurs when there is a multiple bond (double bond or triple bond)
between carbon atoms. A hydrocarbon that contains a double bond between carbon atoms is
called an **alkene**. One that contains a triple bond between carbon atoms is called an **alkyne**.

Alkene and alkyne names are based on alkane names. The presence of a double bond is indicated by changing the ending of the name from "-ane" to "-ene." The presence of a triple bond is indicated by changing the ending to "-yne." Table 3 contains some examples.

The location of the double bond is indicated with a number at the beginning of the name; the number indicates the carbon atom at which the multiple bond starts.

Table 3. Naming Alkenes and Alkynes based on Alkane Name

Class of Compound	Change This Suffix	To This Suffix	Example	
alkane	—	—	propane	$CH_3CH_2CH_3$
alkene	-ane	-ene	propene	$CH_2{=}CHCH_3$
alkyne	-ane	-yne	propyne	$CH{\equiv}CCH_3$

Example 4

Give the name of the compound whose structural formula is shown.

$$CH_2CH_2CH_2CH_2CH_3$$
$$|$$
$$CH_3CH_2CH_2CH_2CHC{\equiv}CCH_3$$

1. Locate the longest chain of carbon atoms that contains the multiple bond. In naming alkenes and alkynes, the multiple bond is always in the main chain.

$$CH_2CH_2CH_2CH_2CH_3$$
$$|$$
$$CH_3CH_2CH_2CH_2CHC{\equiv}CCH_3$$

8 carbon atoms

$$CH_2CH_2CH_2CH_2CH_3$$
$$|$$
$$CH_3CH_2CH_2CH_2CHC{\equiv}CCH_3$$

9 carbon atoms

The longest chain that contains the multiple bond contains 9 carbon atoms. Therefore, the name is based on nonane.

Note: Although there is a longer chain of carbon atoms containing 10 atoms, this chain does not include the multiple bond, and is therefore not used to name the compound.

$$CH_2CH_2CH_2CH_2CH_3$$
$$|$$
$$CH_3CH_2CH_2CH_2CHC{\equiv}CCH_3$$

The 10-carbon chain does not contain the triple bond, therefore it is not used in naming the compound.

The multiple bond is a triple bond.
Therefore, the ending is changed from "-ane" to "-yne."

The base name is nonyne.

2. Number the chain. The multiple bond is at the low-number end of the molecule.

$$\overset{5}{C}H_2\overset{4}{C}H_2\overset{3}{C}H_2\overset{2}{C}H_2\overset{1}{C}H_3$$

$$CH_3CH_2CH_2CH_2\overset{|}{C}HC\equiv CCH_3$$
$$\quad\quad\quad\quad\quad\;\;6\;\;7\;\;\;\;89$$

$$\overset{5}{C}H_2\overset{6}{C}H_2\overset{7}{C}H_2\overset{8}{C}H_2\overset{9}{C}H_3$$

$$CH_3CH_2CH_2CH_2\overset{|}{C}HC\equiv CCH_3$$
$$\quad\quad\quad\quad\quad\quad\;4\;\;\;3\;\;\;\;21$$

Here, the triple bond starts
at carbon #7

Here, the triple bond starts
at carbon #2.

The numbering shown above on the right is used, because the multiple bond has a lower number in that arrangement.

A number prefix indicates where the multiple bond starts: 2-nonyne.

3. Identify and locate the branches on the chain.

$$\overset{5}{C}H_2\overset{6}{C}H_2\overset{7}{C}H_2\overset{8}{C}H_2\overset{9}{C}H_3$$

$$(CH_3CH_2CH_2CH_2\overset{|}{C}HC\equiv CCH_3$$
$$\quad\quad\quad\quad\quad\quad\;4\;\;\;3\;\;\;\;21$$

There is a butyl group at carbon 4.

Therefore, the complete name of this molecule is 4-butyl-2-nonyne.

Exercise Four

Name the compounds whose formulas are given below.

A. $\quad\;\;\;CH_3\quad\;\;CH_2$
$$\quad\quad\quad\;\;|\quad\quad\;\;||$$
$$\quad\quad CH_3CHCH_2CCH_2CH_3$$

B.

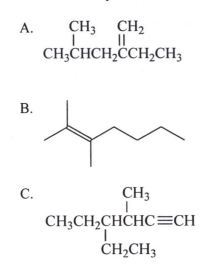

C. $\quad\quad\quad\quad\;\;CH_3$
$$\quad\quad\quad\quad\quad\;\;|$$
$$\quad CH_3CH_2CHCHC\equiv CH$$
$$\quad\quad\quad\quad\quad\;|$$
$$\quad\quad\quad\quad\;\;CH_2CH_3$$

D.

NAMING ALKYL HALIDES

An **alkyl halide** is a compound containing carbon, hydrogen, and a halogen (fluorine, chlorine, bromine, iodine). In naming alkyl halides, the halogen atom is treated as a branch on a hydrocarbon.

Halogen branch prefixes are derived from the name of the halogen by changing "-ine" to "-o."

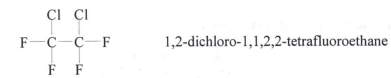

1,2-dichloro-1,1,2,2-tetrafluoroethane

Example 5

Name the molecule whose structural formula is given.

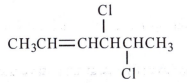

1. The longest carbon chain that contains the C=C double bond is 6 carbon atoms. The name is derived from hexane. The double bond makes it hexene.
2. There is a double bond starting at carbon 2 (when numbering from the left) or carbon 4 (when numbering from the right). The smaller number is used, so the chain is numbered from the left.
    The name contains 2-hexene.
3. There are two chlorine atoms attached to the chain, one at carbon 4 and the other at carbon 5.
    The name of the compound is 4,5-dichloro-2-hexene.

Exercise Five

Name the compounds whose formulas are given below.

A.

$$\begin{array}{ccc} & Cl & Cl \\ & | & | \\ CH_3CH_2 & CHCH_2 & CHCH_3 \end{array}$$

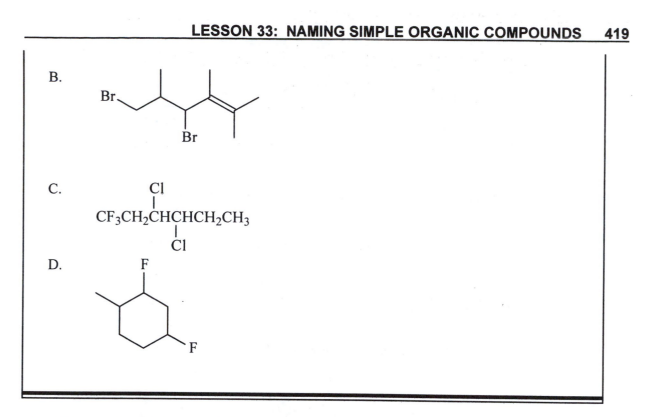

B.

C.

D.

NAMING ALCOHOLS, ALDEHYDES, KETONES, AND ACIDS

For these classes of organic compounds, the name is based on the name of the alkane with a carbon chain of the same length, but the ending is changed to indicate the class of the compound. Below are the functional groups for each of these classes.

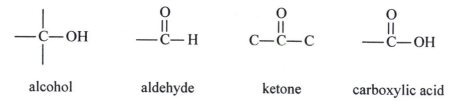

Table 4 contains the names of 3-carbon examples.

Example 6

Name the compounds whose structural formulas are given below.

A.   CH₃   O

1. Find the longest chain that contains the functional group.
   Four carbons, therefore the name is based on butane.
2. The functional group makes this an aldehyde.
   Therefore, this is a butanal.
3. Because the main chain has branches, the atoms in the chain need to be numbered. In

Table 4. Naming of Alcohols, Aldehydes, Ketones, and Carboxylic Acids

Class of Compound	Change -e Suffix To This	Example	
alkane	—	propane	$CH_3CH_2CH_3$
alcohol	-ol	1-propanol	$CH_3CH_2CH_2OH$
aldehyde	-al	propanal	$CH_3CH_2\overset{\overset{\displaystyle O}{\|\|}}{C}H$
ketone	-one	propanone	$CH_3\overset{\overset{\displaystyle O}{\|\|}}{C}CH_3$
carboxylic acid	-oic acid	propanoic acid	$CH_3CH_2\overset{\overset{\displaystyle O}{\|\|}}{C}OH$

aldehydes and carboxylic acids, the carbon in the C=O group is *always* numbered as #1.

4. Indicate the branches and their locations.
        Two methyl groups, both at carbon #3: 3,3-dimethyl-

The name of the compound is 3,3-dimethyl butanal.

B.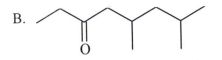

1. The longest chain is 8 carbons; therefore octane.
2. The functional group makes this a ketone.
    Change the ending to indicate a ketone: octanone.
3. Indicate the position along the main chain of the carbon atom in the functional group.
    The possible numbers are 3 and 6. Always use the *lower* number.
        3-octanone
4. Name the branches and number their positions.
        5,7-dimethyl

The name of the compound is 5,7-dimethyl-3-octanone.

Exercise Six

Name the compounds whose formulas are given below.

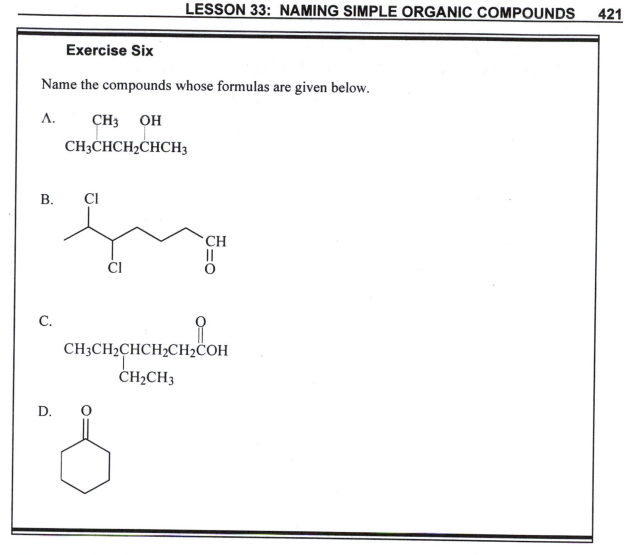

A.
$$CH_3CHCH_2CHCH_3$$
with CH_3 and OH substituents

B.

C.
$$CH_3CH_2CHCH_2CH_2COH$$
with CH_2CH_3

D.

NAMING ETHERS AND AMINES

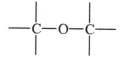

ether
functional group

amine
functional group

In naming ethers and amines, the carbon chains that are bonded to the oxygen or nitrogen are named followed by either "ether" or "amine."

Example 7

Name the following compounds.

A. $CH_3CH_2\!-\!O\!-\!CH_2CH_2CH_3$

1. This contains an oxygen atom with single bonds to two alkane branches. Therefore it is an ether.
2. One alkane branch is an ethyl group. The other is a propyl group.

$$CH_3CH_2—O—CH_2CH_2CH_3$$

ethyl   propyl

3. Therefore, the name is:

ethyl propyl ether

B.

$$CH_3CH_2CH—N—CH_3$$

with H on N and CH₃ below CH.

1. This contains a nitrogen atom with single bonds, one to hydrogen and two to alkane branches. Therefore, it is an amine.
2. The alkane branches are a methyl group and a 2-butyl group.

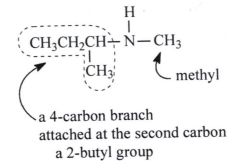

methyl

a 4-carbon branch
attached at the second carbon
a 2-butyl group

3. The name is:

2-butyl methyl amine

Exercise Seven

Name the compounds whose formulas are shown below.

A. $CF_3—O—CH_2CH_3$

B. $H—N—CH_2CH_3$
         $|$
         CH_3

C. $CH_3CH—O—CH_3$
         $|$
         CH_3

NAMING ESTERS

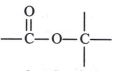

ester functional group

The name of an ester has two parts. One part comes from the name of the carboxylic acid from which it can be made. The other part of an ester's name comes from the remaining hydrocarbon.

Example 8

Name the ester whose structure is given.

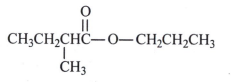

1. Divide the structure after the oxygen, leaving the oxygen attached to the C=O group.

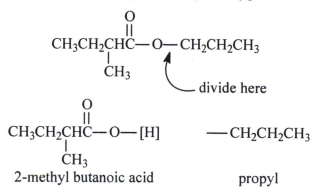

2. Change the ending of the acid from "-ic acid" to "-ate."
    2-methyl-butanoate
3. Put the name of the alkyl group in front of this.
    The name of this ester is propyl 2-methyl-butanoate.

Exercise Eight

Name the compounds whose formulas are shown below.

A.
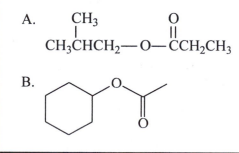

B.

NAMING AROMATIC COMPOUNDS

Aromatic compounds contain a ring of six carbon atoms, and each of these six is bonded to only three other atoms (two from the ring and one other).

The simplest aromatic hydrocarbon is called benzene, and its structure can be represented by any of the following formulas.

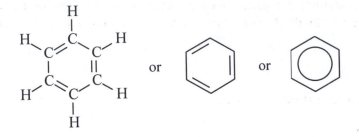

Compounds related to benzene can be formed by replacing one or more of the hydrogens in benzene with other groups, such as halogens or alkane branches. The locations of these replacements are indicated by numbers placed before the names of the replacement group.

Some molecules contain a benzene ring attached through one of its carbon atoms. This is often called a phenyl group.

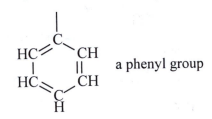

a phenyl group

Example 9

Name the following compounds.

A.

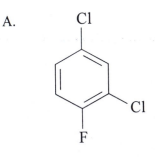

1. The molecule contains a benzene ring. The ring has more than one H replaced, therefore it is not a phenyl group. It is named as a benzene.

2. The location numbers for the attached groups should have the smallest sum.

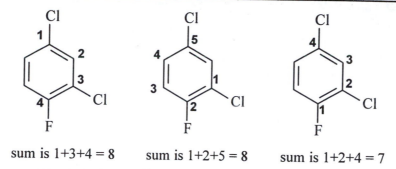

sum is 1+3+4 = 8   sum is 1+2+5 = 8   sum is 1+2+4 = 7

(Numbering each ring in the opposite direction gives higher numbers.)
3. The groups are a fluoro group and two chloro groups.
    The name is:

2,4-dichloro-1-fluorobenzene

B.

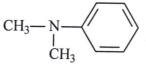

1. The molecule contains a nitrogen atom with single bonds to three other atoms. It is an amine.
2. Two methyl groups are attached to the nitrogen atom. Also attached is a benzene ring with only one H atom replaced; this is a phenyl group.

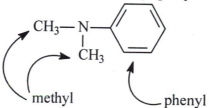

methyl   phenyl

3. The name is:

dimethyl phenyl amine

Exercise Nine

Name the following compounds.

A.

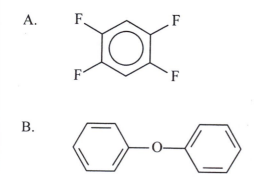

B.

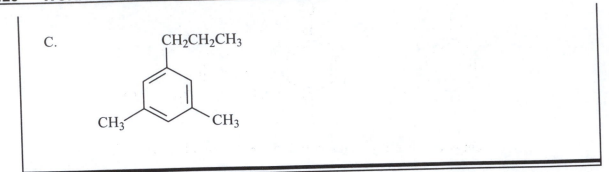

C.

COMMON NAMES

There are many organic compounds that have commonly used names which do not follow the systematic naming system. The only way to learn these names is to memorize them. Table 5 contains some of the most common names.

Often the common names are used in naming compounds with similar molecular structures.

Example 10

Draw a structural formula for each compound named below.

A. dichloroacetic acid.

The structure of acetic acid is $CH_3-\overset{\overset{\displaystyle O}{\displaystyle \|}}{C}-OH$

"dichloro" means that two of the hydrogen atoms attached to carbon have been replaced by chlorine atoms.

The structure of dichloroacetic acid is $Cl-\overset{\overset{\displaystyle H}{\displaystyle |}}{\underset{\displaystyle |}{\underset{\displaystyle Cl}{C}}}-\overset{\overset{\displaystyle O}{\displaystyle \|}}{C}-OH$

B. isopropyl formate

The name indicates that this is an ester.
    "isopropyl" is the hydrocarbon group in isopropyl alcohol $CH_3\overset{\overset{\displaystyle |}{}}{C}HCH_3$

"formate" is derived from formic acid $H-\overset{\overset{\displaystyle O}{\displaystyle \|}}{C}-O-$

Then, the structure of isopropyl formate is

$$H-\overset{\overset{\displaystyle O}{\displaystyle \|}}{C}-O-\overset{\overset{\displaystyle CH_3}{\displaystyle |}}{C}HCH_3$$

Table 5. Common Names of Some Organic Compounds

Formula	Systematic Name	Common Name
$CHCl_3$	trichloromethane	chloroform
CH_3OH	methanol	methyl alcohol
$H-\overset{\overset{\textstyle O}{\|\|}}{C}-H$	methanal	formaldehyde
$H-\overset{\overset{\textstyle O}{\|\|}}{C}-OH$	methanoic acid	formic acid
$CH_3-\overset{\overset{\textstyle O}{\|\|}}{C}-OH$	ethanoic acid	acetic acid
CH_3CH_2OH	ethanol	ethyl alcohol
$CH_3-\overset{\overset{\textstyle O}{\|\|}}{C}-H$	ethanal	acetaldehyde
$H-C{\equiv}C-H$	ethyne	acetylene
$CH_2{=}CH_2$	ethene	ethylene
$CH_3\overset{\overset{\textstyle O}{\|\|}}{C}CH_3$	propanone	acetone
$CH_3\overset{\overset{\textstyle OH}{\|}}{C}HCH_3$	2-propanol	isopropyl alcohol *or* isopropanol
$CH_3\overset{\overset{\textstyle OH}{\|}}{\underset{\underset{\textstyle CH_3}{\|}}{C}}CH_3$	2-methyl-2-propanol	tertiary butyl alcohol *or* *t*-butyl alcohol *or* *t*-butanol

---

Exercise Ten

Draw a structural formula for each compound named below.

A. hexafluoroacetone

B. tertiary butyl acetate

C. trichloroethylene

Note: Naming Compounds with More than One Functional Group

There are many compounds that have more than one functional group. Some of these have already been described, such as those containing halogens and double bonds.

A more complicated situation arises when a compound contains two functional groups, both of which are indicated by a suffix in the name. To deal with these situations, there are rules that assign the priorities of the functional groups. For example, ketones are of higher priority than alcohols, and a molecule that contains both an alcohol and a ketone functional group is named as a ketone with an alcohol group.

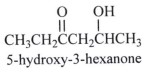
5-hydroxy-3-hexanone

The carboxylic acid functional group has a higher priority than a ketone.

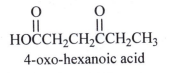
4-oxo-hexanoic acid

These rules that determine the priorities of the functional groups are beyond what is usually needed in General Chemistry, and they are not dealt with here.

PROBLEMS

1. Draw structural formulas for the compounds named below.

   (a) 3-chloro-2-methylbutanal
   (b) cyclopentyl propanoate
   (c) ethyl 3-hexyl amine
   (d) 1,2-dimethylcyclohexene

2. Name the following compounds.

(a)

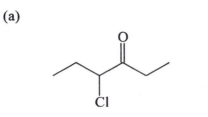

(b)

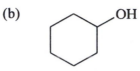

(c)   CH₃CHOCH₂CH₃
        |
        CH₂CH₃

(d)

$$CH_3CH_2CHOCCH_2CH_3$$
        | ‖
        CH₃   O

(e)

$$\overset{O}{\overset{\|}{HOCCH_2CH_2CCl_3}}$$

3. Draw a structural formula for each of the compounds named below.

(a) pentanoic acid
(b) 4-methylcyclohexanone
(c) 1,4-dichlorobenzene
(d) butyl dimethyl amine
(e) octyl acetate

4. Give the systematic name for each of the following.

(a) O=CH
         |
    CH₃CHCH₂CH₃

(b)

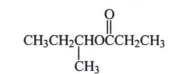

(c)   O   Cl
       ‖ |
    CH₃CCH₂CHCH₃

(d) CH₃CH₂OCH₂CH₃

REACTIONS OF ORGANIC COMPOUNDS

This lesson deals with

1. Predicting the major product of the halogenation of an alkane.
2. Drawing the structure for the product of the reaction of an alkyl halide with aqueous base.
3. Predicting the products of the oxidation of alcohols.
4. Drawing the structure for the product of the reaction of an alcohol and a carboxylic acid.
5. Identifying the reactants needed to produce a particular ester.
6. Drawing structures for the products of a reaction between ammonia and an alkyl halide.
7. Drawing the structure for the product of the reaction of an amine and a carboxylic acid.
8. Identifying the reactants needed to produce a particular amide.
9. Drawing the structure of a compound produced by the reaction of a Grignard reagent with an aldehyde or ketone.

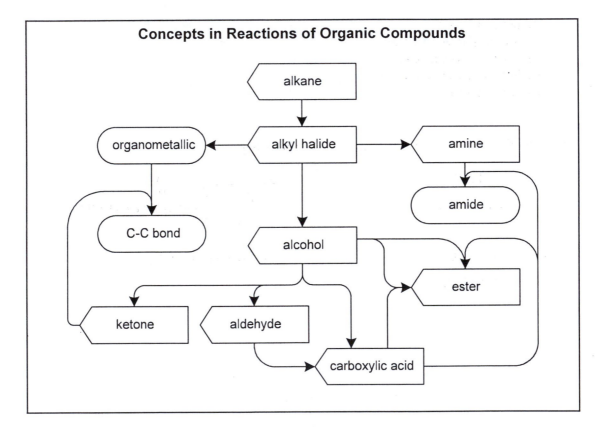

ALKYL HALIDES

An alkyl halide is a substituted hydrocarbon, in which one or more hydrogen atoms have been replaced by halogen atoms.

Formation of alkyl halides

Reaction of a hydrocarbon with a halogen

$$CH_4 + Cl_2 \longrightarrow CH_3Cl + HCl$$

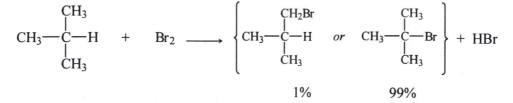

This product ratio is not what would be expected based on the number of replaceable hydrogen atoms. There are 6 terminal hydrogen atoms and 2 central hydrogen atoms. Based on this, we would expect 3 times as much 1-bromopropane as 2-bromopropane.

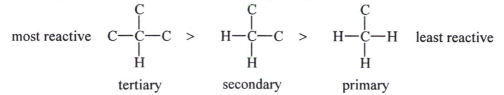

Carbon atoms can be classified by the number of other carbon atoms to which they are bonded. Carbon atoms bonded to three other carbon atoms are called tertiary carbon atoms. Those bonded to two other carbon atoms are called secondary carbons. A carbon atom that is bonded to only one other carbon atom is called a primary carbon atom.

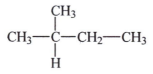

Hydrogen attached to a tertiary carbon is more reactive than hydrogen attached to a secondary carbon, which is more reactive than hydrogen attached to a primary carbon.

Example 1

Draw structural formulas for the products of the monochlorination (*i.e.*, replacement of *one* hydrogen atom by a chlorine atom) of 2-methylbutane and indicate which is the major product.

$$CH_3-\underset{\underset{H}{|}}{\overset{\overset{CH_3}{|}}{C}}-CH_2-CH_3$$

The 2-methylbutane molecule contains hydrogen atoms in 4 different locations:

the 6 hydrogens of the methyl groups on carbon 2   (primary)
the hydrogen on carbon 2   (tertiary)
the 2 hydrogens on carbon 3   (secondary)
the 3 hydrogens of carbon 4   (primary)

Therefore, there are 4 different monochlorination products:

CH₂Cl
|
CH₃—C—CH₂—CH₃
|
H

CH₃
|
CH₃—C—CH₂—CH₃
|
Cl

CH₃
|
CH₃—C—CHCl—CH₃
|
H

CH₃
|
CH₃—C—CH₂—CH₂Cl
|
H

The major product is the one obtained by replacing the tertiary hydrogen.

Therefore, 2-chloro-2-methylbutane is the major product.

Exercise One

A. The octane rating of a gasoline is based on a scale on which iso-octane has a rating
of 100. Iso-octane is 2,2,4-trimethylpentane. How many different products can be
obtained from the monobromination (replacement of *one* hydrogen atom by a
bromine atom) of 2,2,4-trimethylpentane? Draw their structures.

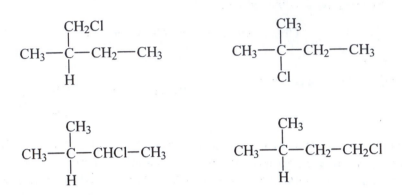

B. Which of the products in A is the major product?

C. Which of the products in A is produced in the smallest amount?

D. What is the major product of the monochlorination of 2,3-dimethylbutane?

REACTIONS OF ALKYL HALIDES

With hydroxide ions

primary alkyl halide:

$$HO^- + CH_3CH_2Br \longrightarrow CH_3CH_2OH + Br^-$$
primary alcohol

secondary alkyl halide

$$HO^- + CH_3\overset{\overset{\displaystyle Cl}{|}}{C}HCH_3 \longrightarrow CH_3\overset{\overset{\displaystyle OH}{|}}{C}HCH_3 + Cl^-$$
secondary alcohol

tertiary alkyl halide

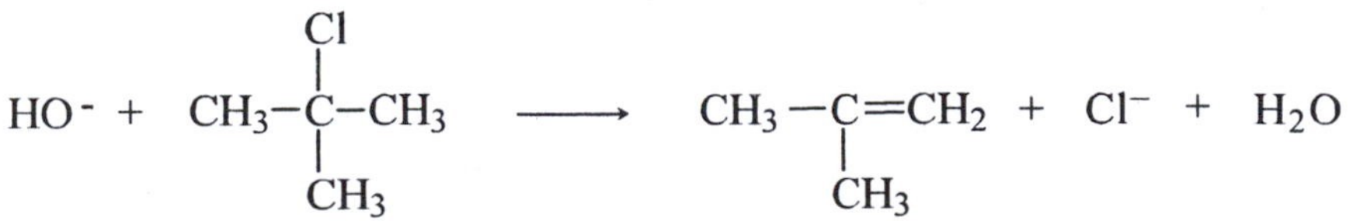

$$HO^- + CH_3-\overset{\overset{\displaystyle Cl}{|}}{\underset{\underset{\displaystyle CH_3}{|}}{C}}-CH_3 \longrightarrow CH_3-\overset{\overset{\displaystyle}{\underset{\underset{\displaystyle CH_3}{|}}{C}}}=CH_2 + Cl^- + H_2O$$
alkene

Tertiary alcohols cannot be made directly from tertiary alkyl halide, because the product is an alkene instead. The halide ion is removed from its carbon atom, a hydrogen atom is removed from an adjacent carbon atom, and a double bond forms between the two carbon atoms.

Example 2

Write an equation for the reaction between 2-chlorobutane and 0.2 M NaOH(aq).

First, draw the structure of 2-chlorobutane to see if it is a primary, secondary, or tertiary alkyl halide.

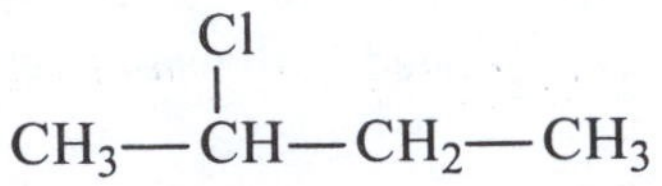

$$\begin{array}{c} Cl \\ | \\ CH_3{-}CH{-}CH_2{-}CH_3 \end{array}$$

2-chlorobutane

It is a secondary alkyl halide, so a secondary alcohol is the product.

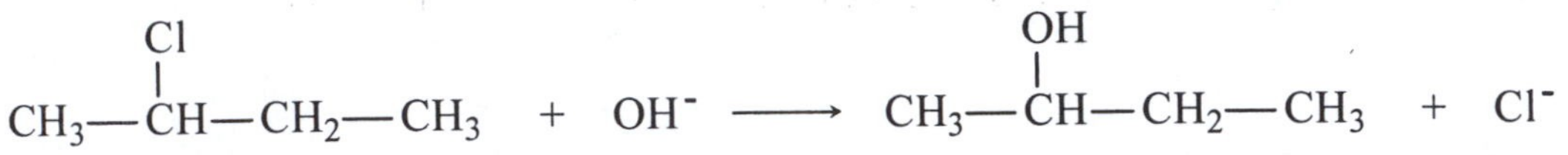

$$\begin{array}{c} Cl \\ | \\ CH_3{-}CH{-}CH_2{-}CH_3 \end{array} + OH^- \longrightarrow \begin{array}{c} OH \\ | \\ CH_3{-}CH{-}CH_2{-}CH_3 \end{array} + Cl^-$$

Exercise Two

A. Write an equation for the reaction of 2-chloro-2,3,3-trimethylbutane with NaOH(aq).

B. Write the structure for the product of the reaction of 1,4-dichlorobutane with NaOH(aq).

C. An alcohol can be made by the reaction of an alkyl halide with NaOH(aq). An alkyl halide can be made by the halogenation of an alkane.

   1. What alkane can be used to make 1-butanol?

   2. What other alcohol could be made from this alkane?

3. Which alcohol would be the easier one to make from the alkane?

OXIDATION OF ALCOHOLS

Many alcohols react with inorganic oxidizing agents, such as dichromate ions ($Cr_2O_7^{2-}$) dissolved in aqueous sulfuric acid (H_2SO_4).

The nature of the reaction depends on whether the alcohol is a primary, secondary, or tertiary alcohol.

Primary alcohol

$$CH_3CH_2CH_2CH_2OH \xrightarrow{\ H^+/Cr_2O_7^{2-}\ } CH_3CH_2CH_2\overset{\displaystyle O}{\overset{\|}{C}}H$$

1-butanol	butanal
a primary alcohol	an aldehyde
	50% yield

Yield is low because aldehydes also react with oxidizing agents.

$$CH_3CH_2CH_2\overset{\displaystyle O}{\overset{\|}{C}}H \xrightarrow{\ H^+/Cr_2O_7^{2-}\ } CH_3CH_2CH_2\overset{\displaystyle O}{\overset{\|}{C}}OH$$

butanal	butanoic acid

Oxidation of a primary alcohol yields an aldehyde and a carboxylic acid. Significant amounts of aldehyde are obtained from the oxidation of primary alcohols only for aldehydes with low boiling points. The aldehyde can be isolated by boiling it off from the reaction mixture before it is oxidized to a carboxylic acid. In general, only aldehydes with 5 or fewer carbon atoms have sufficiently low boiling points to be isolated by this method.

Secondary alcohol

$$CH_3\overset{\displaystyle OH}{\overset{|}{C}}HCH_3 \xrightarrow{\ H^+/Cr_2O_7^{2-}\ } CH_3\overset{\displaystyle O}{\overset{\|}{C}}CH_3$$

2-propanol	acetone (propanone)
a secondary alcohol	a ketone

The oxidation of a secondary alcohol produces a ketone.

Tertiary alcohol:

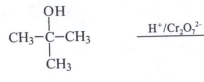

$$\text{CH}_3-\overset{\overset{\displaystyle \text{OH}}{|}}{\underset{\underset{\displaystyle \text{CH}_3}{|}}{\text{C}}}-\text{CH}_3 \xrightarrow{\ \text{H}^+/\text{Cr}_2\text{O}_7{}^{2-}\ } \qquad \text{No Reaction}$$

2-methyl-2-propanol
a tertiary alcohol

A tertiary alcohol does not react with dichromate oxidizing agent.

More potent oxidizing agents can react with tertiary alcohols, but these break bonds between carbon atoms and yield a variety of products.

Example 3

Identify the products of each of the following reactions.

A.

$$\text{CH}_3\text{CH}_2\overset{\overset{\displaystyle \text{CH}_3}{|}}{\text{CHOH}} \xrightarrow{\ \text{H}^+/\text{Cr}_2\text{O}_7{}^{2-}\ }$$

The organic reactant is a secondary alcohol.
The product of oxidation will be a ketone.
The product is

$$\text{CH}_3\text{CH}_2\overset{\overset{\displaystyle \text{CH}_3}{|}}{\text{C}}{=}\text{O}$$

B.

$$\text{CH}_3\overset{\overset{\displaystyle \text{CH}_3}{|}}{\text{CH}}\text{CH}_2\text{OH} \xrightarrow{\ \text{H}^+/\text{Cr}_2\text{O}_7{}^{2-}\ }$$

The reactant is a primary alcohol.
The products are an aldehyde and a carboxylic acid.
This alcohol contains only 4 carbons, so the aldehyde can be isolated.
The products are

$$\text{CH}_3\overset{\overset{\displaystyle \text{CH}_3}{|}}{\text{CH}}{-}\overset{\overset{\displaystyle \text{O}}{\|}}{\text{CH}} \qquad\qquad \text{CH}_3\overset{\overset{\displaystyle \text{CH}_3}{|}}{\text{CH}}{-}\overset{\overset{\displaystyle \text{O}}{\|}}{\text{C}}{-}\text{OH}$$

C.

$$\text{CH}_3\overset{\overset{\displaystyle \text{CH}_3}{|}}{\text{CH}}{-}\overset{\overset{\displaystyle \text{CH}_3}{|}}{\underset{\underset{\displaystyle \text{OH}}{|}}{\text{C}}}{-}\text{CH}_2{-}\text{CH}_2\text{OH} \xrightarrow{\ \text{H}^+/\text{Cr}_2\text{O}_7{}^{2-}\ }$$

This organic reactant contains 2 alcohol groups.
One is a tertiary alcohol and will not react with $\text{H}^+/\text{Cr}_2\text{O}_7{}^{2-}$.
The other is a primary alcohol and will be oxidized to a carboxylic acid. (Any aldehyde formed would contain 7 carbon atoms and have a boiling point too high to be isolated from the mixture and be oxidized to the acid.)

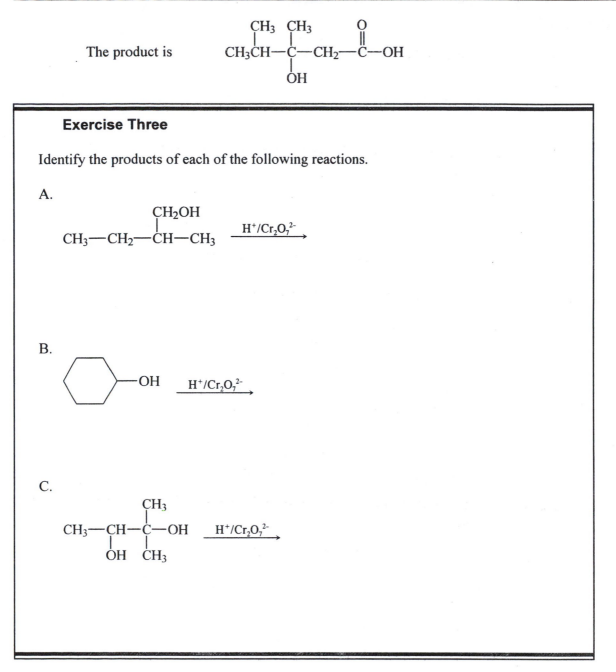

The product is

$$CH_3 \overset{\overset{\displaystyle CH_3}{|}}{CH} - \overset{\overset{\displaystyle CH_3}{|}}{\underset{\underset{\displaystyle OH}{|}}{C}} - CH_2 - \overset{\overset{\displaystyle O}{||}}{C} - OH$$

Exercise Three

Identify the products of each of the following reactions.

A.

$$CH_3 - CH_2 - \overset{\overset{\displaystyle CH_2OH}{|}}{CH} - CH_3 \quad \xrightarrow{\;H^+/Cr_2O_7^{2-}\;}$$

B.

⬡—OH $\xrightarrow{\;H^+/Cr_2O_7^{2-}\;}$

C.

$$CH_3 - \overset{}{\underset{\underset{\displaystyle OH}{|}}{CH}} - \overset{\overset{\displaystyle CH_3}{|}}{\underset{\underset{\displaystyle CH_3}{|}}{C}} - OH \quad \xrightarrow{\;H^+/Cr_2O_7^{2-}\;}$$

REDUCTION OF ALDEHYDES AND KETONES

Aldehydes can be made by oxidizing a primary alcohol. Ketones can be formed by oxidizing a secondary alcohol. These reactions can be reversed by using a strong reducing agent. Such strong reducing agents include $NaBH_4$ and $LiAlH_4$.

Example 4

A. What is the product of the following reaction?

$$\overset{\displaystyle O}{\underset{\displaystyle CH_3CH_2CH}{||}} \xrightarrow{\quad LiAlH_4 \quad}$$

The LiAlH₄ reduces the aldehyde to a primary alcohol, by adding 2 hydrogen atoms. The product is

$$\underset{\displaystyle H}{\overset{\displaystyle O-H}{\underset{|}{\overset{|}{CH_3CH_2CH}}}}$$

which is the primary alcohol, 1-propanol, usually written as $CH_2CH_2CH_2OH$.

B. What compound can be reduced to 2-butanol by LiAlH₄?

The structure of 2-butanol is

$$\overset{\displaystyle OH}{\underset{\displaystyle CH_3CHCH_2CH_3}{|}}$$

We want to make this by reducing some other compound. Reduction is the reverse of oxidation. If 2-butanol were oxidized, the product would be

$$\overset{\displaystyle O}{\underset{\displaystyle CH_3CCH_2CH_3}{||}}$$

which is 2-butanone. This compound can be reduced to 2-butanol by LiAlH₄.

Exercise Four

A. What is the product of the reaction of LiAlH₄ with cyclohexanone?

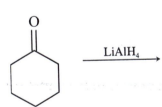

B. What compound can be reduced with LiAlH₄ to produce propylene glycol?

$$\overset{\displaystyle OH}{\underset{\displaystyle CH_3CHCH_2OH}{|}}$$

REACTIONS OF CARBOXYLIC ACIDS

A carboxylic acid reacts with an alcohol forming an ester and water.

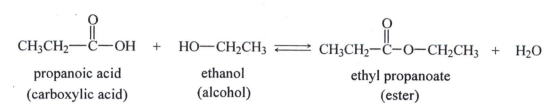

propanoic acid   ethanol   ethyl propanoate
(carboxylic acid)   (alcohol)   (ester)

This reaction is an equilibrium (an ester reacts with water to form an alcohol and a carboxylic acid). The formation of the ester can be promoted by removing the water. This is done by adding a substance with a great affinity for water, such as sulfuric acid.

Example 5

A. Draw the structure for the product of the reaction between acetic acid and 2-butanol.

First draw the structures of the reactants.

acetic acid   2-butanol

In the reaction, the OH of the acetic acid and the H of the alcohol combine to form water, and the remaining fragments of the acid and the alcohol combine:

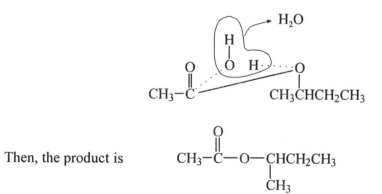

Then, the product is

$$CH_3-\overset{\displaystyle O}{\overset{\|}{C}}-O-\underset{\displaystyle CH_3}{\underset{|}{C}}HCH_2CH_3$$

B. Identify the carboxylic acid and the alcohol from which the following ester can be formed.

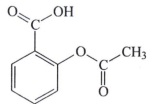

First, identify the portions of the molecule that came from the alcohol and from the acid.

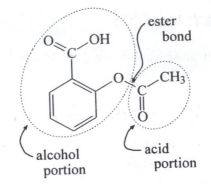

Then, write the complete formulas for the carboxylic acid and alcohol.

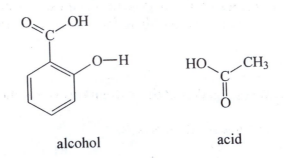

alcohol   acid

Exercise Five

A. Draw structures for the carboxylic acid and the alcohol needed to make the ester with the following structure.

$$CH_3\overset{O}{\overset{\|}{C}}CH_2CH_2-O-\overset{O}{\overset{\|}{C}}H$$

B. Write a chemical equation to illustrate how benzoic acid reacts with 2-methyl-2-propanol.

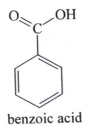

benzoic acid

C. Draw the structure of the ester formed by the reaction between salicylic acid and methanol.

salicylic acid

AMINES AND AMIDES

Amines are compounds of nitrogen. They are related to ammonia, but have one or more of the hydrogen atoms replace by alkyl carbon atoms.

$$H-\underset{\underset{H}{|}}{N}-H \qquad H-\underset{\underset{H}{|}}{N}-CH_3 \qquad CH_3-\underset{\underset{H}{|}}{N}-CH_3 \qquad CH_3-\underset{\underset{CH_3}{|}}{N}-CH_3$$

ammonia	methylamine	dimethylamine	trimethylamine
	primary amine	secondary amine	tertiary amine

Amines can be prepared by reactions between alkyl halides and ammonia.

$$CH_3CH_2Br + NH_3 \longrightarrow CH_3CH_2NH_2 + HBr$$

The primary amine can react with more alkyl halide to form secondary and tertiary amines.

$$CH_3CH_2Br + CH_3CH_2NH_2 \longrightarrow (CH_3CH_2)_2NH + HBr$$

$$CH_3CH_2Br + (CH_3CH_2)_2NH \longrightarrow (CH_3CH_2)_3N + HBr$$

The primary amine can be made the major product by using a large excess of ammonia over alkyl halide.

An amine reacts with a carboxylic acid to form a salt:

$$CH_3CH_2-\overset{\overset{O}{\|}}{C}-OH \ + \ H_2N-CH_2CH_3 \longrightarrow CH_3CH_2-\overset{\overset{O}{\|}}{C}-O^- \ H_3\overset{+}{N}-CH_2CH_3$$

ionic salt

However, if this salt is heated to 225°C, the salt loses water, and an amide is formed.

$$CH_3CH_2-\overset{O}{\overset{\|}{C}}-O^- \ \overset{+}{H_3N}-CH_2CH_3 \ \xrightarrow{\ 225°C\ } \ CH_3CH_2-\overset{O}{\overset{\|}{C}}-\overset{H}{\overset{|}{N}}-CH_2CH_3 \ + \ H_2O$$

Many important polymers are amides, such as proteins and nylon.

Example 6

A. Draw the structures of the amines that can result from the reaction of ammonia with 2-bromobutane.

The reactants are

$$H-\overset{H}{\overset{|}{N}}-H \qquad CH_3\overset{}{\underset{Br}{\overset{|}{C}H}}CH_2CH_3$$

ammonia   2-bromobutane

Three products are possible.

replace 1 hydrogen of NH_3:
$$H-\overset{H}{\overset{|}{N}}-H$$
$$CH_3CHCH_2CH_3$$

replace 2 hydrogens:
$$H-\overset{CH_3}{\overset{|}{N}}-\overset{}{C}HCH_2CH_3$$
$$CH_3CHCH_2CH_3$$

replace all 3 hydrogens:
$$CH_3CH_2\overset{CH_3}{\overset{|}{C}H}-N-\overset{CH_3}{\overset{|}{C}H}CH_2CH_3$$
$$CH_3CHCH_2CH_3$$

B. Draw the structure of the amide that can be formed from 2-butylamine and propanoic acid.

$$CH_3\overset{NH_2}{\overset{|}{C}H}CH_2CH_3 \qquad CH_3CH_2\overset{O}{\overset{\|}{C}}-OH$$

2-butylamine   propanoic acid

The –OH from propanoic acid combines with a hydrogen from the amine nitrogen, and the amine nitrogen attaches to the carbon from which the –OH is released.

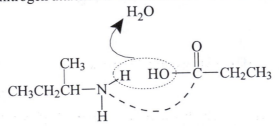

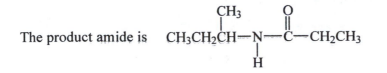

The product amide is CH_3CH_2CH $-$ N $-$ C $-$ CH_2CH_3

with CH_3 above the CH, O double-bonded above the C, and H below the N.

Exercise Six

A. 1. Draw structures for the three products of the reaction between ammonia and bromoethane.

2. Which of these is the major product if an excess of ammonia is used?

3. Which of these is the major product if an excess of bromoethane is used.

B. Draw the structure of the amide that can be made from the reaction of benzoic acid with dimethylamine. The structure of benzoic acid is in exercise 5.

C. Draw structures for the amine and the carboxylic acid that combine to form the following molecule.

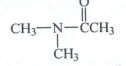

$$CH_3-N-\overset{\overset{\displaystyle O}{\|}}{C}CH_3$$
$$|$$
$$CH_3$$

FORMATION OF CARBON-CARBON BONDS

One method of producing carbon-carbon bonds involves organometallic compounds.

An organometallic compound is one that contains a carbon atom bonded directly to a metal atom.

In the Grignard reaction, an alkyl halide reacts with magnesium metal to produce an organomagnesium compound.

$$CH_3CH_2CH_2Br + Mg \longrightarrow CH_3CH_2CH_2—Mg—Br$$

The bond between metal and carbon is very reactive. It will react with an aldehyde or ketone to form a bond between two carbon atoms and a bond between magnesium and oxygen.

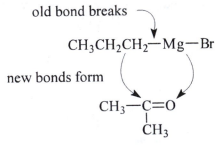

The intermediate product is

$$CH_3CH_2CH_2$$
$$|$$
$$CH_3—C—O–Mg—Br$$
$$|$$
$$CH_3$$

This intermediate product reacts readily with water, producing an alcohol and HOMgBr.

$$CH_3CH_2CH_2$$
$$CH_3{-}\overset{\displaystyle |}{\underset{\displaystyle |}{C}}{-}O{-}Mg{-}Br \;+\; H_2O \;\longrightarrow\; CH_3{-}\overset{\displaystyle |}{\underset{\displaystyle |}{C}}{-}O{-}H \;+\; HO{-}Mg{-}Br$$
$$CH_3 \qquad\qquad\qquad\qquad\qquad CH_3$$

The reaction of a ketone with an organomagnesium compound produces a tertiary alcohol. The reaction of an aldehyde with an organomagnesium compound produces a secondary alcohol.

Example 7

Write equations to show how the primary alkyl halide, 1-bromopropane, can be converted to the secondary alcohol, 3-hexanol.

$$CH_3CH_2CH_2{-}Br \qquad\qquad CH_3CH_2CH_2\overset{\overset{\textstyle OH}{\textstyle |}}{C}HCH_2CH_3$$

1-bromopropane   3-hexanol

The steps in the process can be determined by working backward, from the product, 3-hexanol, to the starting material, 1-bromopropane. When 3-hexanol is formed from an organomagnesium compound, the C–C bond that forms is one of those to the alcohol C. There are two such bonds. One of them attaches a 4-carbon chain to a 2-carbon chain. The other attaches a 3-carbon chain to a 3-carbon chain. This second one is the one that can be formed from two propane units.

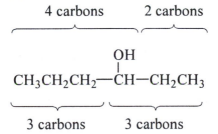

The organomagnesium compound would be $CH_3CH_2CH_2MgBr$. The aldehyde would be propanal.

The organomagnesium compound is made by the reaction of 1-bromopropane with magnesium.

$$CH_3CH_2CH_2Br \;+\; Mg \;\longrightarrow\; CH_3CH_2CH_2MgBr$$

Propanal can be made by oxidizing 1-propanol with $Cr_2O_7{}^{2-}$, and 1-propanol can be made from 1-bromopropane by treating it with NaOH.

$$CH_3CH_2CH_2Br \;+\; NaOH \;\longrightarrow\; CH_3CH_2CH_2OH \;+\; NaBr$$

$$CH_3CH_2CH_2OH \;\xrightarrow{\;Cr_2O_7{}^{2-}\;}\; CH_3CH_2\overset{\overset{\textstyle O}{\textstyle \|}}{C}H$$

Exercise Seven

A. Write an equation for the reaction of 2-bromobutane with magnesium metal.

B. Write the formula of the ketone that could be formed from 2-bromobutane.

C. Draw the structure of the alcohol that can be made from the reaction between the products of the reactions in A and B.

PROBLEMS

1. What is the carbon compound produced, if any, by the reaction of dichromate ions $(Cr_2O_7^{2-})$ dissolved in aqueous sulfuric acid (H_2SO_4) with each of the following.

    (a) 2-butanol   (b) hexanal   (c) 3-octanone   (d) 2-methyl-2-pentanol

2. Describe how to make each of the following from the appropriate alkyl bromide.

    (a) 2-butanol   (b) 3-octanone   (c) ethyl acetate   (d) triethyl amine

3. Draw a structural formula for the major carbon-containing product(s) of each of the following reactions.

    (a)
    $$CH_3CH_2CH_2\overset{\displaystyle O}{\overset{\displaystyle \|}{C}}OH \;+\; CH_3\overset{\displaystyle OH}{\overset{\displaystyle |}{C}}HCH_3 \;\xrightarrow{\;H_2SO_4\;}$$

    (b)
    $$CH_3CH_2CH_2\overset{\displaystyle O}{\overset{\displaystyle \|}{C}}H \;\xrightarrow{\;H^+/Cr_2O_7^{2-}\;}$$

    (c)
    $$CH_3CH_2CH_2\overset{\displaystyle O}{\overset{\displaystyle \|}{C}}H \;\xrightarrow{\;LiAlH_4\;}$$

MOLECULAR STRUCTURES AND ISOMERS

This lesson deals with

1. Identifying formulas that represent structural isomers.
2. Identifying formulas that represent geometric isomers.
3. Drawing formulas to represent structural and geometric isomers.
4. Identifying stereogenic centers in structural formulas.
5. Specifying the R or S configuration at a stereogenic center.
6. Calculating the specific rotation of a substance from optical rotation data.
7. Identifying structures of diastereomers.

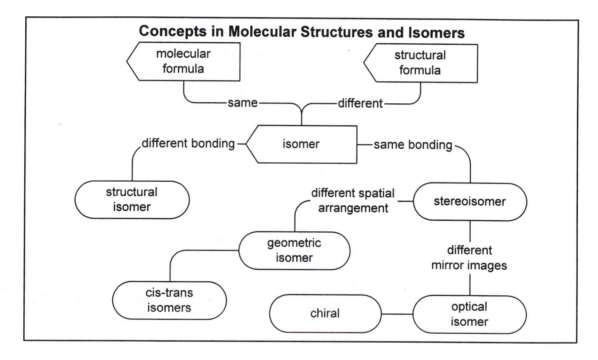

ISOMERS

Isomers are different compounds that have the same molecular formula.

STRUCTURAL ISOMERS

Isomers with different structural formulas are **structural isomers**.

$$CH_3CH_2CH_2CH_2OH \qquad CH_3\overset{\overset{\displaystyle OH}{|}}{C}HCH_2CH_3 \qquad CH_3CH_2{-\!\!-}O{-\!\!-}CH_2CH_3$$

All have the same molecular formula: $C_4H_{10}O$

Structural isomers must have
1. the same molecular formula
2. different structural formulas

Example 1

Determine whether each set of structural formulas represents isomers.

A.

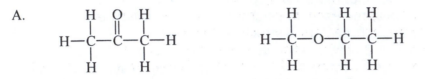

molecular formulas:

$$C_3H_6O \qquad\qquad C_3H_8O$$

The molecular formulas are different. Therefore, these are **not** isomers.

B.

Both have same molecular formula: $C_3H_6Cl_2$

However, the parts of a molecule connected by a single bond between carbon atoms can rotate with respect to each other. The carbon atom at the left can rotate, changing the positions of the two hydrogen atoms and chlorine atom. The same can happen at the right end of the molecule. Therefore, the two structures above represent the same molecule.

Exercise One

Determine whether each of the following sets of formulas represent structural isomers, the same molecule, or nonisomeric molecules.

A. CH_3OCH_3 CH_3CH_2OH

B. $CH_3CH_2CH_2Br$ $CH_2BrCH_2CH_3$

C.

$$CH_3CH{=}CHCH_2\overset{\overset{\displaystyle O}{\|}}{C}H \qquad CH_3CH_2\overset{\overset{\displaystyle O}{\|}}{C}CH_2CH_3$$

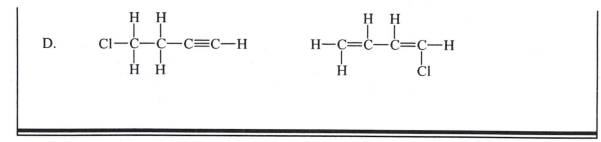

D.

GEOMETRIC ISOMERS

Geometric isomers have molecules whose atoms are bonded together in the same order, but the atoms have different spatial arrangements.

The different spatial arrangements are a result of the lack of free rotation around a bond in the molecules.

Free rotation is prevented by either

    1. a double bond

or 2. a ring of carbon atoms

Carbon-carbon double bond:

sp^2 + sp^2 sigma (σ) bond
p + p pi (π) bond

The π bond has two lobes, above and below the sigma bond. These lobes prevent rotation along the sigma bond and lock the other bonds into a fixed position.

W, X, Y, and Z represent atoms or groups of atoms

For geometrical isomers to exist, W must be different from X **and** Y must be different from Z

Geometric isomers:

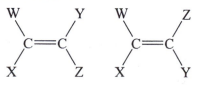 In both structures, the left carbon atom is bonded to W and X, and the right carbon is bonded to Y and Z. The arrangement of W and X is the same in both, but the arrangement of Y and Z is reversed.

Same molecules:

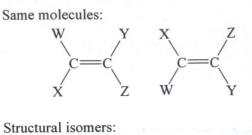

In both structures, the left carbon atom is bonded to W and X, and the right carbon is bonded to Y and Z. The arrangement of W and X is reversed in both, as is the arrangement of Y and Z.

Structural isomers:

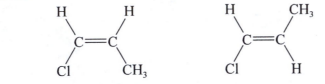

The atoms in these have a different bonding order. In the structure on the left, W is attached to a carbon also bonded to X, while in the other structure W is attached to a carbon also bonded to Z.

Example 2

Determine whether each of the following sets of formulas represent structural or geometric isomers.

A.

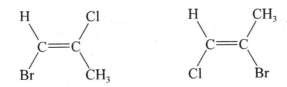

1. These have the same molecular formula: C_3H_5Cl
2. The bonding order of atoms is the same: **not** structural isomers.
3. Two different groups bonded to each double-bonded carbon atom.
4. In each structure, the double-bonded carbon on the left has the same arrangement of attached groups, but on the right, the groups are reversed.

Therefore, these are geometric isomers.

B.

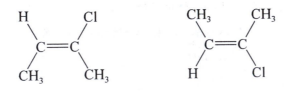

1. These have the same molecular formula.
2. The bonding order is different, they are different structures.

Therefore, these are structural isomers.

C.

1. Same molecular formula.
2. Same structure.
3. Different groups attached to each double-bonded carbon atom.
4. Groups are reversed at *both* ends.

Therefore, these are the same molecule.

Exercise Two

Determine whether each of the following sets of formulas represents structural isomers, geometric isomers, the same molecule, or nonisomeric molecules.

A.

$$
\begin{array}{ccc}
\text{H} & & \text{CH}_3 \\
& \text{C}=\text{C} & \\
\text{CH}_3 & & \text{CH}_2\text{CH}_3
\end{array}
\qquad
\begin{array}{ccc}
\text{H} & & \text{CH}_2\text{CH}_3 \\
& \text{C}=\text{C} & \\
\text{CH}_3 & & \text{CH}_3
\end{array}
$$

B.

$$
\begin{array}{ccc}
\text{H} & & \text{H} \\
& \text{C}=\text{C} & \\
\text{CH}_3 & & \text{CH}_2\text{CHCH}_3 \\
& & \quad\quad | \\
& & \quad\quad \text{Br}
\end{array}
\qquad
\begin{array}{ccc}
\text{CH}_3 & & \text{H} \\
& \text{C}=\text{C} & \\
\text{H} & & \text{CHCH}_2\text{CH}_3 \\
& & \quad | \\
& & \quad \text{Br}
\end{array}
$$

C.

$$
\begin{array}{ccc}
\text{CH}_3\text{CH}_2 & & \text{CH}_3 \\
& \text{C}=\text{C} & \\
\text{H} & & \text{CH}_3
\end{array}
\qquad
\begin{array}{ccc}
\text{H} & & \text{CH}_3 \\
& \text{C}=\text{C} & \\
\text{CH}_3\text{CH}_2 & & \text{CH}_3
\end{array}
$$

D.

$$
\begin{array}{ccc}
\text{CH}_2=\text{CH} & & \text{CH}_3 \\
& \text{C}=\text{C} & \\
\text{H} & & \text{H}
\end{array}
\qquad
\begin{array}{ccc}
\text{CH}_3\text{CH}_2 & & \text{CH}_3 \\
& \text{C}=\text{C} & \\
\text{H} & & \text{H}
\end{array}
$$

E.

$$
\begin{array}{ccc}
\text{CH}_3 & & \text{H} \\
& \text{C}=\text{C} & \\
\text{Cl} & & \text{CH}_2\text{CH}_3
\end{array}
\qquad
\begin{array}{ccc}
\text{CH}_3\text{CH}_2 & & \text{CH}_3 \\
& \text{C}=\text{C} & \\
\text{H} & & \text{Cl}
\end{array}
$$

CHIRALITY

An object that is different from its mirror image is said to be **chiral**.

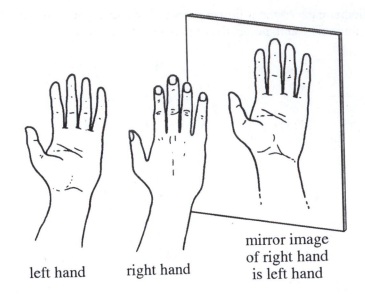

left hand   right hand   mirror image
of right hand
is left hand

The mirror image of a right hand is the left hand. The right hand is chiral (and so is the left hand).

An object whose mirror image is the same as the object is said to be **achiral**.

Achiral objects have a plane of symmetry. A plane of symmetry is an imaginary plane dividing the object into two halves that are mirror images of each other.

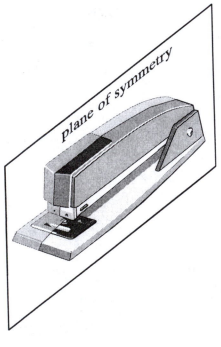

For example, a stapler has a plane of symmetry dividing it vertically front to back. The part of the stapler on one side of this plane is the mirror image of the part on the other side.

A stapler is achiral (not chiral), because it has a plane of symmetry.

Chiral objects have no such plane of symmetry.

Exercise Three

Determine whether each of the following objects is chiral or achiral.

A.

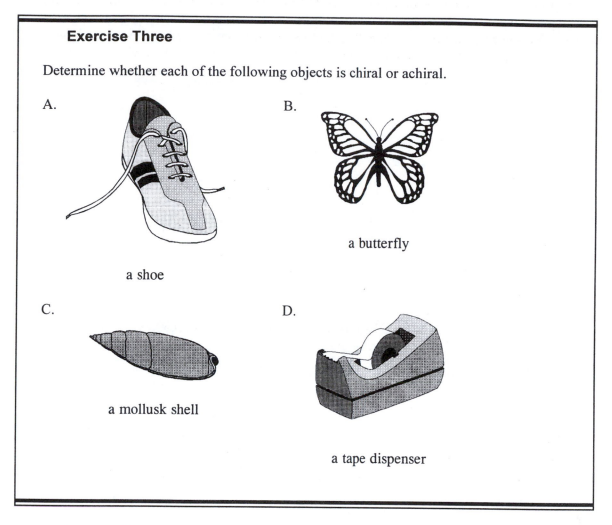

a shoe

B.

a butterfly

C.

a mollusk shell

D.

a tape dispenser

ENANTIOMERS

Enantiomers are isomers whose molecules are mirror images of each other but cannot be superimposed. Such molecules are chiral.

Consider CHFClBr.

The mirror image of a molecule of CHFClBr cannot be superimposed on the molecule. When the carbon atom and any two of its bonded atoms are in the same positions in the molecule as in its mirror image, the other two bonded atoms are in reversed positions. The mirror image is different from the molecule. Therefore, the molecule is chiral.

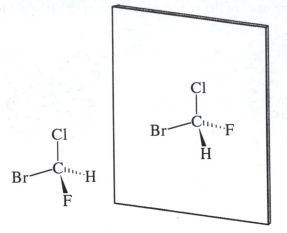

When representing the spatial arrangement of atoms in a molecule, lines represent bonds that connect atoms in the plane of the paper, a filled triangle represents a bond to an atom in front of the paper, a dashed line or dashed triangle represents a bond to an atom behind the paper.

A molecule of an organic compound can be chiral when a carbon atom is bonded to four different atoms or groups of atoms. Such a carbon atom is called a **chiral center** or a **stereogenic center**. A pair of chiral compounds are often called **optical isomers**, because they rotate the plane of polarized light, as described later in this lesson.

Example 3

Find the stereogenic centers in the following molecules

A. 2-butanol

$$CH_3-\overset{\overset{\displaystyle OH}{|}}{\underset{\underset{\displaystyle H}{|}}{C}}-CH_2CH_3$$

The second carbon atom is bonded to 4 different groups; it is a stereogenic center.

B. 2,3-dichlorobutane

$$CH_3-\overset{\overset{\displaystyle Cl}{|}}{\underset{\underset{\displaystyle H}{|}}{C}}-\overset{\overset{\displaystyle Cl}{|}}{\underset{\underset{\displaystyle H}{|}}{C}}-CH_3$$

The second and third carbon atoms are both bonded to 4 different groups; both of these carbon atoms are stereogenic centers.

Exercise Four

Which of the following molecules contain a stereogenic center? For those that do, circle the stereogenic center.

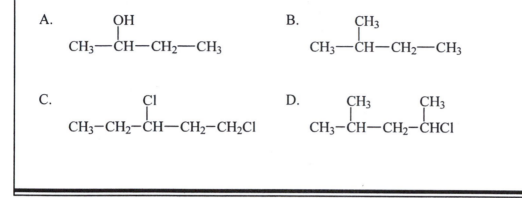

A.
$$OH$$
$$CH_3—CH—CH_2—CH_3$$

B.
$$CH_3$$
$$CH_3—CH—CH_2—CH_3$$

C.
$$Cl$$
$$CH_3—CH_2—CH—CH_2—CH_2Cl$$

D.
$$CH_3 \quad CH_3$$
$$CH_3—CH—CH_2—CHCl$$

SPECIFYING ARRANGEMENTS AT STEREOGENIC CENTERS

At a carbon stereogenic center, two possible arrangements of the four different groups are possible. These are specified with the **R,S convention**.

1. The four groups are numbered 1 through 4 according to decreasing atomic number. The highest atomic number is assigned 1 and the lowest 4.

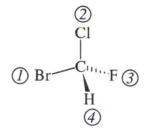

If two attached atoms have the same atomic number, then consider the atoms attached to them.

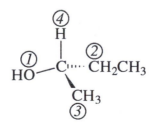

2.   The molecule is oriented with group 4 behind the stereogenic center carbon.

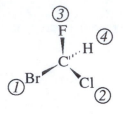

3.   If groups 1, 2, and 3 are arranged clockwise, then the arrangement is *R*. If they are counterclockwise, then the arrangement is *S*.

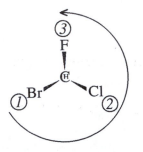

In this example, the arrangement in counterclockwise, so it is the *S* configuration.

Exercise Five

A.   Indicate whether the arrangement of the groups on the stereogenic center is *R* or *S* in each of the following molecules.

1.   2.

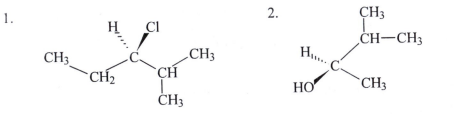

B.   Draw a structure for the following molecules. Be sure to indicate the spacial arrangement of the groups bonded to the stereogenic center.

1.   (*R*)-1-bromo-1-chloropropane

2. (*S*)-2-chlorobutane

OPTICAL ACTIVITY

Chiral molecules show optical activity. An optically active substance rotates the orientation of a plane of polarized light as it passes through. One isomer of a chiral pair rotates the plane in one direction and the other isomer rotates it in the other direction. For this reason, the isomers are often called optical isomers.

In polarized light, the light waves oscillate in only one plane (for example, the xz plane in the figure below) rather than in all 3 dimensions.

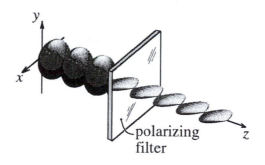

A non-polarized light wave traveling in the z direction vibrates in the x and y directions. A polarized light wave traveling in the z direction vibrates in only one plane.

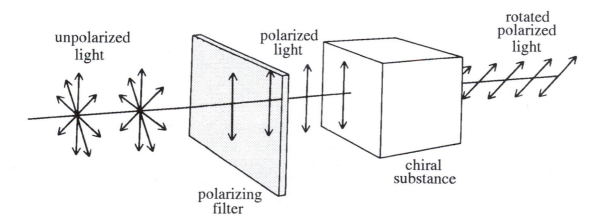

When a plane polarized light wave travels through an optically active substance, the plane of its polarization rotates.

A **polarimeter** measures the rotation of a plane of polarized light.

The rotation is represented by α. The value of α may be positive or negative, and it depends on temperature and on the wavelength of the light.

Measurements are usually made at 25°C and with the yellow light from a sodium lamp (the "D line" of sodium at 589.3 nm). Such a rotation is represented as α_D^{25}.

The degree of rotation is proportional to the concentration of the chiral substance and to the distance the polarized light travels through the substance. The **specific rotation**, represented by $[\alpha]$, is defined as

$$[\alpha]_D^{25} = \frac{\alpha_D^{25}}{l\,c}$$

l is the length of the path of the light through the chiral substance, in dm.
c is the concentration of the solution, in g/mL. (For pure liquids, this concentration is the same as the density of the liquid.)

Example 4

The specific rotation of sucrose (table sugar) is $+66°$ mL-g^{-1} dm^{-1}. How much will a plane of polarized light be rotated when it passes through 10 cm of a solution containing a tablespoon (12 grams) of sucrose in a cup (125 mL) of water?

$$[\alpha]_D^{25} = \frac{\alpha_D^{25}}{l\,c}$$

α_D^{25} is the rotation of a particular solution

$[\alpha]_D^{25} = +66°$ mL g^{-1} dm^{-1}
$l = 10$ cm $= 1$ dm
$c = 12$ g $/ 125$ mL $= 0.1$ g/mL

$$\alpha_D^{25} = [\alpha]_D^{25}\, l\, c$$
$$= (66°\ \text{mL g}^{-1}\ \text{dm}^{-1})\,(1\ \text{dm})\,(0.1\ \text{g mL}^{-1})$$
$$= 7°$$

Substances that rotate a plane of polarized light in a clockwise direction are called **dextrorotatory**. Those that rotate the plane counterclockwise are called **levorotatory**. When one enantiomer of a pair has a specific rotation of $[\alpha]$, then the other has a specific rotation of $-[\alpha]$. There is no relationship between the R and S configurations and the direction in which a chiral substance rotates a plane of polarized light.

Exercise Six

What is the specific rotation of a substance of which a solution containing 0.28 g in 75 mL of solution rotates a plane of polarized light by an angle of −13° in a 5 cm cell?

DIASTEREOMERS

Diastereomers are stereoisomers that are not enantiomers.

Diastereomers exist when the molecules contain more than one stereogenic center.

There are four isomers of 2-chloro-3-bromobutane:

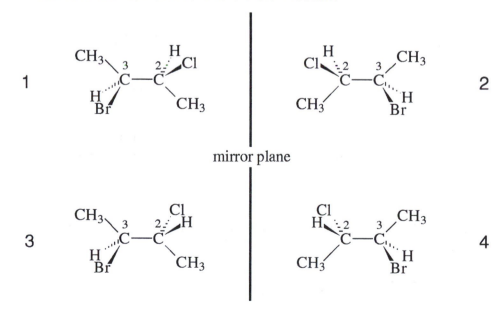

These can be identified by their *R,S* configurations.

Structure	Configuration of carbon 2	Configuration of carbon 3
1	S	S
2	R	R
3	R	S
4	S	R

A diagram shows the enantiomer and diastereomer relationships among these four isomers.

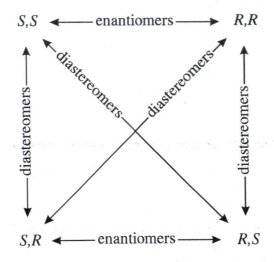

S,S ⟵——— enantiomers ———⟶ R,R

S,R ⟵——— enantiomers ———⟶ R,S

If the two stereogenic centers in a molecule are bonded to identical groups, then the *R,S* and *S,R* confingurations are identical, rather than enantiomers.

Rotating the molecule on the right by an angle of 180° around the bond between carbon 2 and carbon 3 puts it into exactly the same configuration as the molecule on the left.

Example 5

Determine whether the following pair of structures represent enantiomers, diastereomers, or neither type of isomer.

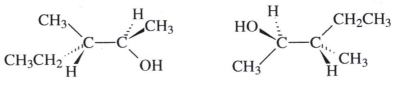

1. The two molecules have the same molecular formula and the same geometric structure. Therefore, they may represent enantiomers, diastereomers, or the same molecule.
2. In the structure on the left, the stereogenic center bonded to CH_3CH_2 is in the S configuration, and the other stereogenic center is in the R configuration.
3. In the structure on the right, the stereogenic center bonded to CH_3CH_2 is in S configuration, and the other stereogenic center is in the R configuration.
4. Because both stereogenic centers have the same configuration, these structures represent the same molecule.

Exercise Seven

For each of the following pairs of structures, indicate whether they represent enantiomers, diastereomers, or neither type of isomer.

A.

B.

C.

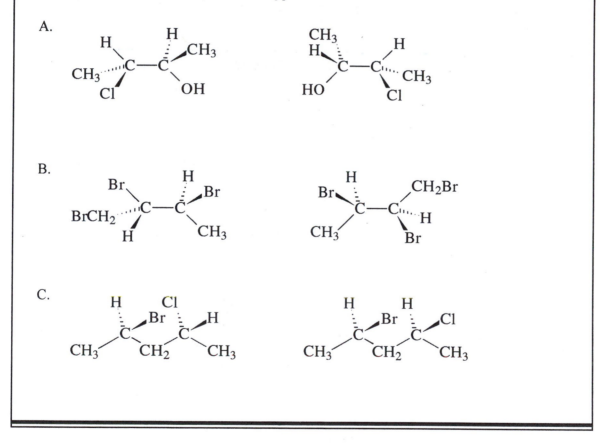

PROBLEMS

Indicate whether each of the following has geometric isomers, optical isomers, both, or neither.

1. 2-butanol	2. 2-pentene	3. 3,4-dimethyl-2-hexene
4. 2-methyl-2-butene	5. 1,2-dichloropropene	6. 2-pentanone

POLYMERS

This lesson deals with

1. Identifying repeating sections in the structural formula of a polymer molecule.
2. Drawing a structural formula for a section of addition polymer molecule, given the monomer.
3. Drawing a structural formula for a section of a polyester, given the monomers.
4. Drawing a structural formula for a section of a polyamide, given the monomers.
5. Drawing a structural formula for a section of an addition copolymer, given the monomers.
6. Drawing a structural formula for the monomer(s) of a polymer, given the structural formula of a section of polymer molecule.

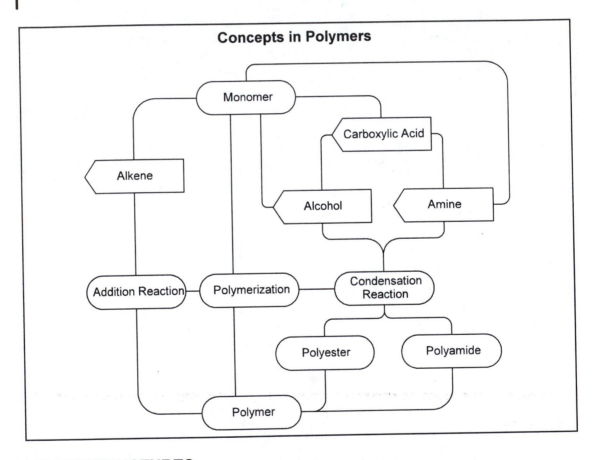

POLYMER STRUCTURES

Many of the materials we encounter every day are polymers. There are both natural and synthetic polymers. These are just a few polymeric materials:

silk   cotton   nylon   protein   polyester   Teflon

Polymers are materials composed of very large molecules that have a distinctive structure:

> The molecules of a polymer contain a long chain of atoms, and this chain is composed of smaller repeating sections.

A common polymer is polyvinyl chloride (PVC). It's used in making raincoats, "vinyl" siding, food packaging, and pipes for plumbing, among many other things. Its molecules contain a chain of thousands of carbon atoms, with hydrogen attached to every carbon and chlorine attached to every other carbon. A section of a polyvinyl chloride molecule has the following structure:

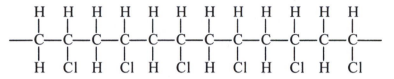

In this structure the repeating unit is

Other polymers have more complex repeating units. Below is the structure of polyethylene terephthalate, from which 2-liter soft-drink containers are made.

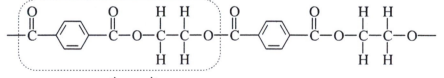

repeating unit

In many polymers, the repeating units are not identical. There are variations from one unit to the next. These variations occur in the atoms that are attached to the structural backbone of the polymer chain, not in the backbone itself. Proteins are such polymers. Below is a structure for a typical segment of a protein molecule.

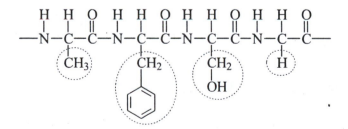

In this structure, the groups that vary are circled. For this polymer, the repeating unit can be represented as

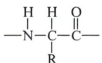

where R represents any one of over 20 different attached groups.

Exercise One

In each of the following structures of polymer segments, identify the repeating unit by circling it.

A. polystyrene

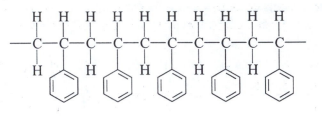

B. nylon 6,6

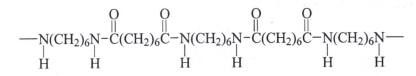

C. cotton

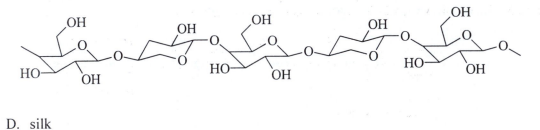

D. silk

POLYMER FORMATION – ADDITION REACTION

Polymer molecules have a repeating structure because they are formed by reactions that attach many smaller molecules together. The smaller molecules that combine to form polymers are called *monomers*. The reaction in which monomers are combined to form a polymer is called *polymerization*.

Polymers are classified by the type of polymerization reaction that forms them. There are two general types of polymerization reactions:

Addition reactions: all atoms in the monomers are incorporated into the polymer molecules.

Condensation reactions: some of the atoms from the monomers are released into other, small molecules.

Addition polymerization

Monomers that undergo self addition polymerization contain multiple bonds between carbon atoms, such as alkenes. Alkenes have double bonds between carbon atoms, such as in chloroethene (commonly called vinyl chloride).

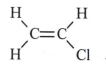

During the polymerization reaction, the individual monomer molecules are linked together when electrons from the double bond rearrange and form new bonds between adjacent monomers.

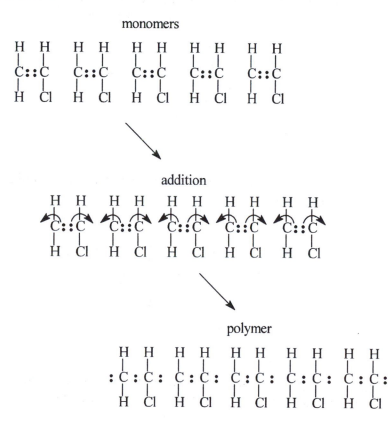

Using the more common bond-line representation for this polymer structure gives the following representation of a segment of the polymer chain.

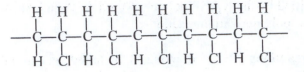

Example 1

Draw a bond-line structure for a segment of the addition polymer molecule that is formed from the monomer below. Include four monomer units in the polymer structure.

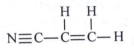

As the instructions indicate, this monomer forms an addition polymer. Therefore, the C=C double bond is the part of the molecule that reacts to form the chain. The N≡C bond is not involved in forming the chain, it is one of the attached groups on the polymer chain.

Start by drawing four of the monomers side-by-side with the C=C bonds arranged horizontally, and the attached groups above and below the chain.

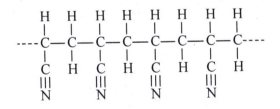

Then, change the C=C double bonds to single bonds, and attach the resulting C–C groups to each other with single bonds.

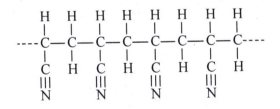

The dashed lines used at the ends of the structure above are sometimes used to indicate that the structure continues with many repetitions of the indicated pattern.

Exercise Two

For each of the following monomers, draw a bond-line structure for a segment of the addition polymer molecule that is formed from the monomer. Include five monomer units in the polymer structure.

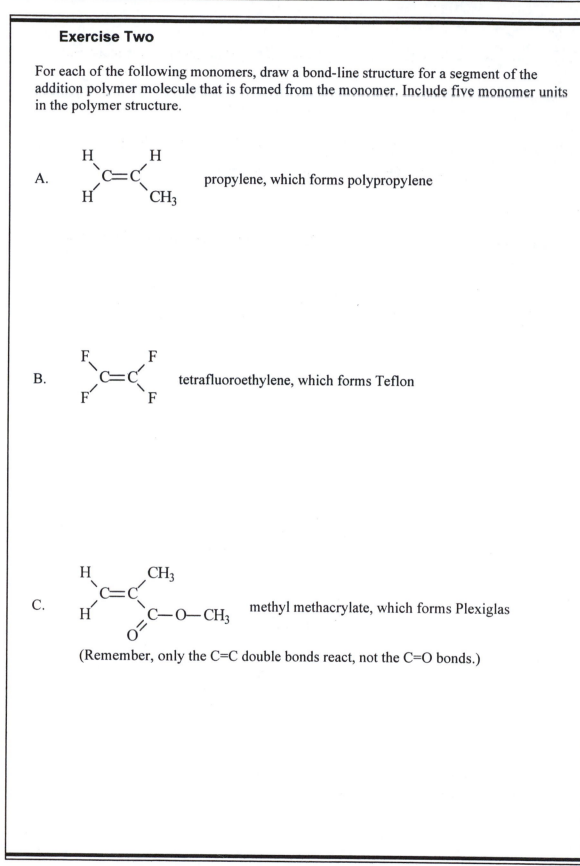

A.   propylene, which forms polypropylene

B.   tetrafluoroethylene, which forms Teflon

C.   methyl methacrylate, which forms Plexiglas

(Remember, only the C=C double bonds react, not the C=O bonds.)

POLYESTERS

Many polymers are formed by reactions in which not all of the atoms in the monomers are included in the polymers. Some of the atoms are ejected in small molecules, such as water or HCl. These reactions are called *condensation reactions*, and the polymers formed in these reactions are called *condensation polymers*.

There are several types of condensation reactions, but the two most common are

ester formation and **amide formation**

Ester formation

An ester is a compound formed by the condensation reaction between an alcohol and a carboxylic acid.

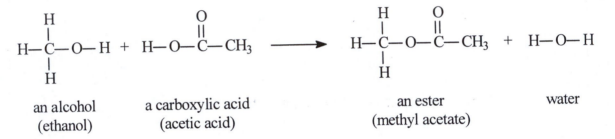

| an alcohol | a carboxylic acid | an ester | water |
| (ethanol) | (acetic acid) | (methyl acetate) | |

The group of atoms characteristic of an ester is

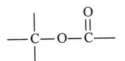

If the alcohol contains two alcohol functional groups, it can form two ester bonds, one with each –OH group. Such is the case with ethylene glycol:

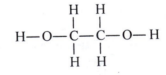

If the carboxylic acid contains two acid groups, it can form two ester bonds, one with each carboxyl group. Such is the case with malonic acid:

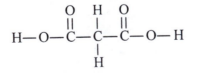

If both the alcohol and the acid contain two functional groups, then the product is a chain of atoms held together by ester groups.

Example 2

Draw a structural formula for a segment of the polymer chain that forms from the polymerization of the monomers below.

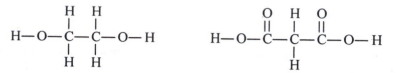

The monomer on the left contains alcohol functional groups. The monomer on the right contains carboxylic acid functional groups. Therefore, these can combine to form ester bonds. An alcohol group reacts with a carboxylic acid group to form an ester group. A water molecule is released, which makes this a condensation reaction.

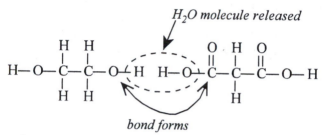

The ester that forms is

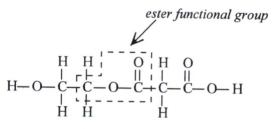

A similar reaction occurs at the right end of the new molecule,

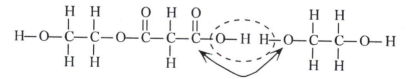

which produces another ester bond.

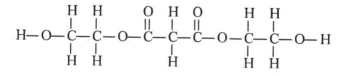

This repeats until all the alcohol and acid groups have reacted and a large polymer molecule is formed, with the structure

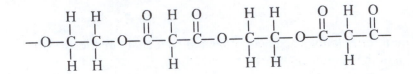

The structure shown above represents two repeating sections with a carboxylic acid unit and an alcohol unit in each repeating section. This gives a total of four monomer units in the represented section.

EXERCISE THREE

Draw a structural formula for a segment of the polymer molecule that forms from the monomers whose structures are below. Use at least four monomer units in the polymer structure.

O H H O   H   CH₃
‖ | | ‖ | |
HO—C—C═C—C—OH   HO—C—C—OH
                             | |
                             H   H

Draw a structural formula for the repeating unit of this polymer.

POLYAMIDES

A second type of condensation polymers is the polymer in which the monomers are joined by amide bonds. An amide bond forms between an amine and a carboxylic acid. (First a salt forms, but after heating the salt, an amide forms. See pages 441-442.)

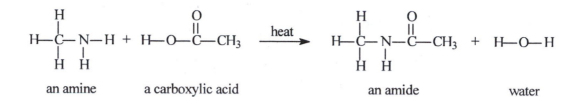

The group of atoms characteristic of an amide is

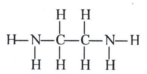

If the amine contains two amine functional groups, it can form two amide bonds, one with each –NH group. This is the case with ethylene diamine:

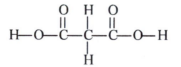

If the carboxylic acid contains two acid groups, it can form two ester bonds, one with each carboxyl group. This is the case with malonic acid:

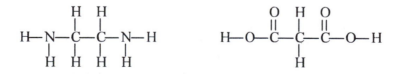

If both the amine and the acid contain two functional groups, then the product is a chain of atoms held together by ester groups.

Example 3

Draw a structural formula for a segment of the polymer chain that forms from the polymerization of the monomers below.

The monomer on the left contains amine functional groups. The monomer on the right contains carboxylic acid functional groups. Therefore, these can combine to form amide bonds. An amine group reacts with a carboxylic acid group after heating to form an amide group. A water molecule is released, which makes this a condensation reaction.

H₂O molecule released

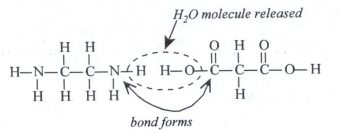

The amide that forms is

amide functional group

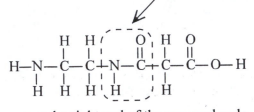

A similar reaction occurs at the right end of the new molecule,

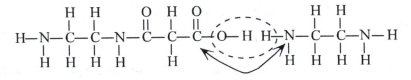

which produces another amide bond.

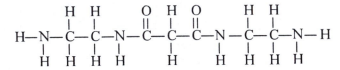

This repeats until all the amine and acid groups have reacted and a large polymer molecule is formed, with the structure

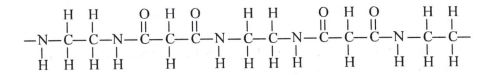

The structure shown above represents five monomer units, two carboxylic acid units and three amine units.

Many synthetic polyamides are sold under the trade name of nylon.

EXERCISE FOUR

Draw a structural formula for a segment of the polymer molecule that forms from the monomers whose structures are below. Use at least four monomer units in the polymer structure.

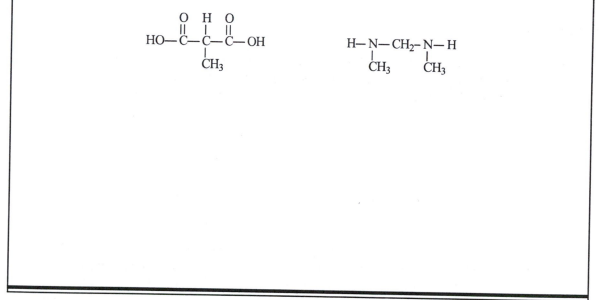

ADDITION COPOLYMERS

A *copolymer* is a polymer that is made from two or more different monomers. The polyesters and polyamides whose formation is described above are copolymers, because they are both made from two different monomers. A polyester is made from a di-alcohol and a di-acid. A polyamide is made from a di-amine and a di-acid. The addition polymers described above are **not** copolymers, because only one kind of monomer is used in forming them. For example, polyvinylchloride is not a copolymer, because all of its monomer units are the same, namely vinyl chloride.

Addition polymers can be made from two or more different monomers. The process in which a polymer is formed from more than one monomer is called *copolymerization*. There are various arrangements of the monomers in the polymer.

When the monomers are arranged in an alternating pattern in the copolymer, then the copolymer is called an *alternating copolymer*. If the monomers are represented by A and B, then the structure of an alternating copolymer can be represented as

—A—B—A—B—A—B—A—B—A—B—

If the arrangement of the monomers is random, then the polymer is called a *random copolymer*. This can be represented as

—A—A—B—A—B—B—B—A—A—B—A—B—B—A—

In a *block copolymer*, the monomers are arranged in segments of like monomers. This can be represented as

—A—A—A—A—A—A—B—B—B—B—B—B—B—B— A—A—A—

Example 4

The major ingredient of some brands of plastic food wrap, such as Saran Wrap, is an alternating copolymer made from the two monomers whose structures are below. Draw a structural formula for a segment of this polymer, showing at least four monomer units.

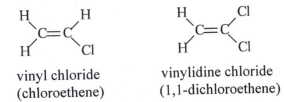

vinyl chloride	vinylidine chloride
(chloroethene)	(1,1-dichloroethene)

To draw a structure of a segment of the polymer chain, follow the same procedure as for any addition polymer. An addition polymer forms by joining the monomers at the C=C double bond. In this case there are two different molecules containing C=C double bonds. The structure should show four monomer units in an alternating arrangement. Put the required four monomer units in a row.

$$
\begin{array}{cccccccc}
H & H & H & Cl & H & H & H & Cl \\
| & | & | & | & | & | & | & | \\
C & = C & C & = C & C & = C & C & = C \\
| & | & | & | & | & | & | & | \\
H & Cl & H & Cl & H & Cl & H & Cl
\end{array}
$$

Then, change the double bonds to single bonds and connect the adjacent monomer molecules with C–C single bonds.

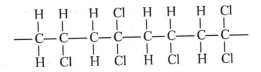

EXERCISE FIVE

An adhesive used in a packaging tape is a random addition copolymer of ethylene and vinyl acetate. Draw a structural formula for a segment of the copolymer, showing at least six monomer units and both monomers in the structure.

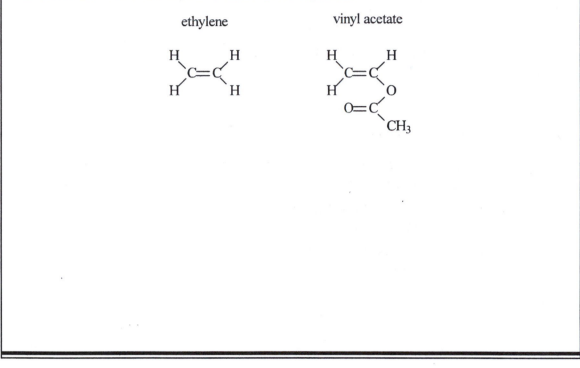

ethylene   vinyl acetate

IDENTIFYING MONOMERS FROM THE POLYMER STRUCTURE

So far we have been looking at how polymer molecules are assembled from their monomers. Now, we will examine the reverse process, determining the structures of the monomers that are used to produce a given polymer structure.

The process involves several steps. (Examples follow.)

1. Identify the type of polymer: addition or condensation.
2. If the polymer is an addition polymer, divide the chain into two-carbon segments with their attached groups and restore the double bonds between the two carbon atoms. This completes the process for addition polymers.
3. If the polymer is a condensation polymer, determine whether it is a polyester or polyamide.
4. If it is a polyester, divide the chain between the C doubled bonded to O and the adjacent O. Add an –OH group to the C and an –H to the O. Do this for every such C and O. This completes the process for a polyester.
5. If it is a polyamide, divide the chain between the C doubled bonded to O and the adjacent N. Add an –OH group to the C and an –H to the N. Do this for every such C and N. This completes the process for a polyamide.

Example 5

Draw structural formulas for the monomers used in making each of the following polymers.

A.

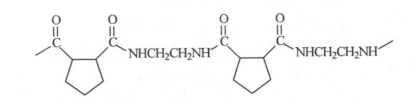

1. Is this an addition or a condensation polymer?

   The chain of addition polymers is made up of only carbon atoms, so this is not an addition polymer.

   The chain contains C and N atoms, so it is not a polyester, whose chain contains C and O atoms.

   The chain contains C and N atoms, and there are amide groups in the chain, so this is a polyamide.

   amide group

   Because it is a polyamide, go to step 5.

5. Divide the chain between the C doubled bonded to O and the adjacent N.

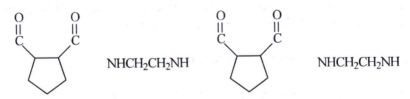

   Add an –OH to each C where a bond was broken, and an –H to each N where a bond was broken.

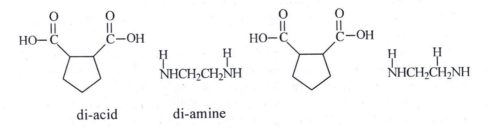

   di-acid   di-amine

   This produces two copies of the structure for the di-acid and two for the di-amine that are used in making this polymer.

B.

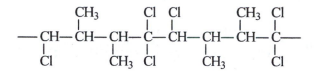

1. Is this an addition or a condensation polymer?

    The chain of this polymer is made up of only carbon atoms, so this is an
    addition polymer.

2. Divide the chain into two-carbon segments with their attached groups.

$$
\underset{\underset{Cl}{|}}{\overset{\overset{CH_3}{|}}{CH}}{-}CH \quad \underset{\underset{CH_3}{|}\ \underset{Cl}{|}}{\overset{\overset{Cl}{|}}{CH}}{-}\overset{|}{C} \quad \underset{\underset{CH_3}{|}}{\overset{\overset{Cl}{|}}{CH}}{-}CH \quad \underset{\underset{Cl}{|}}{\overset{\overset{CH_3}{|}\ \overset{Cl}{|}}{CH}}{-}C
$$

    Restore the double bonds between the two carbon atoms.

$$
\underset{\underset{Cl}{|}}{\overset{\overset{CH_3}{|}}{CH}}{=}CH \quad \underset{\underset{CH_3}{|}\ \underset{Cl}{|}}{\overset{\overset{Cl}{|}}{CH}}{=}C \quad \underset{\underset{CH_3}{|}}{\overset{\overset{Cl}{|}}{CH}}{=}CH \quad \underset{\underset{Cl}{|}}{\overset{\overset{CH_3}{|}\ \overset{Cl}{|}}{CH}}{=}C
$$

    This results in the structures for two monomers, each of which is repeated twice.
    Because there are two different monomers, the polymer is a copolymer. From the
    segment shown, it appears to be an alternating copolymer.

EXERCISE SIX

Draw structural formulas for the monomers used in making each of the following
polymers.

A.

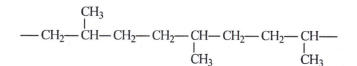

B.

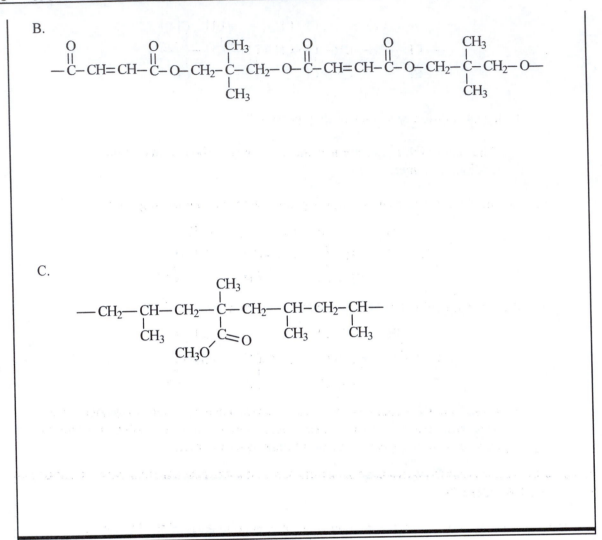

C.

PROBLEMS

1. For each of the following monomers or sets of monomers, draw a structural formula for a segment of the polymer that can be formed from them. Include at least four monomers in the formula.

(a)

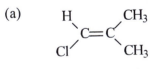

(b)

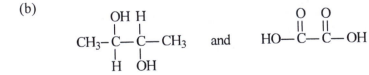

(c)

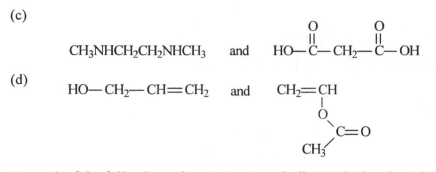

CH₃NHCH₂CH₂NHCH₃   and   HO—C—CH₂—C—OH

(d)

HO—CH₂—CH=CH₂   and   CH₂=CH

2. For each of the following polymer structures, indicate whether the polymer is an addition polymer or condensation polymer. If it is a condensation polymer, indicate if it is a polyester or polyamide.

(a)

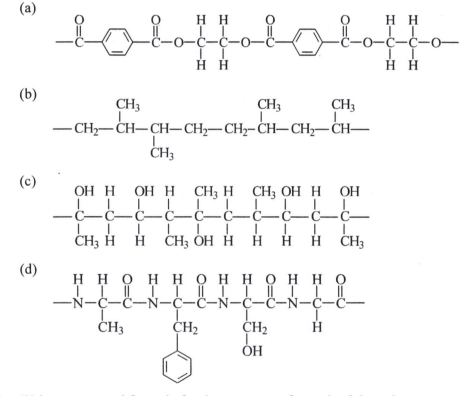

(b)

(c)

(d)

3. Write a structural formula for the monomers for each of the polymer structures in Problem 2.

4. A terpolymer is a copolymer made from three different monomers. Draw a structural formula for a segment of a terpolymer chain made from the following three monomers. Show at least four monomer segments in the polymer structure, and include all three of the monomers.

SIGNIFICANT FIGURES AND EXPONENTIAL NOTATION

This lesson deals with:

1. Determining the significant figures in a number.
2. Expressing a number in exponential notation.
3. Multiplying and dividing powers of ten.
4. Rounding off to a given number of significant figures.
5. Multiplying and dividing numbers in exponential notation.
6. Adding and subtracting numbers written in exponential notation.

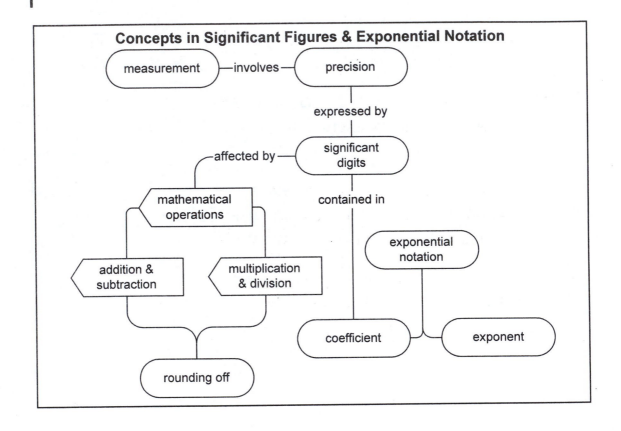

SIGNIFICANT FIGURES

Significant figures are the digits of a number whose values are certain.

You may have encountered a town population sign such as Mudville's, and wondered about the final digit, the "1," and whether that digit is still correct. Maybe today the population is 7,303. Maybe tomorrow it will be 7,304. If you did, you were wondering about significant figures.

MUDVILLE
POP. 7,301

Numbers that are a result of a measurement have a limited number of significant figures. The more precise the measurement the greater the number of significant figures in the result. A reported measurement should indicate the precision of the measurement by the number of figures used to report it.

Numbers that are definitions have an unlimited number of significant figures. For example, 1 foot is exactly 12 inches. Both of these numbers, 1 and 12, have unlimited significant figures.

Guidelines for Recognizing Significant Figures

1. All non-zero digits are significant.
    243 has three significant figures.
    1.287 has four significant figures.

2. Zeros are significant only

    a. when they occur between two non-zero digits.
        207 has three significant figures.
        1.032 has four significant figures.

    b. when they appear after a non-zero digit AND to the left of a decimal point.
        200. has three significant figures.
        200 has only one.

    c. when they occur to the right of a decimal point and to the right of a non-zero digit.
        2.20 has three significant figures.
        0.03400 has four significant figures (the underlined ones: 0.03<u>400</u>).
        Note: Zeros to the left of all non-zero digits in a number are *not* significant.

Exercise One

Determine the number of significant figures in each of the following numbers.

A. 23.4 _____   C. 24700 _____   E. 10000 _____

B. 17.03 _____   D. 0.0020 _____   F. 1.0005 _____

EXPONENTIAL NOTATION

How can we write two hundred with two significant figures?

   200 has 1 significant figure
   200. has 3 significant figures
   200.0 has 4 significant figures

Exponential notation

$$C \times 10^n$$

C is a number between one and ten, called the coefficient.
n is an integer, either positive or negative, called the exponent.

Two hundred may be written as

2×10^2	one significant figure
2.0×10^2	two significant figures
2.00×10^2	three significant figures
2.000×10^2	four significant figures
etc.	

The coefficient tells the number of significant figures.
The exponent indicates the location of the decimal point.

Example 1

Write 628000 in exponential notation.

How many significant figures? There are three.

C is 6.28

$$6\,2\,8\,0\,0\,0 = 6.28 \times 10^5$$
$$5\ \ 4\ \ 3\ \ 2\ \ 1$$

The decimal point is moved five places.
When the decimal point is moved to the left, the exponent is positive.

Example 2

Write 0.0026 in exponential notation.

How many significant figures? There are two

C = 2.6

$$0.0\,0\,2\,6 = 2.6 \times 10^{-3}$$
$$1\ \ 2\ \ 3$$

When the decimal point is moved to the right, the exponent is negative.

Exercise Two

Convert each of the following numbers to exponential notation, using the proper number of significant figures.

A. 34200 = _____

B. 0.000729 = _____

C. 40.20 = _____

D. 0.000000006024 = _____

E. 200000 = _____

F. 0.00300 = _____

CHANGING THE POWER OF TEN

Sometimes it is necessary to express a number in exponential notation with a specific exponent, e.g, when changing the units of measurement.

Example 3

A. Express 6.59×10^6 with an exponent of 4. That is,

$$6.59 \times 10^6 = ? \times 10^4$$

The exponent decreases, so the coefficient must increase. Move decimal point two places to the right.

$$6.59 \times 10^6 = 659 \times 10^4$$

B. Express 7.6×10^{-3} with a coefficient of 0.76

$$7.6 \times 10^{-3} \text{ converts to } 0.76 \times 10^?$$

The coefficient decreases, so the exponent must increase. Add 1 to the exponent.

$$7.6 \times 10^{-3} = 0.76 \times 10^{-2}$$

Exercise Three

A. Rewrite 3.27×10^4 as a number with seven as the exponent of ten.

B. Rewrite 6.33×10^{-2} as a number with -4 as the exponent of ten.

C. Rewrite 1.064×10^{-3} as a number with the coefficient equal to 106.4.

ROUNDING OFF

The result of a mathematical operation with numbers that contain significant figures must be given with the appropriate number of significant figures. This often means that some of the digits displayed by a calculator must be discarded. Discarding insignificant digits is called rounding off.

Guidelines for Rounding Off

1. If the digit following the last significant digit is less than five, the preceding digit number remains unchanged.

    2.673 becomes 2.67 to three significant figures
    5.826 becomes 5.8 to two significant figures

2. If the digit following the last significant digit is more than five, the preceding digit is increased by one.

    0.3148 becomes 0.315 to three significant figures
    2.379 becomes 2.4 to two significant figures

3. If the digit following the last significant digit is exactly five, and the preceding digit is even, it remains unchanged.

    6.25 becomes 6.2 to two significant figures
    3.7858 becomes 3.78 to three significant figures

    If the digit following the last significant digit is exactly five, and the preceding digit is odd, it is increased by 1.

    0.235 becomes 0.24 to two significant figures
    12.655 becomes 12.66 to four significant figures

Exercise Four

Round off the given number to the indicated number of significant figures:

A. 23.62 to three significant figures   _____

B. 23.62 to two significant figures _____

C. 0.365 to two significant figures _____

D. 1.296 to three significant figures   _____

E. 7.206 to two significant figures _____

F. 0.355 to two significant figures _____

SIGNIFICANT FIGURES IN MULTIPLICATION AND DIVISION

In multiplication and division, the result of the calculation has the same number of significant figures as the original number that has the fewest.

3.7×10^4 has two significant figures
7.94×10^{-2} has three significant figures

The answer must have only two significant figures.

$$(3.7 \times 10^4)(7.94 \times 10^{-2}) \quad = 2.9378 \times 10^3 \text{ (calculator result)}$$
$$= 2.9 \times 10^3 \text{ (rounded to 2 significant figures)}$$

Example 4

Perform the following calculation and express the result to the appropriate number of significant figures.

$$\frac{3.52 \times 10^5}{(2.6 \times 10^7)(8.535 \times 10^{-2})}$$

Use calculator to get numerical result.

$$\frac{3.52 \times 10^5}{(2.6 \times 10^7)(8.535 \times 10^{-2})} = 1.58622865 \times 10^{-1} \quad (\text{caculator result})$$

The calculator does not take significant figures into account, so we must do so. The mathematical operation involves only multiplication and division. Therefore, the number of significant figures in the result is equal to the number of significant figures in the number having the fewest. The smallest number is two significant figures in 2.6×10^7.

The result should be rounded to 1.6×10^{-1}.

Exercise Five

Perform the following calculations and express the results to the proper number of significant figures.

A. $(3.24 \times 10^6)(4.87 \times 10^{-2}) =$

B. $\dfrac{2.67 \times 10^{-2}}{9.2 \times 10^4} =$

C. $\dfrac{(3.6 \times 10^4)(2.8 \times 10^{-2})}{(6.68 \times 10^5)} =$

D. $\dfrac{(8.36 \times 10^{-7})}{(2.1 \times 10^6)(1 \times 10^5)} =$

SIGNIFICANT FIGURES IN ADDITION AND SUBTRACTION

In addition and subtraction, the result is rounded off so every digit is the result of an addition or subtraction of a significant digit. Therefore, the result contains the same number of significant figures to the right of the decimal point as the addend which has the fewest digits to the right of the decimal point.

Example 5

Add these numbers and report the result to the proper number of significant figures.
$$(5.72 \times 10^{-2}) + (6.48 \times 10^{-4}) + (3.7916 \times 10^{-1})$$

A calculator gives the sum as 4.37008×10^{-1}.

To determine the number of significant figures in the answer, it is necessary to convert all numbers to the same exponent.

$$
\begin{array}{rl}
0.572 & \times\ 10^{-1} \\
0.00648 & \times\ 10^{-1} \\
3.7916 & \times\ 10^{-1} \\
\hline
4.37008 & \times\ 10^{-1}
\end{array}
$$

The last two digits in the coefficient of the sum, zero and eight, are not significant, because the addends above them do not all have significant digits in these places. Our answer can have only three significant figures after the decimal point, because the first number has only that many. It limits the number of significant digits in the sum.

The result of the addition is
$$4.370 \times 10^{-1}$$

Example 6

Give the result of the subtraction below to the proper number of significant figures.
$$(3.628 \times 10^5) - (7.84 \times 10^6)$$

Convert both numbers to the same exponent.

$$
\begin{array}{rl}
3.628 & \times\ 10^5 \\
-78.4 & \times\ 10^5 \\
\hline
-74.772 & \times\ 10^5
\end{array}
$$

The second number has only one digit to the right of the decimal point. Therefore, the result of the subtraction can have only one. The result of the subtraction is

$$-74.8 \times 10^5.$$

Exercise Six

Perform the following calculations and give the result to the proper number of significant figures.

A. $(3.64 \times 10^{-3}) + (8.29 \times 10^{-5}) =$

B. $(3.289 \times 10^{-2}) - (6.6 \times 10^{-3}) =$

C. $(6.74 \times 10^{7}) + (4.28 \times 10^{6}) - (2.432 \times 10^{6}) =$

D. $(2.69 \times 10^{4}) + (3.27 \times 10^{2})(4.6 \times 10^{2}) =$

ANSWERS

LESSON 1: CHEMICAL SYMBOLS AND FORMULAS: MOLAR MASS CALCULATIONS

Exercise One

A.

name of atom	symbol	number of protons	number of neutrons
chlorine-35	^{35}Cl	17	18
carbon-14	^{14}C	6	8
strontium-90	^{90}Sr	38	52
oxygen-18	^{18}O	8	10
cobalt-60	^{60}Co	27	33
uranium-235	^{235}U	92	143

B.

name of element	atomic number	symbol	name of element	atomic number	symbol
oxygen	8	O	antimony	51	Sb
carbon	6	C	sulfur	16	S
phosphorus	15	P	magnesium	12	Mg
manganese	25	Mn	bromine	35	Br
calcium	20	Ca	argon	18	Ar
strontium	38	Sr	copper	29	Cu
lead	82	Pb	tin	50	Sn
nickel	28	Ni	bismuth	83	Bi
sodium	11	Na	cobalt	27	Co
potassium	19	K	barium	56	Ba
plutonium	94	Pu	gold	79	Au
mercury	80	Hg	iodine	53	I

Exercise Two

$$46.00 \text{ g O} \left(\frac{1 \text{ mol O}}{16.00 \text{ g O}} \right) = 2.875 \text{ mol O}$$

$$\tfrac{1}{2} (2.875 \text{ mol O}) = 1.438 \text{ mol S}$$

$$1.438 \text{ mol S} \left(\frac{32.06 \text{ g S}}{1 \text{ mol S}} \right) = 46.10 \text{ g S}$$

Exercise Three

A. HCl hydrogen chloride (hydrochloric acid if dissolved in water)
NH_3 ammonia
H_2SO_4 sulfuric acid
$NaHCO_3$ sodium hydrogen carbonate (sodium bicarbonate)

B. sodium chloride NaCl   sulfur trioxide SO_3
gaseous hydrogen H_2 methane CH_4
carbon dioxide CO_2 nitric acid HNO_3

C. O_3 three $Ba_3(PO_4)_2$ eight Na_3PO_4 four $Al_2(SO_4)_3$ twelve

D. H_2O three $Ca(OH)_2$ five $CHCl_3$ five $NaAl(SO_4)_2$ twelve

Exercise Four

A. NaCl $22.99 + 35.45 = 58.44$ g/mol   oxygen gas $2 \times 16.00 = 32.00$ g/mol
CO_2 $12.01 + 2 \times 16.00 = 44.01$ g/mol   KCN $39.10 + 12.01 + 14.01 = 65.12$ g/mol
ammonium chloride $14.01 + (4 \times 1.008) + 35.45 = 53.49$ g/mol
$K_2Cr_2O_7$ $(2 \times 39.10) + (2 \times 52.00) + (7 \times 16.00) = 294.2$ g/mol

B. (1) H_2O

2 H: 2.02 g/mol
O: 16.00 g/mol
18.02 g/mol

$\%H = \dfrac{2.02}{18.02} \times 100 = 11.2\%$

$\%O = \dfrac{16.00}{18.02} \times 100 = 88.8\%$

(2) NH_4NO_3

2 N: 28.02 g/mol
4 H: 4.03 g/mol
3 O: 48.00 g/mol
80.05 g/mol

$\%N = \dfrac{28.02}{80.05} \times 100 = 35.00\%$

$\%H = \dfrac{4.032}{80.05} \times 100 = 5.037\%$

$\%O = \dfrac{48.00}{80.05} \times 100 = 59.96\%$

(3) $Ca_3(PO_4)_2$

3 Ca: 120.24 g/mol
2 P: 61.94 g/mol
8 O: 128.00 g/mol
310.18 g/mol

$\%Ca = \dfrac{120.24}{310.18} \times 100 = 38.76\%$

$\%P = \dfrac{61.94}{310.18} \times 100 = 19.97\%$

$\%O = \dfrac{128.00}{310.18} \times 100 = 41.27\%$

C. (1) C in 5.4 g of CO_2

C: 12.01 g/mol
2 O: 32.00 g/mol
44.01 g/mol

$\% C = \dfrac{12.01}{44.01} \times 100 = 27.29\% \text{ C}$

grams C = $(0.2729)(5.4 \text{ g } CO_2) = 1.5 \text{ g C}$

(2) Cl in 1.00 kg of NaCl

Na: 22.99 g/mol
Cl: 35.45 g/mol
58.44 g/mol

$\% Cl = \dfrac{35.45}{58.44} \times 100 = 60.66\% \text{ Cl}$

kg Cl = $(0.6066)(1.00 \text{ kg } CO_2) = 0.607 \text{ kg Cl}$

(3) Na in 78 g of $NaHCO_3$

Na: 22.99 g/mol
H: 1.01 g/mol
C: 12.01 g/mol
3 O: 48.00 g/mol
84.01 g/mol

$\% Na = \dfrac{22.99}{84.01} \times 100 = 27.36\% \text{ Na}$

grams Na = $(0.2736)(78 \text{ g } NaHCO_3) = 21 \text{ g Na}$

Exercise Five

A. (1) The percent composition indicates that 100 grams contains 40.0 g C, 53.3 g O, and 6.67 g H.

(2)

$$40.0 \text{ g C} \left(\frac{1 \text{ mol C}}{12.01 \text{ g C}} \right) = 3.33 \text{ mol C}$$

$$53.3 \text{ g O} \left(\frac{1 \text{ mol O}}{16.00 \text{ g O}} \right) = 3.33 \text{ mol O}$$

$$6.67 \text{ g H} \left(\frac{1 \text{ mol H}}{1.008 \text{ g H}} \right) = 6.62 \text{ mol H}$$

(3)

$$\frac{3.33 \text{ mol C}}{3.33 \text{ mol C}} = 1 \text{ mol C/mol C}$$

$$\frac{3.33 \text{ mol O}}{3.33 \text{ mol C}} = 1 \text{ mol O/mol C}$$

$$\frac{6.62 \text{ mol H}}{3.33 \text{ mol C}} = 1.99 \text{ mol H/mol C}$$

1.99 is very close to 2 – round off to 2.

(4)   The empirical formula is COH_2.

B.   (1,2)

$$85.66 \text{ g C} \left(\frac{1 \text{ mol C}}{12.01 \text{ g C}} \right) = 7.132 \text{ mol C}$$

$$14.34 \text{ g H} \left(\frac{1 \text{ mol H}}{1.008 \text{ g H}} \right) = 14.23 \text{ mol H}$$

(3) $\dfrac{7.132 \text{ mol C}}{7.132 \text{ mol C}} = 1 \text{ mol C/mol C}$ $\dfrac{14.23 \text{ mol H}}{7.132 \text{ mol C}} = 2 \text{ mol H/mol C}$

(4)   The empirical formula is CH_2. Empirical formula mass is $12.01 + 2(1.008) = 14.03$.

$$\frac{70.13 \text{ g/mol}}{14.03 \text{ g/formula}} = 4.988 \text{ formula/mol} \quad (\text{round off to 5 })$$

Molecular formula is C_5H_{10}.

Exercise Six

Find moles in 100 g.

$$26.58 \text{ g K} \left(\frac{1 \text{ mol K}}{39.10 \text{ g K}} \right) = 0.6798 \text{ mol K}$$

$$35.35 \text{ g Cr} \left(\frac{1 \text{ mol Cr}}{52.00 \text{ g Cr}} \right) = 0.6798 \text{ mol Cr}$$

$$38.07 \text{ g O} \left(\frac{1 \text{ mol O}}{16.00 \text{ g O}} \right) = 2.379 \text{ mol O}$$

Find relative moles.

$$\frac{0.6798 \text{ mol K}}{0.6798 \text{ mol K}} = 1 \text{ mol K/mol K} \qquad \frac{0.6798 \text{ mol Cr}}{0.6798 \text{ mol K}} = 1 \text{ mol Cr/mol K}$$

$$\frac{2.379 \text{ mol O}}{0.6798 \text{ mol K}} = 3.50 \text{ mol O/mol K}$$

Not all are whole numbers. The third ratio is 3.5; so multiply all by 2 to get whole numbers. The empirical formula is $K_2Cr_2O_7$.

Exercise Seven

The molar mass of boron: $(0.1978)(10.01 \text{ g/mol}) + (0.8022)(11.01 \text{ g/mol}) = 10.81 \text{ g/mol}$

PROBLEMS

1.   (a) HNO_3 (b) I_2 (c) O_3 (d) NaOH   (e) CH_4
     (f) SO_2 (g) HCN   (h) $HC_2H_3O_2$ (i) H_2O_2 (j) Na_2CO_3

2.   (a) 98.08 g/mol   (b) 60.06 g/mol   (c) 520.9 g/mol
     (d) 159.8 g/mol   (e) 84.01 g/mol   (f) 17.03 g/mol

3. 5.03% Be   10.04% Al   31.35% Si   53.58% O

4. (a) 699.4 kg Fe   (b) 1429 kg hematite

5. $C_6H_8O_6$

LESSON 2: WRITING AND BALANCING CHEMICAL EQUATIONS

Exercise One

A. $NH_3(g) + O_2(g) \longrightarrow NO(g) + H_2O(g)$

B. $Na_2CO_3(aq) + HCl(aq) \longrightarrow NaCl(aq) + H_2O(l) + CO_2(g)$

C. $KClO_3(s) \longrightarrow KCl(s) + O_2(g)$

D. $SO_2(g) + H_2O(l) \longrightarrow H_2SO_3(aq)$

Exercise Two

A. balanced   B. not balanced   C. balanced
D. balanced   E. not balanced   F. balanced

Exercise Three

A. $PCl_5(s) + 4 H_2O(l) \longrightarrow H_3PO_4(aq) + 5 HCl(aq)$

B. $2 NO(g) + Cl_2(g) \longrightarrow 2 NOCl(g)$

C. $Al_2O_3(s) + 6 HCl(aq) \longrightarrow 2 AlCl_3(aq) + 3 H_2O(l)$

D. $P_4O_{10}(s) + 6 H_2O(g) \longrightarrow 4 H_3PO_4(l)$

Exercise Four

A. $3 AsH_3(g) + 4 KClO_3(s) \longrightarrow 4 KCl(s) + 3 H_3AsO_4(s)$

B. $C_3H_8(g) + 5 O_2(g) \longrightarrow 3 CO_2(g) + 4 H_2O(l)$

C. $3 SiF_4(g) + 4 H_2O(l) \longrightarrow H_4SiO_4(aq) + 2 H_2SiF_6(aq)$

D. $Sb_2S_3(s) + 3 Fe(s) \longrightarrow 2 Sb(s) + 3 FeS(s)$

Exercise Five

A. $2 C_2H_6(g) + 7 O_2(g) \longrightarrow 4 CO_2(g) + 6 H_2O(g)$

B. $2 H_2O(l) + 2 F_2(g) \longrightarrow 4 HF(aq) + O_2(g)$

C. $I_2(s) + 3 Cl_2(g) \longrightarrow 2 ICl_3(s)$

D. $Al_2O_3(s) + 3 C(s) + 3 Cl_2(g) \longrightarrow 2 AlCl_3(s) + 3 CO(g)$

E. $Ca(s) + 2 H_2O(l) \longrightarrow Ca(OH)_2(aq) + H_2(g)$

F. $N_2(g) + 3 H_2(g) \longrightarrow 2 NH_3(g)$

G. $12 HNO_3(aq) + P_4O_{10}(s) \longrightarrow 6 N_2O_5(g) + 4 H_3PO_4(aq)$

H. $4 NH_3(g) + 5 O_2(g) \longrightarrow 4 NO(g) + 6 H_2O(g)$

I. $2 Na(s) + 2 H_2O(l) \longrightarrow 2 NaOH(aq) + H_2(g)$

Exercise Six

A. 1. $SO_2(g) + {}^1/_2 O_2(g) \longrightarrow SO_3(g)$

2. $NH_3(g) \longrightarrow {}^1/_2 N_2(g) + {}^3/_2 H_2(g)$

3. ${}^1/_2 CH_4(g) + O_2(g) \longrightarrow {}^1/_2 CO_2(g) + H_2O(l)$

4. $Al(s) + 3 HCl(g) \longrightarrow AlCl_3(aq) + {}^3/_2 H_2(g)$

B. 1. Sixteen moles of hydrogen sulfide gas react with eight moles of sulfur dioxide gas to form three moles of solid sulfur and sixteen moles of liquid water.

AND

Sixteen molecules of hydrogen sulfide gas react with eight molecules of sulfur dioxide gas to form three molecules of solid sulfur and sixteen molecules of liquid water.

2. One mole of ammonia gas reacts with five-fourths moles of oxygen gas to form one mole of nitric oxide gas and three-halves moles of liquid water.

PROBLEMS

1. (a) $HCl(aq) + NaHCO_3(s) \longrightarrow CO_2(g) + NaCl(aq) + H_2O(l)$
    (b) $CH_4(g) + 4 Cl_2(g) \longrightarrow CCl_4(l) + 4 HCl(g)$
    (c) $SiO_2(s) + 3 C(s) \longrightarrow SiC(s) + 2 CO(g)$

2. (a) $P_2O_5(s) + 3 H_2O(l) \longrightarrow 2 H_3PO_4(l)$
    (b) $4 KClO_3(s) \longrightarrow 3 KClO_4(s) + KCl(s)$
    (c) $2 KClO_3(s) \longrightarrow 2 KCl(s) + 3 O_2(g)$
    (d) $2 Al(s) + 3 H_2SO_4(aq) \longrightarrow Al_2(SO_4)_3(s) + 3 H_2(g)$
    (e) $UF_4(g) + 2 Mg(s) \longrightarrow U(s) + 2 MgF_2(s)$
    (f) $2 N_2H_4(g) + N_2O_4(g) \longrightarrow 3 N_2(g) + 4 H_2O(g)$

LESSON 3: USING THE MOLE CONCEPT

Exercise One

$$50.0 \text{ g CaBr}_2 \times \frac{1 \text{ mol CaBr}_2}{199.9 \text{ g CaBr}_2} = 0.250 \text{ mol CaBr}_2$$

Exercise Two

A.

$$160 \text{ g H}_2O \times \frac{1 \text{ mol H}_2O}{18.02 \text{ g H}_2O} \times \frac{6.022 \times 10^{23} \text{ molecules H}_2O}{1 \text{ mol H}_2O} = 5.3 \times 10^{24} \text{ molecules H}_2O$$

B.

$$5.3 \times 10^{24} \text{ molecules H}_2O \times \frac{3 \text{ atoms}}{1 \text{ molecule H}_2O} = 1.6 \times 10^{25} \text{ atoms}$$

Exercise Three

A.

$$0.50 \text{ mol Al} \times \frac{2 \text{ mol Al}_2O_3}{4 \text{ mol Al}} = 0.25 \text{ mol Al}_2O_3$$

B.

$$0.38 \text{ mol O}_2 \times \frac{2 \text{ mol Al}_2O_3}{3 \text{ mol O}_2} = 0.25 \text{ mol Al}_2O_3$$

C. They are the same, because 0.50 mol Al reacts with 0.38 mol O_2.

Exercise Four

$$1.0 \text{ g (NH}_4)_2Cr_2O_7 \times \frac{1 \text{ mol (NH}_4)_2Cr_2O_7}{252.1 \text{ g (NH}_4)_2Cr_2O_7} \times \frac{4 \text{ mol H}_2O}{1 \text{ mol (NH}_4)_2Cr_2O_7}$$

$$\times \frac{6.022 \times 10^{23} \text{ molecules H}_2O}{1 \text{ mol H}_2O} = 9.6 \times 10^{21} \text{ molecules H}_2O$$

PROBLEMS

1. 68.2 g CH_4
2. 2.41×10^{23} molecules HBr
3. 0.45 mol O_2
4. a. 79.8 g O_2 b. 54.9 g CO_2 c. 44.9 g H_2O
    d. 99.8 g substance   e. 99.8 g substance
    f. The masses are equal. This illustrates the Law of Conservation of Mass. In any chemical reaction, the total mass of the products is equal to the total mass of the reactants.

LESSON 4: LIMITING REACTANTS AND PERCENT YIELD

Exercise One

The limiting reactant can be found in either of two ways.

1. How many moles of O_2 are needed to react with 5 moles of C_3H_6?

$$5.0 \text{ mol } C_3H_6 \left(\frac{9 \text{ mol } O_2}{2 \text{ mol } C_3H_6} \right) = 22.5 \text{ mol } O_2$$

Do we have 22.5 mol O_2? No. Therefore, O_2 is the limiting reactant.

or

2. How many moles of C_3H_6 are needed to react with 12 moles of O_2?

$$12.0 \text{ mol } O_2 \left(\frac{2 \text{ mol } C_3H_6}{9 \text{ mol } O_2} \right) = 2.7 \text{ mol } C_3H_6$$

Do we have 2.7 moles of C_3H_6? Yes. Therefore, O_2 is the limiting reactant.

Exercise Two

Given: 5.0 g H_2 and 30.0 g N_2
Which is the limiting reactant? First, find the moles of reactants.

$$5.0 \text{ g } H_2 \left(\frac{1 \text{ mol } H_2}{2.016 \text{ g } H_2} \right) = 2.5 \text{ mol } H_2 \qquad 30.0 \text{ g } N_2 \left(\frac{1 \text{ mol } N_2}{28.02 \text{ g } N_2} \right) = 1.07 \text{ mol } N_2$$

How many moles of N_2 are needed to react with 2.5 moles of H_2?

$$2.5 \text{ mol } H_2 \left(\frac{1 \text{ mol } N_2}{3 \text{ mol } H_2} \right) = 0.83 \text{ mol } N_2$$

We have more than enough N_2, so H_2 is the limiting reactant.

$$2.5 \text{ mol } H_2 \left(\frac{2 \text{ mol } NH_3}{3 \text{ mol } H_2} \right) \left(\frac{17.03 \text{ g } NH_3}{1 \text{ mol } NH_3} \right) = 28 \text{ g } NH_3$$

Exercise Three

Given: 4.55 g Al and 9.62 g NiO
Which is the limiting reactant? First, find the moles of reactants.

$$4.55 \text{ g Al} \left(\frac{1 \text{ mol Al}}{26.98 \text{ g Al}} \right) = 0.169 \text{ mol Al} \qquad 9.62 \text{ g NiO} \left(\frac{1 \text{ mol NiO}}{74.69 \text{ g NiO}} \right) = 0.129 \text{ mol NiO}$$

How many moles of NiO are needed to react with 0.169 mol Al?

$$0.169 \text{ mol Al} \left(\frac{3 \text{ mol NiO}}{2 \text{ mol Al}} \right) = 0.254 \text{ mol NiO}$$

We do not have enough NiO, so NiO is the limiting reactant. Aluminum is in excess.
Find the mass of Al that reacts.

$$0.129 \text{ mol NiO} \left(\frac{2 \text{ mol Al}}{3 \text{ mol NiO}} \right) \left(\frac{26.98 \text{ g Al}}{1 \text{ mol Al}} \right) = 2.32 \text{ g Al react}$$

The amount remaining is 4.55 g − 2.32 g = 2.23 g Al remain.

Exercise Four

A.

$$\text{percent yield} = \frac{\text{actual yield}}{\text{maximum yield}} \times 100 = \frac{15.0 \text{ g}}{19.0 \text{ g}} \times 100 = 78.9\%$$

B.

$$\text{percent yield} = \frac{\text{actual yield}}{\text{maximum yield}} \times 100\%$$

actual yield = 4.0 mol NaCl

What is maximum yield?

Given reactants: 5.0 mol NaOH and 6.0 mol HCl. Which is the limiting reactant?

How many moles of HCl are needed to react with 5.0 mol NaOH?

$$5.0 \text{ mol NaOH} \left(\frac{1 \text{ mol HCl}}{1 \text{ mol NaOH}} \right) = 5.0 \text{ mol HCl}$$

We have enough HCl, so NaOH is the limiting reactant.

$$5.0 \text{ mol NaOH} \left(\frac{1 \text{ mol NaCl}}{1 \text{ mol NaOH}} \right) = 5.0 \text{ mol NaCl}$$

$$\text{percent yield} = \frac{4.0 \text{ mol NaCl}}{5.0 \text{ mol NaCl}} \times 100\% = 80\% \text{ yield}$$

PROBLEMS

1. oxygen
2. H_2S is limiting; SO_2 is in excess. 40.5 g S produced. 5.4 g SO_2 remain.
3. 32%
4. 100%
5. 40%

LESSON 5: WRITING NET IONIC EQUATIONS

Exercise One

A. ionic, K^+ and OH^-
B. not ionic
C. ionic, Co^{2+} and $2 Cl^-$
D. ionic, Na^+ and HCO_3^-
E. not ionic
F. ionic, Ag^+ and NO_3^-
G. not ionic
H. ionic, $2 NH_4^+$ and $Cr_2O_7^{2-}$

Exercise Two

A. soluble, guidelines 1 and 9
B. insoluble, guideline 8
C. insoluble, guideline 6
D. soluble, guideline 4
E. soluble, guideline 2
F. soluble, guideline 2
G. soluble, guideline 1
H. insoluble, guideline 7
I. insoluble, guideline 6
J. insoluble, guideline 9

Exercise Three

A. strong electrolyte, H^+ (aq) + NO_3^- (aq), $i = 2$
B. weak electrolyte, NH_3 (aq), $1 < i < 2$
C. strong electrolyte, $3 Na^+$ (aq) + PO_4^{3-} (aq), $i = 4$
D. non-electrolyte, CH_3OH (aq), $i = 1$
E. weak electrolyte, H_3PO_4 (aq), $1 < i < 2$
F. strong electrolyte, H^+ (aq) + HSO_4^- (aq), $i = 2$

Exercise Four

A. 1. H^+ (aq) + OH^- (aq) $\longrightarrow$ H_2O (l)
   2. Ag^+ (aq) + Cl^- (aq) $\longrightarrow$ AgCl (s)
   3. H_2S (aq) + $2 NH_3$ (aq) $\longrightarrow$ $2 NH_4^+$ (aq) + S^{2-} (aq)
   4. PO_4^{3-} (aq) + H_2O (l) $\longrightarrow$ HPO_4^{2-} (aq) + OH^- (aq)
   5. $2 H^+$ (aq) + $CaCO_3$ (s) $\longrightarrow$ CO_2 (g) + H_2O (l) + Ca^{2+} (aq)

B. 1. Pb^{2+} (aq) + $2 Cl^-$ (aq) $\longrightarrow$ $PbCl_2$ (s)
   2. OH^- (aq) + NH_4^+ (aq) $\longrightarrow$ NH_3 (g) + H_2O (l)

Exercise Five

A. MnS is insoluble.
   $$Mn^{2+} \text{ (aq)} + S^{2-} \text{ (aq)} \longrightarrow MnS \text{ (s)}$$
B. $FeBr_3$ and NH_4NO_3 are both soluble.
C. $Ni(OH)_2$ is insoluble.
   $$Ni^{2+} \text{ (aq)} + 2 OH^- \text{ (aq)} \longrightarrow Ni(OH)_2 \text{ (s)}$$
D. $Mg_3(PO_4)_2$ is insoluble.
   $$3 Mg^{2+} \text{ (aq)} + 2 PO_4^{3-} \text{ (aq)} \longrightarrow Mg_3(PO_4)_2 \text{ (s)}$$

PROBLEMS

1. (a) complete ionic:
$$Pb^{2+}(aq) + 2\ ClO_3^-(aq) + 2\ Na^+(aq) + 2\ Br^-(aq) \longrightarrow PbBr_2(s) + 2\ Na+(aq) + 2\ ClO_3^-(aq)$$
net ionic:
$$Pb^{2+}(aq) + 2\ Br^-(aq) \longrightarrow PbBr_2(s)$$

   (b) complete ionic:
$$2\ Na^+(aq) + 2\ OH^-(aq) + Cu^{2+}(aq) + SO_4^{2-}(aq) \longrightarrow 2\ Na^+(aq) + SO_4^{2-}(aq) + Cu(OH)_2(s)$$
net ionic:
$$2\ OH^-(aq) + Cu^{2+}(aq) \longrightarrow Cu(OH)_2(s)$$

   (c) complete ionic:
$$3\ H_2S(g) + 2\ Cr^{3+}(aq) + 6\ Cl^-(aq) \longrightarrow Cr_2S_3(s) + 6\ H^+(aq) + 6\ Cl^-(aq)$$
net ionic:
$$3\ H_2S(g) + 2\ Cr^{3+}(aq) \longrightarrow Cr_2S_3(s) + 6\ H^+(aq)$$

   (d) complete ionic:
$$CaCO_3(s) + 2\ HCH_3CO_2(aq) \longrightarrow Ca^{2+}(aq) + 2\ CH_3CO_2^-(aq) + H_2O(l) + CO_2(g)$$
net ionic:
$$CaCO_3(s) + 2\ HCH_3CO_2(aq) \longrightarrow Ca^{2+}(aq) + 2\ CH_3CO_2^-(aq) + H_2O(l) + CO_2(g)$$

2. (a) complete:
$$FeCl_3(aq) + 3\ NaOH(aq) \longrightarrow Fe(OH)_3(s) + 3\ NaCl(aq)$$
complete ionic:
$$Fe^{3+}(aq) + 3\ Cl^-(aq) + 3\ Na^+(aq) + 3\ OH^-(aq) \longrightarrow Fe(OH)_3(s) + 3\ Na^+(aq) + 3\ Cl^-(aq)$$
net ionic:
$$Fe^{3+}(aq) + 3\ OH^-(aq) \longrightarrow Fe(OH)_3(s)$$

   (b) complete:
$$Pb(NO_3)_2(aq) + K_2SO_4(aq) \longrightarrow PbSO_4(s) + 2\ KNO_3(aq)$$
complete ionic:
$$Pb^{2+}(aq) + 2\ NO_3^-(aq) + 2\ K^+(aq) + SO_4^{2-}(aq) \longrightarrow PbSO_4(s) + 2\ K^+(aq) + 2\ NO_3^-(aq)$$
net ionic:
$$Pb^{2+}(aq) + SO_4^{2-}(aq) \longrightarrow PbSO_4(s)$$

   (c) complete:
$$Hg(NO_3)_2(aq) + K_2CO_3(aq) \longrightarrow HgCO_3(s) + 2\ KNO_3(aq)$$
complete ionic:
$$Hg^{2+}(aq) + 2\ NO_3^-(aq) + 2\ K^+(aq) + CO_3^{2-}(aq) \longrightarrow HgCO_3(s) + 2\ K^+(aq) + 2\ NO_3^-(aq)$$
net ionic:
$$Hg^{2+}(aq) + CO_3^{2-}(aq) \longrightarrow HgCO_3(s)$$

   (d) complete:
$$2\ H_3PO_4(aq) + 3\ Sr(OH)_2(aq) \longrightarrow 6\ H_2O(l) + Sr_3(PO_4)_2(s)$$
complete ionic:
$$2\ H_3PO_4(aq) + 3\ Sr^{2+}(aq) + 6\ OH^-(aq) \longrightarrow 6\ H_2O(l) + Sr_3(PO_4)_2(s)$$
net ionic:
$$2\ H_3PO_4(aq) + 3\ Sr^{2+}(aq) + 6\ OH^-(aq) \longrightarrow 6\ H_2O(l) + Sr_3(PO_4)_2(s)$$

3. (a) net ionic equation: $3\ Cu^{2+}(aq) + 2\ H_3PO_4(aq) \longrightarrow Cu_3(PO_4)_2(s) + 6\ H^+(aq)$
      spectator ions: $H^+(aq)$ and $Cl^-(aq)$

   (b) net ionic equation: $2\ Ag^+(aq) + S^{2-}(aq) \longrightarrow Ag_2S(s)$
      spectator ions: $K^+(aq)$ and $NO_3^-(aq)$

   (c) no reaction

   (d) net ionic equation: $Ni^{2+}(aq) + SO_4^{2-}(aq) + Ba^{2+}(aq) + 2\ OH^-(aq) \longrightarrow Ni(OH)_2(s) + BaSO_4(s)$
      spectator ions: none

4. (a) any one of these: $AgNO_3(aq)$ $AgCH_3CO_2(aq)$
      with any one of these: $NaCl(aq)$ $BaCl_2(aq)$ $HCl(aq)$ $NH_4Cl(aq)$ and others

   (b) any one of these: $CaCl_2(aq)$ $Ca(NO_3)_2(aq)$ $CaBr_2(aq)$ and others
      with any one of these: $NaF(aq)$ $KF(aq)$ $NH_4F(aq)$ and others [note: *not* $HF(aq)$]

   (c) any one of these: $HCl(aq)$ $HNO_3(aq)$ $HClO_3(aq)$ and others
      with any one of these: $Na_2CO_3(aq)$ $K_2CO_3(aq)$ $(NH_4)_2CO_3(aq)$ and others

   (d) this one: $HF(aq)$
      with any one of these: $NaOH(aq)$ $KOH(aq)$ and others [note: *not* $Ba(OH)_2(aq)$]

LESSON 6: MOLARITY CALCULATIONS

Exercise One

$$\text{moles of solute} = 2.42 \text{ g RbCl} \left(\frac{1 \text{ mol RbCl}}{120.92 \text{ g RbCl}} \right) = 0.0200 \text{ mol RbCl}$$

$$\text{liters of solution} = 250.0 \text{ mL} \left(\frac{1 \text{ L}}{1000 \text{ mL}} \right) = 0.2500 \text{ L}$$

$$\text{molarity} = \frac{0.0200 \text{ mol RbCl}}{0.2500 \text{ L}} = 0.0800 \text{ M RbCl}$$

Exercise Two

$$\text{moles NaOH} = (2.00 \text{ moles NaOH/L}) \times (0.2500 \text{ L})$$
$$= 0.500 \text{ moles NaOH}$$

$$\text{grams NaOH} = 0.500 \text{ mol NaOH} \left(\frac{40.00 \text{ g NaOH}}{1 \text{ mol NaOH}} \right) = 20.0 \text{ g NaOH}$$

Exercise Three

$$\text{to prepare the solution, need: moles H}_2\text{SO}_4 = (3.0 \text{ mol H}_2\text{SO}_4\text{/L}) \times (0.5000 \text{ L}) = 1.5 \text{ mol H}_2\text{SO}_4$$

$$\text{get 1.5 mol H}_2\text{SO}_4 \text{ from:} \quad 1.5 \text{ mol H}_2\text{SO}_4 \left(\frac{1 \text{ L}}{18 \text{ mol H}_2\text{SO}_4} \right) = 0.083 \text{ L of concentrated H}_2\text{SO}_4$$

Dilute 83 mL of concentrated H_2SO_4 to 500.0 mL.

Exercise Four

For a dilution of a solution with solvent, we may use the relationship $M_1V_1 = M_2V_2$

$$M_1 = 0.800 \text{ M} \qquad\qquad M_2 = 0.150 \text{ M}$$
$$V_1 = ? \qquad\qquad\qquad V_2 = 250.0 \text{ mL}$$

$$V_1 = \frac{M_2 V_2}{M_1} = \frac{(0.150 \text{ M})(250.0 \text{ mL})}{0.800 \text{ M}} = 46.9 \text{ mL}$$

To prepare the solution, dilute 46.9 mL of 0.800 M $CuSO_4$ to 250.0 mL with water.

Exercise Five

$$\text{Molarity} = \frac{\text{moles of solute}}{\text{liters of solution}} = \frac{\text{mol CuSO}_4}{\text{L solution}}$$

L solution = 0.1000 L + 0.5000 L = 0.6000 L
There is $CuSO_4$ from both solutions.
   From one solution, mol $CuSO_4$ = (0.1000 L) (0.200 mol $CuSO_4$ / L) = 0.0200 mol $CuSO_4$
   From the other solution, mol $CuSO_4$ = (0.5000 L) (0.150 mol $CuSO_4$ / L) = 0.0750 mol $CuSO_4$
   The total mol $CuSO_4$ = 0.0200 mol + 0.0750 mol = 0.0950 mol $CuSO_4$

$$\text{Molarity} = \frac{0.0950 \text{ mol CuSO}_4}{0.6000 \text{ L}} = 0.158 \text{ M CuSO}_4$$

Exercise Six

$$\text{volume of mixture} = 0.5000 \text{ L} + 0.2500 \text{ L} = 0.7500 \text{ L}$$

For Mg^{2+}:

$$\text{mol Mg(NO}_3)_2 = 0.500 \text{ L} \left(\frac{1.00 \text{ mol Mg(NO}_3)_2}{1 \text{ L}} \right) = 0.500 \text{ mol Mg(NO}_3)_2$$

$$\text{mol Mg}^{2+} = 0.500 \text{ mol} \left(\frac{1 \text{ mol Mg}^{2+}}{1 \text{ mol Mg(NO}_3)_2} \right) = 0.500 \text{ mol Mg}^{2+}$$

$$\text{molarity Mg}^{2+} = \frac{0.500 \text{ mol Mg}^{2+}}{0.750 \text{ L}} = 0.667 \text{ M Mg}^{2+}$$

For $Ag^{+:}$

$$\text{mol AgNO}_3 = 0.2500 \text{ L} \left(\frac{2.00 \text{ mol AgNO}_3}{1 \text{ L}} \right) = 0.500 \text{ mol AgNO}_3$$

$$\text{mol Ag}^+ = 0.500 \text{ mol} \left(\frac{1 \text{ mol Ag}^+}{1 \text{ mol AgNO}_3} \right) = 0.500 \text{ mol Ag}^+$$

$$\text{molarity Ag}^+ = \frac{0.500 \text{ mol Ag}^+}{0.750 \text{ L}} = 0.667 \text{ M Ag}^+$$

For NO_3^-:
  Nitrate ions come from both $AgNO_3$ and $Mg(NO_3)_2$.

$$\text{mol NO}_3^- \text{ from AgNO}_3 = 0.500 \text{ mol} \left(\frac{1 \text{ mol NO}_3^-}{1 \text{ mol AgNO}_3} \right) = 0.500 \text{ mol NO}_3^-$$

$$\text{mol NO}_3^- \text{ from Mg(NO}_3)_2 = 0.500 \text{ mol} \left(\frac{2 \text{ mol NO}_3^-}{1 \text{ mol Mg(NO}_3)_2} \right) = 1.00 \text{ mol NO}_3^-$$

$$\text{total NO}_3^- = 0.500 \text{ mol} + 1.00 \text{ mol} = 1.50 \text{ mol NO}_3^-$$

$$\text{molarity NO}_3^- = \frac{1.50 \text{ mol NO}_3^-}{0.750 \text{ L}} = 2.00 \text{ M NO}_3^-$$

Exercise Seven

The steps are: vol NaOH $\xrightarrow{\text{molarity of NaOH}}$ mol NaOH $\xrightarrow{\text{balanced chemical equation}}$ mol H_2SO_4 $\xrightarrow{\text{molarity of } H_2SO_4}$ vol H_2SO_4

$$0.1000 \text{ L NaOH} \left(\frac{0.884 \text{ mol NaOH}}{1 \text{ L}} \right) \left(\frac{1 \text{ mol H}_2SO_4}{2 \text{ mol NaOH}} \right) \left(\frac{1 \text{ L}}{1.024 \text{ mol H}_2SO_4} \right) = 0.0432 \text{ L H}_2SO_4$$

Exercise Eight

$$\text{molarity of H}_2C_2O_4 = \frac{\text{mol H}_2C_2O_4}{\text{L solution}}$$

L solution = 0.05000 L

For mole of $H_2C_2O_4$, the steps are: vol $K_2Cr_2O_7$ $\xrightarrow{\text{molarity of } K_2Cr_2O_7}$ mol $K_2Cr_2O_7$ $\xrightarrow{\text{balanced chemical equation}}$ mol $H_2C_2O_4$

$$0.02844 \text{ L K}_2Cr_2O_7 \left(\frac{0.3242 \text{ mol K}_2Cr_2O_7}{1 \text{ L}} \right) \left(\frac{3 \text{ mol H}_2C_2O_4}{1 \text{ mol K}_2Cr_2O_7} \right) = 0.02766 \text{ mol H}_2C_2O_4$$

$$\text{molarity of H}_2C_2O_4 = \frac{0.02766 \text{ mol H}_2C_2O_4}{0.05000 \text{ L solution}} = 0.5532 \text{ M H}_2C_2O_4$$

Exercise Nine

$$\text{moles of AgNO}_3 = 32.0 \text{ g AgNO}_3 \left(\frac{1 \text{ mol AgNO}_3}{169.9 \text{ g AgNO}_3} \right) = 0.188 \text{ mol AgNO}_3$$

$$\text{mass of solution} = 32.0 \text{ g} + 68.0 \text{ g} = 100.0 \text{ g}$$

$$\text{volume} = 100.0 \text{ g} \left(\frac{1 \text{ mL}}{1.35 \text{ g}} \right) \left(\frac{1 \text{ L}}{1000 \text{ mL}} \right) = 0.07400 \text{ L}$$

$$\text{molarity} = \frac{0.188 \text{ mol}}{0.07400 \text{ L}} = 2.54 \text{ M}$$

PROBLEMS

1. 0.174 g Ca(OH)$_2$	2. 3.30 M NaOH	3. 0.65 L	4. 3.70 mL
5. 0.167 M HCl	6. 68000 g/mol	7. 0.153 M I$_3^-$	8. 0.3637 M NaOH

LESSON 7: MOLE FRACTION AND MOLALITY

Exercise One

$$\text{mol C}_2\text{H}_5\text{OH} = 100.0 \text{ g C}_2\text{H}_5\text{OH} \left(\frac{1 \text{ mol C}_2\text{H}_5\text{OH}}{46.07 \text{ g C}_2\text{H}_5\text{OH}} \right) = 2.171 \text{ mol C}_2\text{H}_5\text{OH}$$

$$\text{mol H}_2\text{O} = 100.0 \text{ g H}_2\text{O} \left(\frac{1 \text{ mol H}_2\text{O}}{18.02 \text{ g H}_2\text{O}} \right) = 5.549 \text{ mol H}_2\text{O}$$

$$X_{\text{C}_2\text{H}_5\text{OH}} = \frac{2.171 \text{ mol C}_2\text{H}_5\text{OH}}{2.171 \text{ mol C}_2\text{H}_5\text{OH} + 5.549 \text{ mol H}_2\text{O}} = 0.2812$$

$$X_{\text{H}_2\text{O}} = \frac{5.549 \text{ mol H}_2\text{O}}{2.171 \text{ mol C}_2\text{H}_5\text{OH} + 5.549 \text{ mol H}_2\text{O}} = 0.7188$$

Check: 0.2812 + 0.7188 = 1.000

Exercise Two

$$\text{mass of ethanol} = (0.710 \text{ g/mL}) \times (100.0 \text{ mL}) = 71.0 \text{ g C}_4\text{H}_9\text{OH}$$

$$\text{mass of water} = (0.990 \text{ g/mL}) \times (100.0 \text{ mL}) = 99.0 \text{ g H}_2\text{O}$$

$$\text{moles of ethanol} = 71.0 \text{ g C}_2\text{H}_5\text{OH} \left(\frac{1 \text{ mol C}_2\text{H}_5\text{OH}}{46.07 \text{ g C}_2\text{H}_5\text{OH}} \right) = 1.54 \text{ mol C}_2\text{H}_5\text{OH}$$

$$\text{moles of H}_2\text{O} = 99.0 \text{ g H}_2\text{O} \left(\frac{1 \text{ mol H}_2\text{O}}{18.02 \text{ g H}_2\text{O}} \right) = 5.49 \text{ mol H}_2\text{O}$$

$$X_{\text{C}_2\text{H}_5\text{OH}} = \frac{1.54 \text{ mol C}_2\text{H}_5\text{OH}}{1.54 \text{ mol C}_2\text{H}_5\text{OH} + 5.49 \text{ mol H}_2\text{O}} = 0.219$$

$$X_{\text{H}_2\text{O}} = \frac{5.49 \text{ mol H}_2\text{O}}{1.54 \text{ mol C}_2\text{H}_5\text{OH} + 5.49 \text{ mol H}_2\text{O}} = 0.781$$

Check: 0.219 + 0.781 = 1.000

Exercise Three

Need:

$$X_{C_2H_5OH} = \frac{\text{mol } C_2H_5OH}{\text{mol } C_2H_5OH + \text{mol } H_2O} = 0.200$$

$$\text{mol } C_2H_5OH = 150.0 \text{ g } C_2H_5OH \left(\frac{1 \text{ mol } C_2H_5OH}{46.07 \text{ g } C_2H_5OH} \right) = 3.256 \text{ mol } C_2H_5OH$$

Let n represent moles of water,

$$\frac{3.256 \text{ mol}}{3.256 \text{ mol} + n} = 0.200$$

$$3.256 \text{ mol} = 0.200 \, (\, 3.256 \text{ mol} + n \,) = 0.651 + 0.200 \, n$$

$$n = \frac{2.605}{0.200} = 13.02 \text{ mol } H_2O$$

From the moles of water, determine grams required.

$$13.02 \text{ mol } H_2O \left(\frac{18.02 \text{ g } H_2O}{1 \text{ mol } H_2O} \right) = 235 \text{ g } H_2O$$

Exercise Four

$$25.3 \text{ g } C_4H_9OH \left(\frac{1 \text{ mol } C_4H_9OH}{74.12 \text{ g } C_4H_9OH} \right) = 0.341 \text{ mol solute}$$

$$752.6 \text{ g } H_2O \left(\frac{1 \text{ kg}}{1000 \text{ g}} \right) = 0.7526 \text{ kg solvent}$$

$$\text{molality} = \frac{0.341 \text{ mol } C_4H_9OH}{0.7526 \text{ kg solvent}} = 0.453 \text{ m } C_4H_9OH$$

Exercise Five

$$\text{molality} = \frac{\text{moles of solute}}{\text{kilograms of solvent}}$$

$$\text{kilograms of solvent} = \frac{\text{moles of solute}}{\text{molality}}$$

$$\text{mol NaCl} = 12.64 \text{ g NaCl} \left(\frac{1 \text{ mol NaCl}}{58.44 \text{ g NaCl}} \right) = 0.2163 \text{ mol NaCl}$$

$$\text{kg } H_2O = \frac{0.2163 \text{ mol NaCl}}{0.500 \text{ mol NaCl} \, / \, \text{kg } H_2O} = 0.433 \text{ kg } H_2O$$

Add 0.433 kg, which is 433 g of water.

Exercise Six

$$\text{kg solvent} = 38.96 \text{ g } CCl_4 \left(\frac{1 \text{ kg}}{1000 \text{ g}} \right) = 0.03896 \text{ kg}$$

$$\text{moles of unknown} = 0.03896 \text{ kg solvent} \left(\frac{0.0856 \text{ mole solute}}{1 \text{ kg solvent}} \right) = 0.00333 \text{ mol unknown}$$

$$\text{molar mass} = \frac{1.08 \text{ g solute}}{0.00333 \text{ mol solute}} = 324 \text{ g/mol}$$

PROBLEMS

1. 1.54 g KOH   2. 130 g/mol   3. 0.058   4. (a) 0.0612   (b) 3.62 m C_2H_5OH
5. 0.291   6. 0.585 m   7. 0.0314   8. 19.2 g   9. 0.023
10. (a) Methanol mole fraction: 0.591, ethanol mole fraction: 0.409. (b) 31.3 m

---

LESSON 8: THERMAL ENERGY IN CHEMICAL REACTIONS

Exercise One

A. q = (4.184 J/g K) (750 g) (359.45 K − 288.55 K)
      = 2.2×10^5 J

B. q = (4.184 J/g K) (150 g) (277.55 K − 297.51 K)
      = -1.3×10^4 J

C.   3260 J = (4.184 J/g K)(135 g)(T_f − 294.55 K)
       5.8 = (T_f − 294.55 K)
       T_f = 300.35 K
       T_f = 27.2°C

Excercise Two

A.   1. endothermic   2. exothermic 3.   exothermic   4. endothermic

B.   A temperature of 25°C, a pressure of 1 atm, and solute concentrations of 1 M.

C.   1.   −(−349 kJ) = +349 kJ
     2.   2(−349 kJ) = −698 kJ
     3.   4(−349 kJ) = −1396 kJ
     4.   −2(−349 kJ) = +698 kJ

Exercise Three

A.   The equation involves $^1/_2$ mole Mg_3N_2.

$$\tfrac{1}{2} \text{ mol } Mg_3N_2 \left(\frac{100.92 \text{ g } Mg_3N_2}{1 \text{ mol } Mg_3N_2} \right) \left(\frac{3.4 \text{ kJ}}{0.50 \text{ g } Mg_3N_2} \right) = 340 \text{ kJ}$$

Because heat is given off, $\Delta H < 0$. Therefore, $\Delta H°_{rxn} = -340$ kJ

B.   The equation involves 1 mole H_2S.

$$1 \text{ mol } H_2S \left(\frac{34.08 \text{ g } H_2S}{1 \text{ mol } H_2S} \right) \left(\frac{4.85 \text{ kJ}}{1.00 \text{ g } H_2S} \right) = 165 \text{ kJ}$$

The reaction is exothermic in the direction the reverse of the way the equation is written. Therefore, $\Delta H°_{rxn}$ = +165 kJ.

C. $0.23 \text{ g Na} \left(\dfrac{1 \text{ mol Na}}{22.99 \text{ g Na}} \right) \left(\dfrac{360 \text{ kJ}}{1 \text{ mol Na}} \right)$ = 3.6 kJ released

Exercise Four

Assume: all heat produced by the reaction is absorbed by the combined solutions,
          the combined solutions have a volume of 300 mL and a mass of 300 grams,
          the heat capacity of the combined solutions is 4.184 J/g K.

q =   (4.184 J/g K) (300 g) (298.95 K − 297.65 K)

$$= \quad 1630 \text{ J released}$$

$$\text{mol NaOH} = \text{mol HF} = 150 \text{ mL} \left(\frac{1 \text{ L}}{1000 \text{ mL}} \right) \left(\frac{0.20 \text{ mol}}{1 \text{ L}} \right) = 0.030 \text{ mol}$$

$$\frac{1630 \text{ J}}{0.030 \text{ mol}} = 53000 \text{ J/mol} = 53 \text{ kJ/mol}$$

Because heat is released by the reaction (the temperature goes up), $\Delta H < 0$. Therefore,

$$\Delta H°_{rxn} = -53 \text{ kJ}$$

Exercise Five

A. 1. $Na(s) + {}^1/_2 N_2(g) + {}^3/_2 O_2(g) \longrightarrow NaNO_3(s)$
   2. $6 C(s) + 3 H_2(g) \longrightarrow C_6H_6(l)$

B. $\Delta H°_f$ is for equation ${}^1/_2 F_2(g) + {}^1/_2 H_2(g) \longrightarrow HF(g)$

$$0.19 \text{ g } F_2 \longrightarrow 360J$$
$${}^1/_2 \text{ mol } F_2 \longrightarrow \text{? kJ}$$

$$\left(\frac{2.68 \text{ kJ}}{0.19 \text{ g } F_2} \right) \left(\frac{38 \text{ g } F_2}{1 \text{ mol } F_2} \right) \left(\frac{\frac{1}{2} \text{ mol } F_2}{\text{equation}} \right) = 268 \text{ kJ for the equation}$$

Because heat is released by the reaction (the temperature goes up), $\Delta H < 0$. Therefore,

$$\Delta H°_{rxn} = \Delta H°_f = -268 \text{ kJ}$$

C.

$$10.0 \text{ g Li} \left(\frac{1 \text{ mol Li}}{6.939 \text{ g Li}} \right) \left(\frac{1 \text{ mol Li}_2O}{2 \text{ mol Li}} \right) \left(\frac{595.8 \text{ kJ}}{1 \text{ mol Li}_2O} \right) = 429 \text{ kJ}$$

Exercise Six

A. $\quad 2 N_2(g) + 4 O_2(g) \longrightarrow 4 NO_2(g) \qquad \Delta H = 2(67.4 \text{ kJ}) = 134.8 \text{ kJ}$

$\quad \underline{4 NO_2(g) + 6 H_2O(l) \longrightarrow 4 NH_3(g) + 7 O_2(g) \qquad \Delta H = -(-1395 \text{ kJ}) = 1395 \text{ kJ}}$

$\quad 2 N_2(g) + 6 H_2O(l) \longrightarrow 4 NH_3(g) + 3 O_2(g) \qquad \Delta H = 134.8 + 1395 \text{ kJ} = 1530 \text{ kJ}$

B. $\Delta H°_{rxn} = 765$ kJ, because the equation's coefficients are half those in part A.

C.

$\quad N_2(g) + 2 O_2(g) \longrightarrow 2 NO_2(g) \qquad\qquad \Delta H = +67.4 \text{ kJ}$

$\quad 2 H_2(g) + O_2(g) \longrightarrow 2 H_2O(l) \qquad\qquad \Delta H = -571.5 \text{ kJ}$

$\quad \underline{2 NO_2(g) + 2 H_2O(l) \longrightarrow N_2H_4(l) + 3 O_2(g) \qquad \Delta H = +554.4 \text{ kJ}}$

$\quad N_2(g) + 2 H_2(g) \longrightarrow N_2H_4(l) \qquad\qquad \Delta H_f = +50.3 \text{ kJ}$

Exercise Seven

A. $\Delta H°_{rxn} = \sum \Delta H°_f \text{ (products)} - \sum \Delta H°_f \text{ (reactants)}$
$= [\, 2 \Delta H°_f(Fe) + 3 \Delta H°_f(CO_2)\,] - [\, \Delta H°_f(Fe_2O_3) + 3 \Delta H°_f(CO)\,]$
$= [\, 2 (0 \text{ kJ}) + 3 (-393.7 \text{ kJ})\,] - [\, (-822.2 \text{ kJ}) + 3 (-110.4 \text{ kJ})\,]$
$= -27.7 \text{ kJ}$

B. $\Delta H°_{rxn} = \sum \Delta H°_f \text{ (products)} - \sum \Delta H°_f \text{ (reactants)}$
$-3535 \text{ kJ} = [\, 5 \Delta H°_f(CO_2) + 6 \Delta H°_f(H_2O)\,] - [\, \Delta H°_f(C_5H_{12}) + 8 \Delta H°_f(O_2)\,]$
$-3535 \text{ kJ} = [\, 5 (-393.7 \text{ kJ}) + 6 (-285.8 \text{ kJ})\,] - [\, \Delta H°_f(C_5H_{12}) + 8 (0 \text{ kJ})\,]$
$-3535 \text{ kJ} = -3683.3 \text{ kJ} - \Delta H°_f(C_5H_{12})$
$\Delta H°_f(C_5H_{12}) = -148 \text{ kJ}$

PROBLEMS

1. -84.8 kJ $\qquad$ 2. 123 kJ $\qquad$ 3. -74.1 kJ $\qquad$ 4. -52 kJ $\qquad$ 5. $+28.4$ kJ
6. -335 kJ $\qquad$ 7. -462 kJ $\qquad$ 8. 27.4°C

LESSON 9: USING THE IDEAL GAS LAW

Exercise One

$P = 200.0$ atm $\qquad$ T : to be found
$V = 50.0$ L $\qquad$ $n = 250.0$ mol

$R = 0.0821$ L-atm/mol-K

$$T = \frac{PV}{nR} = \frac{(200.0 \text{ atm})(50.0 \text{ L})}{(250.0 \text{ mol})(0.0821 \text{ L-atm/mol-K})} = 487 \text{ K}$$

Exercise Two

$P_1 = 25$ atm $\qquad$ $P_2 = ?$ $\qquad\qquad$ $V_1 = V_2$
$T_1 = 22°C$ $\qquad$ $T_2 = -26°C$ $\qquad$ $n_1 = n_2$
$\quad\;\; = 273 + 22$ K $\quad\;\; = 273 - 26$ K
$\quad\;\; = 295$ K $\qquad\;\; = 247$ K

$$\frac{P_1 V_1}{n_1 T_1} = \frac{P_2 V_2}{n_2 T_2} \qquad \frac{P_1}{T_1} = \frac{P_2}{T_2} \qquad \frac{25 \text{ atm}}{295 \text{ K}} = \frac{P_2}{247 \text{ K}} \qquad P_2 = 21 \text{ atm}$$

Exercise Three

$$d = \frac{MP}{RT}$$

For helium: $\quad M = 4.003$ g/mol
$\qquad\qquad\qquad P = 1.0$ atm
$\qquad\qquad\qquad T = 0°C = 273$ K

$$d = \frac{(1.0 \text{ atm})(4.003 \text{ g/mol})}{(0.0821 \text{ L-atm/mol-K})(273 \text{ K})} = 0.18 \text{ g/L}$$

For nitrogen: $\quad M = 28.02$ g/mol

$$P = 55 \text{ torr} \left(\frac{1 \text{ atm}}{760 \text{ torr}} \right) = 0.072 \text{ atm}$$

$$T = 200°C = 473 \text{ K}$$

$$d = \frac{(0.072 \text{ atm})(28.02 \text{ g/mol})}{(0.0821 \text{ L-atm/mol-K})(473 \text{ K})} = 0.052 \text{ g/L}$$

Helium is more dense.

Exercise Four

$$M = \frac{mRT}{PV}$$

$m = 0.527$ g

$T = 98°C + 273 = 371$ K

$$P = 756.3 \text{ torr} \left(\frac{1 \text{ atm}}{760 \text{ torr}} \right) = 0.9951 \text{ atm}$$

$$V = 135 \text{ mL} \left(\frac{1 \text{ L}}{1000 \text{ mL}} \right) = 0.135 \text{ L}$$

$$M = \frac{(0.527 \text{ g})(0.0821 \text{ L-atm/mol-K})(371 \text{ K})}{(0.9951 \text{ atm})(0.135 \text{ L})} = 119 \text{ g/mol}$$

Exercise Five

$$P_{total} = P_{original} + P_{added}$$

Find pressure of gas added:

$$n = 0.500 \text{ mol}$$

$$T = 23°C = 296 \text{ K}$$
$$V = 40.0 \text{ L}$$
$$R = 0.0821 \text{ L-atm/mol-K}$$

$$P = \frac{nRT}{V} = \frac{(0.500 \text{ mol})(0.0821 \text{ L-atm/mol-K})(296 \text{ K})}{40.0 \text{ L}} = 0.304 \text{ atm}$$

$$0.304 \text{ atm}\left(\frac{760 \text{ torr}}{1 \text{ atm}}\right) = 231 \text{ torr}$$

$$P_{total} = 1820 \text{ torr} + 231 \text{ torr} = 2050 \text{ torr}$$

PROBLEMS

1.	19°C	2.	1100 mL	3.	49.1 g/mol	4.	7.95 L
5.	58.2 atm	6.	10.3 L	7.	30°C	8.	32.0 g/mol
9.	60°C						

LESSON 10:   ELECTRONIC STRUCTURE OF ATOMS

Exercise One

A. 2; 2

B. 6; 10

C. 1.   2.   3.   4.

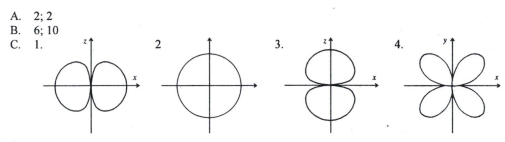

Exercise Two

Be: $1s^2\, 2s^2$

N: $1s^2\, 2s^2\, 2p_x^1\, 2p_y^1\, 2p_z^1$

Na: $1s^2\, 2s^2\, 2p_x^2\, 2p_y^2\, 2p_z^2\, 3s^1$

S: $1s^2\, 2s^2\, 2p_x^2\, 2p_y^2\, 2p_z^2\, 3s^2\, 3p_x^2\, 3p_y^1\, 3p_z^1$

Exercise Three

A. Ti: (1) expanded: $1s^2\, 2s^2\, 2p_x^2\, 2p_y^2\, 2p_z^2\, 3s^2\, 3p_x^2\, 3p_y^2\, 3p_z^2\, 4s^2\, 3d_{xy}^1\, 3d_{xz}^1$

  (2) condensed $1s^2\, 2s^2\, 2p^6\, 3s^2\, 3p^6\, 4s^2\, 3d^2$

  (3) noble-gas notation: [Ar] $4s^2\, 3d^2$

B. As (1) expanded: $1s^2\, 2s^2\, 2p_x^2\, 2p_y^2\, 2p_z^2\, 3s^2\, 3p_x^2\, 3p_y^2\, 3p_z^2\, 4s^2\, 3d_{xy}^2\, 3d_{xz}^2\, 3d_{yz}^2\, 3d_{x^2-y^2}^2\, 3d_{z^2}^2\, 4p_x^1\, 4p_y^1\, 4p_z^1$

  (2) condensed: $1s^2\, 2s^2\, 2p^6\, 3s^2\, 3p^6\, 4s^2\, 3d^{10}\, 4p^3$

  (3) noble-gas notation: [Ar] $4s^2\, 3d^{10}\, 4p^3$

C. Cu (1) expanded: $1s^2\, 2s^2\, 2p_x^2\, 2p_y^2\, 2p_z^2\, 3s^2\, 3p_x^2\, 3p_y^2\, 3p_z^2\, 4s^1\, 3d_{xy}^2\, 3d_{xz}^2\, 3d_{yz}^2\, 3d_{x^2-y^2}^2\, 3d_{z^2}^2$

  (2) condensed: $1s^2\, 2s^2\, 2p^6\, 3s^2\, 3p^6\, 4s^1\, 3d^{10}$

  (3) noble-gas notation: [Ar] $4s^1\, 3d^{10}$

Exercise Four

A. 1. $1s^2\, 2s^2\, 2p^6\, 3s^2\, 3p^6$

  2. $1s^2\, 2s^2\, 2p^6\, 3s^2\, 3p^6$

  3. $1s^2\, 2s^2\, 2p^6$

  4. $1s^2\, 2s^2\, 2p^6$

B. S^{2-} is isoelectronic with K^+.

  Al^{3+} is isoelectronic with F^-.

C. 1. [Ar] $3d^4$

  2. [Kr] $5s^2\, 4d^{10}\, 5p^6$

  3. [Xe] $6s^2\, 5d^{10}\, 4f^{14}$

Exercise Five

A. $3p_y$: n = 3, l = 1, m_l = 0, m_s = +½
    $3p_z$: n = 3, l = 1, m_l = 1, m_s = +½

Note: The m_l values may be 1, 0, or -1, as long as the two given are different, because one electron is in a p_y orbital and the other is in a p_z orbital. The m_s values may be +½ or –½ as long as both are the same (Hund's rule).

B. 5. Br: $1s^2\, 2s^2\, 2p^6\, 3s^2\, 3p^6\, 4s^2\, 3d^{10}\, 4p^5$

PROBLEMS

1. (a) F: $1s^2\, 2s^2\, 2p_x^2\, 2p_y^2\, 2p_z^1$
    (b) Se: $1s^2\, 2s^2\, 2p_x^2\, 2p_y^2\, 2p_z^2\, 3s^2\, 3p_x^2\, 3p_y^2\, 3p_z^2\, 4s^2\, 3d_{xy}^2\, 3d_{xz}^2\, 3d_{yz}^2\, 3d_{x^2-y^2}^2\, 3d_{z^2}^2\, 4p_x^2\, 4p_y^1\, 4p_z^1$
    (c) Ar: $1s^2\, 2s^2\, 2p_x^2\, 2p_y^2\, 2p_z^2\, 3s^2\, 3p_x^2\, 3p_y^2\, 3p_z^2$
    (d) Cr: $1s^2\, 2s^2\, 2p_x^2\, 2p_y^2\, 2p_z^2\, 3s^2\, 3p_x^2\, 3p_y^2\, 3p_z^2\, 4s^1\, 3d_{xy}^1\, 3d_{xz}^1\, 3d_{yz}^1\, 3d_{x^2-y^2}^1\, 3d_{z^2}^1$

2. (a) Ga^{3+}: $1s^2\, 2s^2\, 2p^6\, 3s^2\, 3p^6\, 3d^{10}$
    (b) Zn^{2+}: $1s^2\, 2s^2\, 2p^6\, 3s^2\, 3p^6\, 3d^{10}$
    (c) Se^{2-}: $1s^2\, 2s^2\, 2p^6\, 3s^2\, 3p^6\, 4s^2\, 3d^{10}\, 4p^6$
    (d) Ag^+: $1s^2\, 2s^2\, 2p^6\, 3s^2\, 3p^6\, 3d^{10}\, 4s^2\, 4p^6\, 4d^{10}$

3. (a) Ba: [Xe] $6s^2$
    (b) Sn^{2+}: [Kr] $5s^2\, 4d^{10}$
    (c) Sb: [Kr] $5s^2\, 4d^{10}\, 5p^3$
    (d) Au^{3+}: [Xe] $4f^{14}\, 5d^8$

4. (a) Xe
    (b) Cs^+ or Ba^{2+} or La^{3+} . . .
    (c) Te^{2-} or Sb^{3-} . . .

LESSON 11:   PERIODIC PROPERTIES

Exercise One

A. transition metals: elements in groups IB to VIIIB, such as Fe, Ni.
    alkaline earth metals: elements in group IIA, such as Mg, Ca.
    lanthanides: element nos. 57-71, such as La, Ce.
    halogens: elements in group VIIA, such as Cl, Br.
    inner transition elements: elements no. 57-71 and 89-103, such as La, U.
    alkali metals: elements in group IA, such as Na, K.

B. Mg; Lu; W; He

Exercise Two

A. 1. K_2SO_4 2. Na_2TeO_4 3. Li_2SeO_4

B. 1. $MgWO_4$ 2. SiF_4 3. $NaReO_4$ 4. AsH_3

C. Because fluorine is a non-metal of the second period, it is less likely to be similar to chlorine than is bromine. Therefore, $KBrO_3$ is more likely to be the formula of a compound that exists.

Exercise Three

A. $ZrBr_4(s) + 2\,H_2O(l) \longrightarrow ZrO_2(s) + 4\,HBr(aq)$

B. $2\,RbBrO_3(s) \longrightarrow 2\,RbBr(s) + 3\,O_2(g)$

Exercise Four

A. Ge: $4s^2\, 4p^2$ Te: $5s^2\, 5p^4$
    Ba: $6s^2$ Xe: $5s^2\, 5p^6$

B. Sn; Sc; Na

Exercise Five

A.

	largest	smallest
atomic size	Sn	O
ionization energy	O	Sn
electronegativity	O	Sn

B. ionization energy: F
electronegativity: Mg
atomic size: Sn

Exercise Six

A. melting point estimation: $\dfrac{814°C + 547°C}{2} = 680°C$

boiling point estimation: $\dfrac{1676°C + 1265°C}{2} = 1470°C$

B. boiling point difference = $-55°C - (-17°C) = -38°C$

Because the boiling point decreases by 38°C from SbH_3 to AsH_3, estimate that the boiling point will decrease 38°C from AsH_3 to PH_3:

estimated boiling point of PH_3: $-55°C - 38°C = -93°C$

C. The heat of vaporization increases by $230 - 180 = 50$ kJ/mole from Pb to Sn. Estimate that it will increase by 50 kJ/mole from Sn to Ge:

estimated heat of vaporization of Ge = 230 kJ/mole + 50 kJ/mole = 280 kJ/mol

PROBLEMS

1. iodine

2. $Cd(s) + 2\ HBr(aq) \longrightarrow CdBr_2(aq) + H_2(g)$

3. a. BaSe   b. CaO   c. SrTe   d. MgSe   e. BeS

4. $GeCl_4$ estimated m.p.: $-51.5°C$ (actual value is $-49.5°C$)
$SnCl_4$ estimated b.p.: $110.4°C$ (actual value is $114.1°C$)

5. $2\ K(s) + 2\ H_2O(l) \longrightarrow 2\ KOH(aq) + H_2(g)$

6. a. O < S < Si < Ge   b. Na < Mg < F < Ne
    c. C < B < Al < K   d. K < Li < C < N

7. a. sulfur   b. radium   c. nitrogen   d. copper   e. ruthenium

---

LESSON 12:   LEWIS STRUCTURES AND THE OCTET RULE

Exercise One

A. $5 + 3(1) = 8$ B. $2(4) + 4(1) = 12$
C. $4 + 2(6) = 16$ D. $5 + 3(7) = 26$

Exercise Two

A. 1. $4 + 4(7) = 32$ 2.

$$\begin{array}{c} Cl \\ | \\ Cl-C-Cl \\ | \\ Cl \end{array}$$

3.

$$\begin{array}{c} :\ddot{Cl}: \\ | \\ :\ddot{Cl}-C-\ddot{Cl}: \\ | \\ :\ddot{Cl}: \end{array}$$

4. 32 electrons, same as in 1.

B. $5 + 3(1) = 8$

$$\begin{array}{c} H-\ddot{N}-H \\ | \\ H \end{array}$$

Exercise Three

A. Step 1. $5 + 2(1) + 7 = 14$

Step 2.   H— N— F
                |
                H

Step 3.   H— N̈— F̈ :
                |
                H

Step 4.   This uses 14 electrons.

B. Step 1: $4 + 6 + 4(1) = 14$ electrons

Step 2.
                    H
                    |
            H— C— O— H
                    |
                    H

Step 3.
                    H
                    |
            H— C— Ö— H
                    |
                    H

Step 4.   This uses 14 electrons.

Exercise Four

A. Step 1. $5 + 6 + 7 = 18$ electrons
    Step 2.   N forms the greatest number of bonds. Therefore, it goes in the center.

                    F — N — O

    Step 3.   : F̈— N̈— Ö :

    Step 4.   Step 3 uses 20 electrons, which is 2 too many. Add one bond.
               Fluorine forms only one bond, so added bond must be between N and O.

                    F — N = O

    Give each atom 8 electrons.   : F̈— N̈= Ö

    This uses 18 electrons, which is the correct number.

B. Step 1. $4 + 6 = 10$ electrons   Step 2.   C— O   Step 3.   : C̈— Ö :

    Step 4.   This uses 14 electrons, which is 4 too many. Add 2 more bonds.
                    C ≡ O
               Give each atom 8 electrons.   : C ≡ O :
               This uses 10 electrons, which is the correct number.

Exercise Five

Step 1. $6 + 3(6) = 24$ electrons
Step 2.   Put the unique atom, S, in the center.   O— S— O
                                                          |
                                                          O

Step 3.   Give each atom 8 electrons.
            : Ö— S̈— Ö :
                    |
                  : O :

Step 4.   This uses 26 electrons, which is 2 too many. Add one more bond. The bond can go between the sulfur
           atom and any one of the oxygen atoms. This yields resonance structures.

            O= S— O   O— S— O   O— S= O
                | ‖ |
                O   O   O

Give each atom 8 electrons.

            Ö= S— Ö :   : Ö— S— Ö: : Ö— S= Ö
                | ‖ |
              : O :   : O :   : O :

In each resonance structure, 24 electrons are used, which is the correct number.

Exercise Six

A. Step 1.   6 + 4(6) + 2 = 32 electrons (2 is for the charge)

Step 2.   O—S—O

Step 3.   :O—S—O:

Step 4.   This uses 32 electrons, which is the correct number. Because this is an ion, add square brackets and charge.

$$\left[\begin{array}{c} :\ddot{O}: \\ | \\ :\ddot{O}-S-\ddot{O}: \\ | \\ :\ddot{O}: \end{array} \right]^{2-}$$

B. Step 1.   5 + 2(6) + 1 = 18 electrons (1 is from the charge)

Step 2.   O—N—O   Step 3.   :O—N—O:

Step 4.   This uses 20 electrons, which is 2 too many. Add one bond. Because the bond can be added in two equivalent positions, two resonance structures must be drawn.

O=N—O   ⟷   O—N=O

$\ddot{O}=\ddot{N}-\ddot{O}:$ ⟷ $:\ddot{O}-\ddot{N}=\ddot{O}$

These structures use 18 electrons, which is the correct number. Because they represent ions, square brackets and charge must be added to them.

$$\left[\ddot{O}=\ddot{N}-\ddot{O}: \right]^{-} \quad ⟷ \quad \left[:\ddot{O}-\ddot{N}=\ddot{O} \right]^{-}$$

Exercise Seven

[Na]⁺

A. [Na]⁺ $\left[\begin{array}{c} :\ddot{O}: \\ | \\ :\ddot{O}-P-\ddot{O}: \\ | \\ :O: \end{array} \right]^{3-}$

[Na]⁺

B. $\left[:\ddot{O}-H \right]^{-}$ [Ba]²⁺ $\left[:\ddot{O}-H \right]^{-}$

Exercise Eight

A.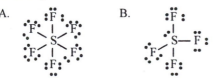

B.   :F:
      |
   F—S—F:
      |
     :F:

Exercise Nine

A.   :Br—O: ⟷   :Br—O:

B.   :O—Cl—O: ⟷   :O—Cl—O: ⟷   :O—Cl—O:

PROBLEMS

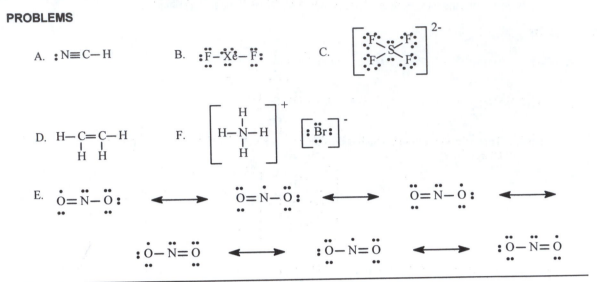

LESSON 13:   OXIDATION NUMBERS & FORMAL CHARGE

Exercise One

A. ox. no. hydrogen = +1   ox. no. chlorine = +5   ox. no. oxygen = –2
B. ox. no. sulfur = +6   ox. no. oxygen = –2
C. ox. no. sodium = +1   ox. no. sulfur = +2   ox. no. oxygen = –2
D. ox. no. mercury = +1
E. ox. no. potassium = +1   ox. no. phosphorus = +5   ox. no. fluorine = –1
F. ox. no. hydrogen = +1   ox. no. carbon = 0   ox. no. oxygen = –2

Exercise Two

A. ox. no. aluminum = +3   ox. no. bromine = –1
B. ox. no. arsenic = +3   ox. no. nitrogen = –3
C. ox. no. carbon = +4   ox. no. chlorine = –1
D. ox. no. sodium = +1   ox. no. carbon = +2   ox. no. nitrogen = –3
E. ox. no. potassium = +1   ox. no. tin = +4   ox. no. sulfur = –2

Exercise Three

A. ox. no. oxygen = +2   ox. no. fluorine = –1
B. ox. no. magnesium = +2   ox. no. chlorine = –1
C. ox. no. magnesium = +2   ox. no. chlorine = +5   ox. no. oxygen = –2
D. ox. no. magnesium = +2   ox. no. hydrogen = –1
E. ox. no. potassium = +1   ox. no. oxygen = – ½

Exercise Four

A. highest: +5; lowest: –3
B. These do not exist: H_2PO_4 Ca_2P
C. These are possible: VO VO_2 V_2O_5
D. ox. no. sodium = +1   ox. no. sulfur = +6   ox. no. oxygen = –1¾
    Applying the rules directly yields an oxidation number of +7 for sulfur, which is too high. Assigning the highest possible number to sulfur, +6, gives –1¾ for oxygen.

Exercise Five

A. 1. Formal charge of all atoms in H_2O is zero.
    2. Formal charge of all atoms in H–C≡N is zero.
    3. Formal charge of oxygen atom on left: zero
        Formal charge of sulfur atom: +1

Formal charge of oxygen atom on right: –1
B.   From left to right: +5   0   0   +5.
The average is 10/4, which is 2.5.

Exercise Six

A.   Ö = N̈ — F̈:   :Ö — N̈ = F̈
       0   0   0   –1   0   +1

Because all formal charges are zero in the structure on the left, it more accurately represents the distribution of electrons in the ONF molecule.

B.   [:Ö–S̈–Ö: with :Ö: top and :O: bottom]²⁻   In this structure, the formal charge of S is +2, and each O is –1.

[structure with charges –1, 2– charge]   In this structure, the O atoms above and below S have a formal charge of –1. The other two O atoms and the S atom have a formal charge of zero. Because the formal charges are lower in this structure, it is a better description of the arrangement of the electrons in the sulfate ion than the first structure. Of course, this is only one of six similar resonance forms.

PROBLEMS

1.   (a)   nitrogen: +1   oxygen: –2
     (b)   nitrogen: +4   oxygen: –2
     (c)   hydrogen: +1   chlorine: +3   oxygen: –2
     (d)   chlorine: +1   oxygen: –2
     (e)   nitrogen: +2   oxygen: –2
     (f)   nitrogen: +5   oxygen: –2
     (g)   nitrogen: –2   hydrogen: +1
     (h)   carbon: –2   hydrogen: +1   oxygen: –2
     (i)   carbon: +2   hydrogen: +1   oxygen: –2
     (j)   nitrogen: –3   hydrogen: +1   iodine: –1
     (k)   lithium: +1   phosphorus: +5   oxygen: –1¾
     (l)   magnesium: +2   chlorine: +1   oxygen: –2

2.   H– C≡N :   H– N≡C :
     formal charges: 0   0   0   0 +1 -1
     The formal charges are nearest zero in the structure on the left, so it is more likely.

3.   -1   0   -1
          :Ö: :O: :Ö:
     H– Ö– N= Ö   ←→   H– Ö– N= Ö :   ←→   H– Ö= N– Ö :
       0   0 +1   0   0   0 +1 -1   0 +1 +1 -1
     In the structure at the right, the formal charges are further from zero than in the other two structures. Therefore, it contributes the least to the bonding.

LESSON 14:   GEOMETRY OF SIMPLE MOLECULES

Exercise One

A.   The Cl–Si–Cl bond angle is 109.5°.

B. The geometry of CO_2 is linear.   O—C—O

Exercise Two

A. I—P—I   Geometry of molecule: pyramidal, like NH_3.
                 Predicted I—P—I bond angles: 109°

B. Cl—S—Cl   Geometry of molecule: non-linear, like H_2O.
                 Predicted Cl—S—Cl bond angles: 109°

C. Cl—C—H   Geometry of molecule: tetrahedral, like CH_4.
                 Predicted Cl—C—Cl and Cl—C—H bond angles: 109°

Exercise Three

A. O=N—F   Three electron groups on central atom, 2 atoms bonded.
                 Geometry: non-linear. Bond angles: about 120°.

B. H—C≡N   Two electron groups on central atom, 2 atoms bonded.
                 Geometry: linear. Bond angles: 180°.

Exercise Four

A. $\left[\begin{array}{c} O \\ O-S-O \\ O \end{array}\right]^{2-}$ Four electron groups on central atom, 4 atoms bonded.
                 Geometry: tetrahedral. Bond angles: 109°.

B. $\left[\begin{array}{c} O-S-O \\ O \end{array}\right]^{2-}$ Four electron groups on central atom, 3 atoms bonded.
                 Geometry: pyramidal. Bond angles: about 109°.

C. $\left[O-N=O \right]^{-}$ Three electron groups on central atom, 2 atoms bonded.
                 Geometry: non-linear. Bond angle: about 120°.

D. $\left[\begin{array}{c} H \\ H-N-H \\ H \end{array}\right]^{+}$ Four electron groups on central atom, 4 atoms bonded.
                 Geometry: tetrahedral. Bond angles: 109°.

Exercise Five

A. (SF₆ structure)   Six electron groups on central atom, 6 atoms bonded.
                 Geometry: octahedral. Bond angles: 90°.

B. F—Br—F with F   Five electron groups on central atom, 3 atoms bonded.
                 Geometry: T-shaped. Bond angles: about 90°.

C. (XeF₄ structure)   Six electron groups on central atom, 4 atoms bonded.
                 Geometry: square. Bond angles: 90°.

D. $\left[I-I-I \right]^{-}$ Five electron groups on central atom, 2 atoms bonded.
                 Geometry: linear.   Bond angels: 180°.

PROBLEMS

1. (a) non-linear (bent)   (b) linear
    (c) non-linear (bent)   (d) non-linear (bent)

2. (a) triangular   (b) triangular pyramidal
    (c) tetrahedral   (d) triangular pyramidal

3. (a) 120 degrees   (b) 109 degrees
    (c) 120 degrees   (d) 180 degrees

LESSON 15:   VALENCE BONDS AND HYBRID ORBITALS

Exercise One

A. $1s$ from H + $4p_z$ from Br
B. $5p_z$ from I + $3p_z$ from Cl

Exercise Two

Each H–C bond is formed by overlap of $1s$ from H and sp^3 from C.

Exercise Three

A. Each O–H bond is formed by overlap of sp^3 from O and $1s$ from H.
B. Because the H–S–H bond angle is close to the angle of the p orbitals, the orbitals of sulfur do not hybridize. One S–H bond is formed by overlap of the $3p_y$ orbital of S with the $1s$ orbital of one H. The other S–H bond is formed by overlap of the $3p_z$ of S with the $1s$ orbital of the other H.

Exercise Four

A. sigma: $4p_z$ from Br + sp^3 from C   B. sigma: $1s$ from H + sp^3 from C
C. sigma: $1s$ from H + sp^2 from C   D. sigma: sp^3 from C + sp^2 from C
E. sigma: sp^2 from C + sp^2 from C
    pi: $2p_z$ from C + $2p_z$ from C

Exercise Five

A. two   B. sp
C. 1. sigma: $1s$ from H + sp from C
    2. sigma: sp from C + sp from C
       pi: $2p_y$ from C + $2p_y$ from C
       pi: $2p_z$ from C + $2p_z$ from C
    3. sigma: sp from C + $1s$ from H

Exercise Six

A. Each Se–F bond is formed by overlap of sp^3d^2 of Se with sp^3 (or $2p_z$) of F.
B. Each Sb–I bond is formed by overlap of sp^3d of Sb with $5p_z$ of I.

Exercise Seven

A. H—$\overline{\underline{O}}$—$\overline{\underline{O}}$—H sp^3

B. $\left[I\overline{\underline{O}}-\overline{N}=\overline{\underline{O}} \right]^{-}$ sp^2

C. $I\overline{F}-\overset{\displaystyle I\overline{F}I}{\underset{\displaystyle I\overline{F}I}{Xe}}-\overline{F} I$ sp^3d^2

D. $\left[I\underline{\overline{C}}I-\underline{\overline{I}}-\underline{\overline{C}}lI \right]^{-}$ sp^3d

E. $\left[I\overline{\underline{O}}-\overset{\displaystyle IOI}{\underset{}{C}}-\overline{\underline{O}}I \right]^{2-}$ sp^2

F. $\left[\begin{array}{c} H \\ | \\ H-N-H \\ | \\ H \end{array} \right]^{+}$ sp^3

PROBLEMS

1. A. (a) sp (b) linear   (c) $1s$ from H + sp from C
    (d) σ: sp from C + sp from N
         π: $2p_y$ from C + $2p_y$ from N
         π: $2p_z$ from C + $2p_z$ from N

    B. (a) sp^2 (b) triangular   (c) $1s$ from H + sp^2 from C
    (d) σ: sp^2 from C + sp^2 from O
         π: $2p_y$ from C + $2p_y$ from O

    C. (a) sp (b) linear   (c) $1s$ from H + sp from C
    (d) σ: sp from C + sp from C
         π: $2p_y$ from C + $2p_y$ from C
         π: $2p_z$ from C + $2p_z$ from C

2. $1s$ from H with $4p$ from Br; a σ bond

3. (a) sp^3d (b) sp^3 (c) sp^2 (d) sp^3

LESSON 16: COLLIGATIVE PROPERTIES OF SOLUTIONS

Exercise One

A. $50.0 \text{ g NaCl}\left(\dfrac{1 \text{ mol NaCl}}{58.44 \text{ g NaCl}}\right) = 0.8556 \text{ mol NaCl}$

$1.00 \text{ kg H}_2\text{O}\left(\dfrac{1000 \text{ g}}{1 \text{ kg}}\right)\left(\dfrac{1 \text{ mol H}_2\text{O}}{18.02 \text{ g H}_2\text{O}}\right) = 55.49 \text{ mol H}_2\text{O}$

For NaCl, $i = 2$.

$X_{\text{H}_2\text{O}} = \dfrac{55.49 \text{ mol H}_2\text{O}}{55.49 \text{ mol H}_2\text{O} + 2(0.8556 \text{ mol NaCl})} = 0.9701$

$P_{\text{H}_2\text{O}} = (0.9701)(19.8 \text{ mm Hg}) = 19.2 \text{ mm Hg}$

B. 1. $150.0 \text{ g C}_6\text{H}_6\left(\dfrac{1 \text{ mol C}_6\text{H}_6}{78.11 \text{ g C}_6\text{H}_6}\right) = 1.920 \text{ mol C}_6\text{H}_6$

       $150.0 \text{ g C}_5\text{H}_{12}\left(\dfrac{1 \text{ mol C}_5\text{H}_{12}}{72.15 \text{ g C}_5\text{H}_{12}}\right) = 2.079 \text{ mol C}_5\text{H}_{12}$

       For both C_6H_6 and C_5H_{12}, $i = 1$.

       $X_{\text{C}_6\text{H}_6} = \dfrac{1.920 \text{ mol C}_6\text{H}_6}{1.920 \text{ mol C}_6\text{H}_6 + 2.079 \text{ mol C}_5\text{H}_{12}} = 0.4801$

       $P_{\text{C}_6\text{H}_6} = (0.4801)(77.4 \text{ mm Hg}) = 37.2 \text{ mm Hg}$

   2. $X_{\text{C}_5\text{H}_{12}} = \dfrac{2.079 \text{ mol C}_5\text{H}_{12}}{1.920 \text{ mol C}_6\text{H}_6 + 2.079 \text{ mol C}_5\text{H}_{12}} = 0.5199$

       $P_{\text{C}_5\text{H}_{12}} = (0.5199)(442 \text{ mm Hg}) = 230 \text{ mm Hg}$

   3. $P_{\text{total}} = P_{\text{C}_6\text{H}_6} + P_{\text{C}_5\text{H}_{12}} = 37.2 + 230 = 267 \text{ mm Hg}$

Exercise Two

$$40.0 \text{ g } C_6H_6O_2 \left(\frac{1 \text{ mol } C_6H_6O_2}{110.1 \text{ g } C_6H_6O_2} \right) = 0.363 \text{ mol } C_6H_6O_2$$

$$150.0 \text{ g } CCl_4 \left(\frac{1 \text{ kg}}{1000 \text{ g}} \right) = 0.1500 \text{ kg } CCl_4$$

$$m = \frac{0.363 \text{ mol}}{0.1500 \text{ kg}} = 2.42 \text{ m}$$

For $C_6H_6O_2$, $i=1$. For CCl, $k_b = 5.02$ K/m and $k_f = 29.8$ K/m.

$\Delta T_b = (5.02 \text{ K/m})(2.42 \text{ m})(1) = 12.1 \text{ K} = 12.1°C$ and boiling pt $= 76.8°C + 12.1°C = 88.9°C$

$\Delta T_f = (29.8 \text{ K/m})(2.42 \text{ m})(1) = 72.1 \text{ K} = 72.1°C$ and freezing pt $= -22.8°C - 72.1°C = -94.9°C$

Exercise Three

$\Delta T_f = 5.51°C - (-0.37°C) = 5.88°C = 5.88 \text{ K}$

$k_f = 5.12$ K/m

$i = 1$

$$m = \frac{5.88 \text{ K}}{(5.12 \text{ K/m})(1)} = 1.15 \text{ m}$$

$$\text{moles solute} = 0.0200 \text{ kg solvent} \left(\frac{1.15 \text{ mol solute}}{1 \text{ kg solvent}} \right) = 0.0230 \text{ mole}$$

$$\text{molar mass} = \frac{2.14 \text{ g solute}}{0.0230 \text{ mol solute}} = 93.0 \text{ g/mol}$$

Exercise Four

$\Delta T_f = 0°C - (-0.56°C) = 0.56°C = 0.56 \text{ K}$

$k_f = 1.86$ K/m

$m = 0.10$

$$i = \frac{0.56 \text{ K}}{(1.86 \text{ K/m})(0.10\text{m})} = 3.0$$

Equation (1) is better, because it predicts a mole number of 3.

Exercise Five

For $(NH_4)_2SO_4$, $i = 3$.

$$100.0 \text{ g } (NH_4)_2SO_4 \left(\frac{1 \text{ mol } (NH_4)_2SO_4}{132.1 \text{ g } (NH_4)_2SO_4} \right) = 0.7570 \text{ mol } (NH_4)_2SO_4$$

$$250.0 \text{ mL} \left(\frac{1 \text{ L}}{1000 \text{ mL}} \right) = 0.2500 \text{ L}$$

$$M = \frac{0.7570 \text{ mol}}{0.2500\text{L}} = 3.028 \text{ M}$$

$R = 0.0821$ L-atm/mol-K $\qquad$ $T = 0°C + 273 = 273$ K

$\Pi = (3)(3.028 \text{ mol/L})(0.0821 \text{ L-atm/mol-K})(273 \text{ K}) = 204 \text{ atm}$

Exercise Six

R = 0.0821 L-atm/mol-K

$$\Pi = 7.43 \text{ mm Hg} \left(\frac{1 \text{ atm}}{760 \text{ mm Hg}} \right) = 9.78 \times 10^{-3} \text{ atm}$$

$i = 1$ T = 22°C + 273 = 295 K

$$M = \frac{9.78 \times 10^{-3} \text{ atm}}{(1)(0.0821 \text{ L-atm/mol-K})(295 \text{ K})} = 4.04 \times 10^{-4} \text{ mol/L}$$

$$\text{moles solute} = 0.0500 \text{ L solution} \left(\frac{4.04 \times 10^{-4} \text{ mol solute}}{1 \text{ L solvent}} \right) = 2.02 \times 10^{-5} \text{ mol}$$

$$\text{molar mass} = \frac{2.463 \text{ g solute}}{2.02 \times 10^{-5} \text{ mol solute}} = 1.22 \times 10^{5} \text{ g/mol}$$

PROBLEMS

1. 83.7°C	2. 670 g
3. 42 mm Hg	4. 387 g/mol
5. 4.7 K/m	6. (a) 22.7 torr (b) 101.45°C (c) –5.17°C
7. 13,500 g	8. 108 g/mol
9. –0.37 °C	10. 428 torr

LESSON 17:   EQUILIBRIUM IN REACTIONS OF GASES

Exercise One

A. The equation is not balanced as given. When the equation is balanced as
$$CH_4(g) + 2 H_2S(g) \rightleftharpoons 4 H_2(g) + CS_2(g)$$

$$K = \frac{[H_2]^4 [CS_2]}{[CH_4] [H_2S]^2}$$

B. $K = \dfrac{[CO]^{16} [H_2O]^6}{[Si(CH_3)_4] [CO_2]^{12}}$ C. $K = [CO_2]$ D. $K = \dfrac{[SO_2]^2}{[O_2]^3}$

Exercise Two

1. Write the equilibrium constant expression.

$$K = \frac{[NH_3]^2}{[N_2] [H_2]^3}$$

2. Arrange the data in tabular form under the equation.

	$N_2(g)$	+	$3 H_2(g)$	$\rightleftharpoons$	$2 NH_3(g)$
initial conc	2.5		2.5		0
change in conc	–x		–3x		+2x
equil conc	2.5–x		2.5–3x		2x

3. Find the values of the equilibrium concentrations.
    1.0 mol NH_3 at equilibrium.
      Therefore, $2x = 1.0$ mol/L
          $x = 0.50$ mol/L
        $3x = 1.5$ mol/L
    Then, the equilibrium concentrations are
     $[N_2] = 2.5 - 0.50 = 2.0$ mol/L

$[H_2] = 2.5 - 1.5 = 1.0$ mol/L
$[NH_3] = 1.0$ mol/L

4. Calculate the value of K.

$$K = \frac{[NH_3]^2}{[N_2][H_2]^3} = \frac{(1.0)^2}{(2.0)(1.0)^3} = 0.50$$

Exercise Three

1. Write the equilibrium constant expression.

$$K = \frac{[H_2O][CO]}{[H_2][CO_2]} = 0.77$$

2. Arrange the data in tabular form.

	$H_2(g)$	+	$CO_2(g)$	$\rightleftharpoons$	$H_2O(g)$	+	$CO(g)$
initial moles	6.0		6.0		0		0
initial conc	0.60		0.60		0		0
change in conc	$-x$		$-x$		$+x$		$+x$
equil conc	$0.60-x$		$0.60-x$		x		x

3. Substitute the equilibrium concentrations into the K expression, and solve for x.

$$\frac{(x)(x)}{(0.60-x)(0.60-x)} = 0.77$$

$$\frac{(x)^2}{(0.60-x)^2} = 0.77$$

Take the square root of both sides.

$$\frac{(x)}{(0.60-x)} = 0.877$$

Solve for x.

$$x = 0.877 \, (0.60 - x)$$

$$x = 0.526 - 0.877 \, x$$

$$1.877 \, x = 0.526$$

$$x = \frac{0.526}{1.877} = 0.28 \text{ mol/L}$$

4. Determine the values of the equilibrium concentrations and amounts.
    $[H_2] = [CO_2] = 0.60 - x = 0.60 - 0.28 = 0.32$ mol/L.
    moles of H_2 and moles of $CO_2 = 0.32$ mol/L $\times$ 10.0 L = 3.2 moles

    $[H_2O] = [CO] = 0.28$ mol/L
    moles of H_2O and moles of CO = 0.28 mol/L $\times$ 10.0 L = 2.8 moles

Exercise Four

1. Write the equilibrium constant expression and calculate its value.

$$K = \frac{[HI]^2}{[H_2][I_2]} = \frac{(5.0)^2}{(0.71)(0.71)} = 50$$

2. Summarize the data in tabular form.

	$H_2(g)$	+	$I_2(g)$	$\rightleftharpoons$	2 HI(g)
initial conc	0.71		0.71		5.0
added conc					1.0
change in conc	$+x$		$+x$		$-2x$
equil conc	$0.71+x$		$0.71+x$		$6.0-2x$

3. Substitute the equilibrium concentrations into the K expression and solve for x.

$$\frac{(6.0-2x)^2}{(0.71+x)(0.71+x)} = 50$$

$$\frac{(6.0-2x)^2}{(0.71+x)^2} = 50$$

$$\frac{(6.0-2x)}{(0.71+x)} = 7.1$$

$$6.0-2x = (0.71+x)\,7.1$$
$$6.0-2x = 5.0+7.1\,x$$
$$9.1\,x = 1.0$$
$$x = 0.11 \text{ mol/L}$$

4. Determine the values of the equilibrium concentrations and amounts.

$$[HI] = 6.0 - 2(0.11) = 5.8 \text{ mol/L}$$
$$\text{moles HI} = (5.8 \text{ mol/L})\,(1.00 \text{ L}) = 5.8 \text{ mol}$$
$$[H_2] = [I_2] = 0.71 + 0.11 = 0.82 \text{ mol/L}$$
$$\text{moles } H_2 = \text{moles } I_2 = (0.82 \text{ mol/L})\,(1.00 \text{ L}) = 0.82 \text{ mol}$$

Exercise Five

A. 1. forward   2. reverse
B. 1. forward   2. forward   3. reverse
C. 1. forward   2. no effect

Exercise Six

$K_p = K(RT)^x$

$$\begin{aligned}
x &= 2 - 4 = -2 \\
T &= 400°C = 673 \text{ K} \\
R &= 0.0821 \text{ L-atm/mol-K} \\
K &= 0.50
\end{aligned}$$

$$K_p = (0.50)\,[\,(0.0821)\,(673)\,]^{-2} = 1.6 \times 10^{-4}$$

PROBLEMS

1. $[H_2] = [I_2] = 0.45$ mol/L and $[HI] = 3.10$ mol/L

2. At 100°C, $K = \dfrac{[HBr]^2}{[H_2]\,[Br_2]}$.   At 0°C, $K = \dfrac{[HBr]^2}{[H_2]}$. At -100°C, $K = \dfrac{1}{[H_2]}$.

3. 0.156
4. H_2: 0.57 mol, CO_2: 1.57 mol, H_2O: 0.43 mol, CO: 3.43 mol
5. 1.5 mol PCl_5
6. $K = 34$
7. $[H_2] = [I_2] = 0.218$ mol/L and $[HI] = 1.564$ mol/L
8. 0.26 mol/L added
9. $[PCl_5] = 11.25$ mol/L, $[PCl_3] = [Cl_2] = 0.75$mol/L

LESSON 18:   ACIDS AND BASES

Exercise One

A. $[OH^-] = \dfrac{K_w}{[H^+]} = \dfrac{1.0 \times 10^{-14}}{2.36 \times 10^{-5}} = 4.2 \times 10^{-10}$ M

Because $[H^+] > [OH^-]$, this is an acidic solution.

B.   At 65°C, $K_w = 1.0 \times 10^{-13}$. In pure water, $[H^+] = [OH^-]$.
Therefore, $[H^+][OH^-] = [H^+][H^+] = 1.0 \times 10^{-13}$.
$$[H^+] = 3.2 \times 10^{-7} \text{ M.}$$

Exercise Two

A.   moles of $HClO_4$ = 0.50 g $HClO_4 \left(\dfrac{1 \text{ mol } HClO_4}{100.4 \text{ g } HClO_4} \right)$ = 0.0050 mol $HClO_4$

conc of $HClO_4 = \dfrac{0.0050 \text{ mol } HClO_4}{0.750 \text{ L}} = 6.7 \times 10^{-3}$ M

Because $HClO_4$ is a strong acid, it is completely dissociated into ions.
$$HClO_4 \longrightarrow H^+ + ClO_4^-$$
Therefore, $[H^+] = 6.7 \times 10^{-3}$ M.

$[OH^-] = \dfrac{K_w}{[H^+]} = \dfrac{1.0 \times 10^{-14}}{6.7 \times 10^{-3}} = 1.5 \times 10^{-12}$ M

B. $[H^+] = \dfrac{K_w}{[OH^-]} = \dfrac{1.0 \times 10^{-14}}{1.6 \times 10^{-11}} = 6.3 \times 10^{-4}$ M

Because $[H^+] > [OH^-]$, this solution is an acid.
The concentration of CH_2O_2 is moles of CH_2O_2 divided by liters of solution.

moles of CH_2O_2 = 0.10 g $CH_2O_2 \left(\dfrac{1 \text{ mol } CH_2O_2}{46.03 \text{ g } CH_2O_2} \right)$ = 2.2×10^{-3} mol CH_2O_2

molarity of $CH_2O_2 = \left(\dfrac{2.2 \times 10^{-3} \text{ mol } CH_2O_2}{1.0 \text{ L}} \right)$ = 2.2×10^{-3} M CH_2O_2

Because the concentration of H^+ is less than the concentration of CH_2O_2, not all of the CH_2O_2 has formed ions. Therefore, it is a weak acid.

Exercise Three

A.   pOH = $-\log [OH^-]$ = $-\log (4.2 \times 10^{-5})$ = 4.38
pH = 14.00 − pOH = 14.00 − 4.38 = 9.62

B.   KOH is a strong base. Therefore, $[OH^-]$ is equal to the concentration of KOH.

molarity of KOH = $\dfrac{0.36 \text{ g KOH} \left(\dfrac{1 \text{ mol KOH}}{56.11 \text{ g KOH}} \right)}{0.25 \text{ L}}$ = 2.6×10^{-2} M OH^-

pOH = $-\log (2.6 \times 10^{-2})$ = 1.59
pH = 14.00 − 1.59 = 12.41

Exercise Four

0.36 g NaOH $\left(\dfrac{1 \text{ mol NaOH}}{40.00 \text{ g NaOH}} \right)$ = 9.0×10^{-3} mol NaOH

0.24 L HCl $\left(\dfrac{0.055 \text{ mol HCl}}{1 \text{ L}} \right)$ = 1.3×10^{-2} mol HCl

There is an excess of HCl. Therefore, the mixture will be acidic.
The excess of HCl is 1.3×10^{-2} mol − 9.0×10^{-3} mol = 4×10^{-3} mol

$[H^+] = \dfrac{4 \times 10^{-3} \text{ mol } H^+}{0.24 \text{ L}}$ = 2×10^{-2} M H^+

pH = $-\log (2 \times 10^{-2})$ = 1.7

Exercise Five

A. Cl^- HPO_4^{2-} $CCl_3CO_2^-$
B. H_3PO_4 NH_4^+ HCO_3^-
C. ClO_2^- + H_2O $\rightleftharpoons$ $HClO$ + OH^-
     base   acid   acid   base

       HSO_4^- + H_2O $\rightleftharpoons$ H_3O^+ + SO_4^2
       acid   base   acid   base

       NH_4^+ + NH_2^- $\rightleftharpoons$ NH_3 + NH_3
       acid   base   acid   base
                                    or
                           base   acid

Exercise Six

A. Lewis structures: $[Ag]^+$ $[|C{\equiv}N|]^-$ $[|N{\equiv}C{-}Ag{-}C{\equiv}N|]^-$
     CN^- donates electron pair to Ag^+.
     CN^- is the Lewis base, Ag^+ is the Lewis acid.
B. Lewis structures:

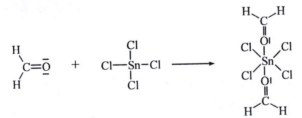

     O atom in CH_3OCH_3 donates a pair of electrons to B atom in BCl_3.
     CH_3OCH_3 is Lewis base, BCl_3 is Lewis acid.
C. Lewis structures:

     The O from H_2CO donates a pair of electrons to the Sn from $SnCl_4$.
     H_2CO is a Lewis base, $SnCl_4$ is a Lewis acid.

PROBLEMS

1. (a) 1.70   (b) 12.30   (c) 12.60
2. (a) 1.5×10^{-7} M   (b) 5×10^{-3} M   (c) 4.2×10^{-6} M   (d) 3.0×10^{-11} M
3. (a) 1.336   (b) 1.165   (c) 7.000   (d) 11.903
4. 7.9 mL
5. (a) OH^-, NH_2^-, $C_2O_4^{2-}$, $Fe(H_2O)_5(OH)^{2+}$ (b) H_3O^+, NH_4^+, HO^-, H_2S, $HN(CH_3)_3^+$
6. (a) base   (b) acid   (c) acid   (d) base

LESSON 19:   IONIZATION OF WEAK ACIDS AND BASES

Exercise One

A. $HF \rightleftharpoons H^+ + F^-$ $K_a = \dfrac{[H^+][F^-]}{[HF]}$

B. $C_5H_5N + H_2O \rightleftharpoons C_5H_5NH^+ + OH^-$ $K_b = \dfrac{[C_5H_5NH^+][OH^-]}{[C_5H_5N]}$

C. $HC_6H_5O \rightleftharpoons H^+ + C_6H_5O^-$ $K_a = \dfrac{[H^+][C_6H_5O^-]}{[HC_6H_5O]}$

D. $CH_3NH_2 + H_2O \rightleftharpoons CH_3NH_3^+ + OH^-$ $K_b = \dfrac{[CH_3NH_3^+][OH^-]}{[CH_3NH_2]}$

Exercise Two

	NH_3	+	H_2O	$\rightleftharpoons$	NH_4^+	+	OH^-
initial conc.	0.50		–		0		0
conc. change	$-x$		–		$+x$		$+x$
equil. conc.	$0.50-x$		–		x		x

$$K_b = \frac{[NH_4^+][OH^-]}{[NH_3]} = 1.8 \times 10^{-5}$$

Check for possible simplifications:

$$400 \times K_b = 400 \times 1.8 \times 10^{-5} = 0.0072$$

This is smaller than 0.50, so assume $0.50 - x = 0.50$.
Then,

$$\frac{(x)(x)}{0.50} = 1.8 \times 10^{-5}$$

$$x^2 = (0.50)(1.8 \times 10^{-5}) = 9.0 \times 10^{-6}$$

$$x = 3.0 \times 10^{-3}$$

$[OH-] = x = 3.0 \times 10^{-3}$ M
$pOH = -\log(3.0 \times 10^{-3}) = 2.52$
$pH = 14.00 - pOH = 14.00 - 2.52 = 11.48$

Exercise Three

	$HOCl$	$\rightleftharpoons$	H^+	+	OCl^-
initial conc.	0.20		0		0
conc. change	$-x$		$+x$		$+x$
equil. conc.	$0.20-x$		x		x

$$K_a = \frac{[H^+][OCl^-]}{[HOCl]}$$

$pH = 4.08$
$[H^+] = 10^{-4.08} = 8.3 \times 10^{-5} = x$

The equilibrium concentrations are:
   $[H^+] = 8.3 \times 10^{-5}$ M
   $[OCl^-] = x = 8.3 \times 10^{-5}$ M
   $[HOCl] = 0.20 - x = 0.20 - 8.3 \times 10^{-5} = 0.20$ M

Then,

$$K_a = \frac{[H^+][OCl^-]}{[HOCl]} = \frac{(8.3 \times 10^{-5})(8.3 \times 10^{-5})}{0.20} = 3.4 \times 10^{-8}$$

Exercise Four

	CH_3NH_2	$+$	H_2O	$\rightleftharpoons$	$CH_3NH_3^+$	$+$	OH^-
initial conc.	0.250		$-$		0		0
conc. change	$-x$		$-$		$+x$		$+x$
equil. conc.	$0.250-x$		$-$		x		x

$$K_b = \frac{[CH_3NH_3^+][OH^-]}{[CH_3NH_2]} = 4.38 \times 10^{-4}$$

Check for possible simplifications:
$$400 \times (4.38 \times 10^{-4}) = 0.175$$

This is less than 0.250, so assume $0.250 - x = 0.250$.
Then, the equilibrium concentrations are:

$[CH_3NH_2] = 0.250$
$[CH_3NH_3^+] = x$
$[OH^-] = x$

$$\frac{(x)(x)}{(0.250)} = 4.38 \times 10^{-4}$$

$$x^2 = (0.250)(4.38 \times 10^{-4}) = 1.095 \times 10^{-4}$$

$$x = 0.0105 = [CH_3NH_3^+]$$

$$\text{percent ionization} = \frac{[CH_3NH_3^+]}{\text{initial conc of } CH_3NH_2} \times 100$$

$$= \frac{0.0105}{0.250} \times 100 = 4.20\%$$

Exercise Five

	$HC_2H_3O_2$	$\rightleftharpoons$	H^+	$+$	$C_2H_3O_2^-$
initial conc.	y		0		0
conc. change	$-x$		$+x$		$+x$
equil. conc.	$y-x$		x		x

$$K_a = \frac{[H^+][C_2H_3O_2^-]}{[HC_2H_3O_2]} = 1.8 \times 10^{-5}$$

$pH = 2.87$
$[H^+] = 10^{-2.87} = 1.35 \times 10^{-3} = x$

The equilibrium concentrations are:
$[HC_2H_3O_2] = y - 1.35 \times 10^{-3}$
$[H^+] = 1.35 \times 10^{-3}$
$[C_2H_3O_2^-] = 1.35 \times 10^{-3}$

Then,

$$\frac{(1.35 \times 10^{-3})(1.35 \times 10^{-3})}{y - 1.35 \times 10^{-3}} = 1.8 \times 10^{-5}$$

$$(1.35 \times 10^{-3})^2 = (y - 1.35 \times 10^{-3})(1.8 \times 10^{-5})$$

$$1.82 \times 10^{-6} = (1.8 \times 10^{-5})y - 2.43 \times 10^{-8}$$

$$(1.8 \times 10^{-5})y = 1.84 \times 10^{-6}$$

$$y = \frac{1.84 \times 10^{-6}}{1.8 \times 10^{-5}} = 0.10$$

The value of y is the concentration of acetic acid, namely 0.10 M.

Exercise Six

Step 1:

$$H_2C_6H_6O_6 \rightleftharpoons H^+ + HC_6H_6O_6$$

	H₂C₆H₆O₆	H⁺	HC₆H₆O₆⁻
initial conc.	0.085	0	0
conc. change	−x	+x	+x
equil. conc.	0.085−x	x	x

$$K_{a1} = \frac{[H^+][HC_6H_6O_6^-]}{[H_2C_6H_6O_6]} = 7.94 \times 10^{-5}$$

$$\frac{(x)(x)}{0.085} = 7.94 \times 10^{-5}$$

$$x^2 = 6.75 \times 10^{-6}$$

$$x = 2.60 \times 10^{-3} = [H^+]$$

$$pH = -\log(2.60 \times 10^{-3}) = 2.58$$

$$[HC_6H_6O_6^-] = x = 2.60 \times 10^{-3}$$

$$[C_6H_6O_6^{2-}] = K_{a2} = 1.62 \times 10^{-12}$$

PROBLEMS

1. pH = 4.72   percent ionization = 3.2 × 10⁻³ %
2. 1.53 × 10⁻³
3. 1.1 × 10–3
4. 0.18
5. 3.4 × 10⁻⁴ M HCl, 9.9 × 10⁻⁴ M HCHO₂
6. pH = 1.28, [HC₂O₄⁻] = 0.0525 M, [C₂O₄²⁻] = 6.40 × 10⁻⁵ M

LESSON 20:   HYDROLYSIS EQUILIBRIA

Exercise One

A.
$$F^- + H_2O \rightleftharpoons HF + OH^-$$
conjugate base   conjugate acid   alkaline pH > 7

B.
$$C_5H_5NH^+ + H_2O \rightleftharpoons C_5H_5N + H_3O^+$$
conjugate acid   conjugate base   acidic pH < 7

C. No hydrolysis. pH = 7

D.
$$C_7H_5O_2^- + H_2O \rightleftharpoons HC_7H_5O_2 + OH^-$$
conjugate base   conjugate acid   alkaline pH > 7

Exercise Two

$$MH^+ \rightleftharpoons M + H^+$$

$$K_a = \frac{[M][H^+]}{[MH^+]}$$

$$K_a = \frac{[M][H^+]}{[MH^+]} \times \frac{[OH^-]}{[OH^-]}$$

$$K_a = \frac{[M]}{[MH^+][OH^-]} \times [H^+][OH^-]$$

$$K_a = \frac{1}{K_b} \times K_w$$

Exercise Three

	$C_2H_3O_2^-$	$+$	H_2O	$\rightleftharpoons$	$HC_2H_3O_2$	$+$	OH^-
initial conc	0.10				0		0
conc change	$-x$				$+x$		$+x$
equil conc	$0.10 - x$				x		x

$$K_b = \frac{[HC_2H_3O_2][OH^-]}{[C_2H_3O_2^-]} = \frac{K_w}{K_a} = \frac{1.0 \times 10^{-14}}{1.8 \times 10^{-5}} = 5.6 \times 10^{-10}$$

Check for simplifications:
$$400 \times K_b = 400 (5.6 \times 10^{-10}) = 2.2 \times 10^{-7}$$
This is much less than 0.10. Assume $0.10 - x = 0.10$.
The equilibrium concentrations are:
$$[C_2H_3O_2^-] = 0.10 \qquad [HC_2H_3O_2] = x \qquad [OH^-] = x$$
Then,

$$\frac{(x)(x)}{0.10} = 5.6 \times 10^{-10}$$
$$x^2 = 5.6 \times 10^{-11}$$
$$x = 7.5 \times 10^{-6}$$
$$pOH = -\log[OH^-] = -\log(7.5 \times 10^{-6}) = 5.12$$
$$pH = 14.00 - pOH = 14.00 - 5.12 = 8.88$$

Exercise Four

	NO_2^-	$+$	H_2O	$\rightleftharpoons$	HNO_2	$+$	OH^-
initial conc	0.045				0		0
conc change	$-x$				$+x$		$+x$
equil conc	$0.045 - x$				x		x

$$K_b = \frac{[HNO_2][OH^-]}{[NO_2^-]} = \frac{K_w}{K_a}$$

$pH = 8.0$; $pOH = 14.0 - 8.0 = 6.0$; $[OH^-] = 1 \times 10^{-6} = x = [HNO_2]$
$[NO_2^-] = 0.045 - 1 \times 10^{-6} = 0.045$

$$K_b = \frac{(1 \times 10^{-6})(1 \times 10^{-6})}{0.045} = 2 \times 10^{-11}$$

$$K_a = \frac{K_w}{K_b} = \frac{1.0 \times 10^{-14}}{2 \times 10^{-11}} = 5 \times 10^{-4}$$

Exercise Five

	NH_4^+	$+$	H_2O	$\rightleftharpoons$	H_3O^+	$+$	NH_3
initial conc.	y				0		0
conc. change	$-x$				$+x$		$+x$
equil. conc.	$y-x$				x		x

$$K_a = \frac{[H^+][NH_3]}{[NH_4^+]} = \frac{K_w}{K_b} = \frac{1.0 \times 10^{-14}}{1.8 \times 10^{-5}} = 5.6 \times 10^{-10}$$

$$pH = 5.00$$
$$[H^+] = 10^{-5.00} = 1.00 \times 10^{-5} = x$$

Then, $[NH_3] = x = 1.00 \times 10^{-5}$ and $[NH_4^+] = y - x = y - 1.00 \times 10^{-5}$.

$$\frac{(1.0 \times 10^{-5})(1.0 \times 10^{-5})}{y - 1.0 \times 10^{-5}} = 5.6 \times 10^{-10}$$

$$y - 1.0 \times 10^{-10} = \frac{(1.0 \times 10^{-5})^2}{5.6 \times 10^{-10}} = 0.18$$

$$y = 0.18 + 1.0 \times 10^{-5} = 0.18 \text{ M}$$

$$\text{moles of } NH_4NO_3 = (0.18 \text{ mol/L})(0.50 \text{ L}) = 0.090 \text{ mol}$$

PROBLEMS

1. 8.52
2. 9.0×10^{-11}
3. 4.89
4. 0.36 gram
5. 11.242
6. 8.79
7. 0.23 M
8. 8.70

LESSON 21: BUFFER SOLUTIONS

Exercise One

A. $H^+ + C_2H_3O_2^- \longrightarrow HC_2H_3O_2$
   $OH^- + HC_2H_3O_2 \longrightarrow C_2H_3O_2^- + H_2O$

B. $H^+ + NH_3 \longrightarrow NH_4^+$
   $OH^- + NH_4^+ \longrightarrow NH_3 + H_2O$

Exercise Two

A. 1. Write the equation for the equilibrium reaction and its K expression.

$$HCHO_2 \rightleftharpoons H^+ + CHO_2^-$$

$$K_a = \frac{[H^+][CHO_2^-]}{[HCHO_2]} = 1.77 \times 10^{-4}$$

2. Summarize the information in a table based on the chemical equation.

	$HCHO_2$	$\rightleftharpoons$	H^+	+	CHO_2^-
initial moles	0.400		0		0.100
initial conc	1.60		0		0.400
conc change	$-x$		$+x$		$+x$
equil conc	$1.60-x$		x		$0.400+x$

3. Test to see if the calculation can be simplified:

$$400 \times K_a = 400 \times (1.77 \times 10^{-4}) = 0.071$$

0.071 is less than both 0.400 and 1.60.

Therefore, assume $1.60 - x = 1.60$ and $0.400 + x = 0.400$.

4. Find the value of x by substituting equilibrium concentrations into K expression.

$$\frac{(x)(0.400)}{(1.60)} = 1.77 \times 10^{-4}$$

$$x = \frac{1.60}{0.400} \times (1.77 \times 10^{-4})$$

$$x = 7.08 \times 10^{-4}$$

$$[H^+] = x = 7.08 \times 10^{-4} \text{ M}$$

$$pH = -\log(7.08 \times 10^{-4}) = 3.150$$

B. 1. Write the equation for the equilibrium reaction and its K expression.

$$NH_3 + H_2O \rightleftharpoons NH_4^+ + OH^-(aq)$$

$$K_b = \frac{[NH_4^+][OH^-]}{[NH_3]} = 1.8 \times 10^{-5}$$

2. Summarize the information in a table based on the chemical equation.

	NH_3	+	H_2O	$\rightleftharpoons$	NH_4^+	+	OH^-
initial moles	0.20				0.15		0
initial conc	0.40				0.30		0
conc change	$-x$				$+x$		$+x$
equil conc	$0.40-x$				$0.30+x$		x

3. Test to see if the calculation can be simplified:

$$400 \times K_b = 400 \times (1.8 \times 10^{-5}) = 7.2 \times 10^{-3}$$

0.0072 is less than both 0.400 and 0.30.

Therefore, assume $0.40 - x = 0.40$ and $0.30 + x = 0.30$.

4. Find the value of x by substituting equilibrium concentrations into K expression.

$$\frac{(x)(0.30)}{(0.40)} = 1.8 \times 10^{-5}$$

$$x = \frac{0.40}{0.30} \times (1.8 \times 10^{-5})$$

$$x = 2.4 \times 10^{-5}$$

$$[OH^-] = x = 2.4 \times 10^{-5} \text{ M}$$

$$pOH = -\log(2.4 \times 10^{-5}) = 4.62$$

$$pH = 14.00 - pOH = 14.00 - 4.62 = 9.38$$

Exercise Three

1. Write the equation for the equilibrium reaction and its K expression.

$$HCHO_2 \rightleftharpoons H^+ + CHO_2^-$$

$$K_a = \frac{[H^+][CHO_2^-]}{[HCHO_2]} = 1.77 \times 10^{-4}$$

2. Summarize the information in a table based on the chemical equation.

	$HCHO_2$	$\rightleftharpoons$	H^+	+	CHO_2^-
initial conc	0.400		0		0.100

Now take into account the buffering action that results when 0.100 mole of NaOH is added.

$$OH^- \quad + \quad HCHO_2 \quad \longrightarrow \quad CHO_2^- \quad + \quad H_2O$$

| 0.100 mol added | 0.100 mol consumed | 0.100 mol formed |

Therefore, as a result of buffering action, the concentration of $HCHO_2$ decreases by 0.100 mol/L, and the concentration of CHO_2^- increases by 0.100 mol/L.

Put these results of buffering action into the table.

	$HCHO_2 \rightleftharpoons$	H^+ +	CHO_2^-
initial conc	0.400	0	0.100
buff action	−0.100	0	+0.100
conc change	−x	+x	+x
equil conc	0.300−x	x	0.200+x

3. Simplifications can still be made.
    Assume $0.300 - x = 0.300$ and $0.200 + x = 0.200$.

4. Find the value of x by substituting equilibrium concentrations into K expression.

$$\frac{(x)(0.200)}{(0.300)} = 1.77 \times 10^{-4}$$

$$x = \frac{0.300}{0.400} \times (1.77 \times 10^{-4}) = 2.66 \times 10^{-4} = [H^+]$$

$$pH = -\log(2.66 \times 10^{-4}) = 3.575$$

Exercise Four

Initial concentrations are as in Exercise Three.
    Take into account the buffering action that results when 0.050 mole of HCl is added.

$$H^+ \quad + \quad CHO_2^- \quad \longrightarrow \quad HCHO_2 \quad + \quad H_2O$$

| 0.050 mol added | 0.050 mol consumed | 0.050 mol formed |

Therefore, as a result of buffering action, the concentration of CHO_2^- decreases by 0.050 mol/L, and the concentration of $HCHO_2$ increases by 0.050 mol/L.

Put these results of buffering action into the table.

	$HCHO_2 \rightleftharpoons$	H^+ +	CHO_2^-
initial conc	0.400	0	0.100
buff action	+0.050	0	−0.050
conc change	−x	+x	+x
equil conc	0.450−x	x	0.050+x

3. Simplifications can still be made.
    Assume $0.450 - x = 0.450$ and $0.050 + x = 0.050$.

4. Find the value of x by substituting equilibrium concentrations into K expression.

$$\frac{(x)(0.050)}{(0.450)} = 1.77 \times 10^{-4}$$

$$x = \frac{0.450}{0.050} \times (1.77 \times 10^{-4}) = 1.6 \times 10^{-3} = [H^+]$$

$$pH = -\log(1.6 \times 10^{-3}) = 2.80$$

Exercise Five

A. 1. Write the equation for the equilibrium reaction and its K expression.

$$NH_3(aq) + H_2O(l) \rightleftharpoons NH_4^+(aq) + OH^-(aq)$$

$$K_b = \frac{[NH_4^+][OH^-]}{[NH_3]} = 1.8 \times 10^{-5}$$

2. Summarize the information in a table based on the chemical equation. The volume of the solution is 0.250 L.

	NH$_3$	+	H$_2$O	$\rightleftharpoons$	NH$_4^+$	+	OH$^-$
initial moles	0.20				0.60		
initial conc	0.27				0.80		

Take into account the buffering action. The added OH$^-$ is consumed by NH$_4^+$.

	OH$^-$	+	NH$_4^+$	$\longrightarrow$	NH$_3$	+	H$_2$O
	0.10 mol added		0.10 mol consumed		0.10 mol formed		

Therefore, as a result of buffering action,

the concentration of NH$_4^+$ decreases by $\dfrac{0.10 \text{ mol}}{0.75 \text{ L}} = 0.13$ mol/L

and

the concentration of NH$_3$ increases by $\dfrac{0.10 \text{ mol}}{0.75 \text{ L}} = 0.13$ mol/L.

	NH$_3$	+	H$_2$O	$\rightleftharpoons$	NH$_4^+$	+	OH$^-$
initial conc	0.27				0.80		0
buff action	+0.13				−0.13		
conc change	−x				+x		+x
at equil	0.40−x				0.67+x		x

3. Simplifications:

$[NH_3] = 0.40 - x = 0.40$
$[NH_4^+] = 0.67 + x = 0.67$
$[OH^-] = x$

4. Solve for x and find pH.

$$\frac{(0.67)(x)}{(0.40)} = 1.8 \times 10^{-5}$$

$$x = \frac{0.40}{0.67} \times (1.8 \times 10^{-5}) = 1.1 \times 10^{-5} = [OH^-]$$

$$pOH = -\log(1.1 \times 10^{-5}) = 4.96$$

$$pH = 14.000 - pOH = 14.000 - 4.96 = 9.04$$

B.   One mole of NaOH is much more base than the 0.1 mole of NH_4^+ can react with. All of the NH_4^+ reacts, and the buffer is destroyed. The excess NaOH increases the pH.

PROBLEMS

1.   a.   2.65   b.   2.59   2.   a.   11.05   b.   10.99

LESSON 22:   SOLUBILITY EQUILIBRIA

Exercise One

A. $MgCO_3 \rightleftharpoons Mg^{2+} + CO_3^{2-}$
     For $MgCO_3$, $K_{sp} = [Mg^{2+}][CO_3^{2-}]$.

B. $Cu(OH)_2 \rightleftharpoons Cu^{2+} + 2\ OH^-$
     For $Cu(OH)_2$, $K_{sp} = [Cu^{2+}][OH^-]^2$

C. $Cu_2S \rightleftharpoons 2\ Cu^+ + S^{2-}$
     For Cu_2S, $K_{sp} = [Cu^+]^2[S^{2-}]$

D. $Bi_2S_3 \rightleftharpoons 2\ Bi^{3+} + 3\ S^{2-}$
     For Bi_2S_3, $K_{sp} = [Bi^{3+}]^2[S^{2-}]^3$

Exercise Two

A.   1. $SrF_2 \rightleftharpoons Sr^{2+} + 2\ F^-$
     2.   For SrF_2, $K_{sp} = [Sr^{2+}][F^-]^2$
     3.

SrF_2	$\longrightarrow$	Sr^{2+}	+	$2\ F^-$
8.9×10^{-4} moles	$\longrightarrow$	8.9×10^{-4} moles	+	$2\ (8.9 \times 10^{-4}$ moles$)$

   $[Sr^{2+}] = 8.9 \times 10^{-4}$ moles per liter
   $[F^-] = 17.8 \times 10^{-4}$ moles per liter
     4. $K_{sp} = [Sr^{2+}][F^-]^2 = (8.9 \times 10^{-4})(17.8 \times 10^{-4})^2 = 2.8 \times 10^{-9}$

B.   1. $Ag_2CO_3 \rightleftharpoons 2\ Ag^+ + CO_3^{2-}$
     2.   For Ag_2CO_3, $K_{sp} = [Ag^+]^2\ [CO_3^{2-}]$
     3. 1.0 g $Ag_2CO_3 \left(\dfrac{1\ mol}{276\ g} \right) = 3.6 \times 10^{-3}$ mol Ag_2CO_3

   $\dfrac{3.6 \times 10^{-3}\ mol\ Ag_2CO_3}{33\ liters} = 1.1 \times 10^{-4}$ mol/L

Ag_2CO_3	$\longrightarrow$	$2\ Ag^+$	+	CO_3^{2-}
1.1×10^{-4} mol	$\longrightarrow$	$2\ (1.1 \times 10^{-4}$ mol$)$	+	1.1×10^{-4} mol

   $[Ag^+] = 2.2 \times 10^{-4}$ M
   $[CO_3^{2-}] = 1.1 \times 10^{-4}$ M
     4. $K_{sp} = [Ag^+]^2\ [CO_3^{2-}] = (2.2 \times 10^{-4})^2\ (1.1 \times 10^{-4}) = 5.3 \times 10^{-12}$

Exercise Three

A.   1. $Cu_2S \rightleftharpoons 2\ Cu^+ + S^{2-}$
     2.   For Cu_2S, $K_{sp} = [Cu^+]^2\ [S^{2-}] = 2.3 \times 10^{-48}$
     3.   Let x = molar solubility

Cu_2S	$\longrightarrow$	$2\ Cu^+$	+	S^{2-}
x mol	$\longrightarrow$	$2\ (x$ mol$)$	+	x mol

   $[Cu^+] = 2x$ $[S^{2-}] = x$
     4. $2.3 \times 10^{-48} = (2x)^2\ (x) = 4x^3$
          $x^3 = 5.75 \times 10^{-49}$
          $x = 8 \times 10^{-17}$ moles per liter

B.   1. $MgCO_3 \rightleftharpoons Mg^{2+} + CO_3^{2-}$

2. For $MgCO_3$, $K_{sp} = [Mg^{2+}][CO_3^{2-}] = 6.8 \times 10^{-6}$
3. Let x = molar solubility of $MgCO_3$

$$MgCO_3 \longrightarrow Mg^+ + CO_3^{2-}$$
$$x \text{ mol} \longrightarrow x \text{ mol} + x \text{ mol}$$

$[Mg^{2+}] = x$ $\qquad [CO_3^{2-}] = x$

4. $6.8 \times 10^{-6} = (x)(x) = x^2$
$x = 2.6 \times 10^{-3}$ moles per liter
Amount that dissolves in 20.0 liters:

$$20.0 \text{ L} \left(\frac{2.6 \times 10^{-3} \text{ mol}}{1 \text{ L}} \right) \left(\frac{84 \text{ g}}{1 \text{ mol}} \right) = 4.4 \text{ g } MgCO_3$$

Exercise Four

A. 1. $PbCl_2 \rightleftharpoons Pb^{2+} + 2 Cl^-$
2. For $PbCl_2$, $K_{sp} = [Pb^{2+}][Cl^-]^2 = 1.2 \times 10^{-5}$
3. Let x = number of moles of $PbCl_2$ to dissolve

$$PbCl_2 \longrightarrow Pb^{2+} + 2 Cl^-$$
$$x \text{ moles} \longrightarrow x \text{ moles} + 2x \text{ moles}$$

But there is also Pb^{2+} from the $Pb(NO_3)_2$, 0.20 mole per liter.
$[Pb^{2+}] = 0.20 + x$
$[Cl^-] = 2x$

4. $1.0 \times 10^{-4} = (0.20 + x)(2x)^2$
Assume $x < 0.20$. Then, $0.20 + x = 0.20$.
$(0.20)(4x^2) = 1.2 \times 10^{-5}$
$x^2 = 1.5 \times 10^{-5}$
$x = 3.9 \times 10^{-3}$ moles $PbCl_2$ dissolve

B. See Exercise 3 B for steps 1 and 2.

3. Let x = moles $MgCO_3$ to dissolve per liter

$$MgCO_3 \longrightarrow Mg^{2+} + CO_3^{2-}$$
$$x \text{ mol} \longrightarrow x \text{ mol} + x \text{ mol}$$

But there is also 1.0 mole CO_3^{2-} per liter of 1.0 M Na_2CO_3.
$[Mg^{2+}] = x$
$[CO_3^{2-}] = 1.0 + x$

4. Assume $x < 1.0$. Then, $1.0 + x = 1.0$
$6.8 \times 10^{-6} = (x)(1.0)$
$x = 6.8 \times 10^{-6}$ mole of $MgCO_3$ per liter

$$20.0 \text{ L} \left(\frac{6.8 \times 10^{-6} \text{ mol}}{1 \text{ L}} \right) \left(\frac{84 \text{ g}}{1 \text{ mol}} \right) = 0.011 \text{ g } MgCO_3$$

Compare this to the answer of Exercise 3 B.

Exercise Five

1. $Fe(OH)_3 \rightleftharpoons Fe^{3+} + 3 OH^-$
2. For $Fe(OH)_3$, $K_{sp} = [Fe^{3+}][OH^-]^3 = 2.6 \times 10^{-39}$
3. Final volume = $100 + 0.05$ mL = 100 mL = 0.1 L
moles $OH^- = (0.00005 \text{ L})(8.0 \text{ mol/L}) = 4 \times 10^{-4}$ mol

$$[OH^-] = \frac{4 \times 10^{-4} \text{ mol}}{0.1 \text{ L}} = 4 \times 10^{-3} \text{ M}$$

$[Fe^{3+}] = 0.20$ M

4. $K = (0.20)(4 \times 10^{-3})^3$
$= (0.20)(64 \times 10^{-9}) = 1.3 \times 10^{-8}$
Compare this to K_{sp} (2.6×10^{-39})
$K > K_{sp}$
A precipitate will form.

PROBLEMS

1. 10.68
2. 1.4×10^{-11} mol/L
3. 9.2×10^{-17}

4. 4.0×10^{-6} mol/L in pure water. 2.5×10^{-8} mol/L at pH = 10.0.

LESSON 23:   RATES OF CHEMICAL REACTIONS

Exercise One

A. 1. 1/2, –1/2, –1/2
      2. 4, –4
      3. 1/2, –1/3

B. (4/6)(0.42 mol/L-min) = 0.28 mol/L-min
    (–7/6)(0.42 mol/L-min) = 0.49 mol/L-min

Exercise Two

A. $\dfrac{d[C_2H_5I]}{dt} = k[C_2H_4]^p[HI]^q$

B. 1. (1) and (2).   A factor of 4.0/2.0 = 2. $2 = 2^1$.   Exponent of $[C_2H_4]$ is 1.
      2. (1) and (3).   A factor of 8.0/4.0 = 2. $2 = 2^1$.   Exponent of $[HI]$ is 1.

C. $\dfrac{d[C_2H_5I]}{dt} = k[C_2H_4][HI]$

    4.0 M-min^{-1} = k (0.20 M) (0.10 M)
    k = 200 M^{-1} min^{-1}

Exercise Three

A. $[A]_t$; $-k$ B. $\ln[A]_t$; $-k$ C. $1/[A]_t$; k
D. second; (3 L/mol – 2 L/mol) / (120 sec – 60 sec) = 0.017 L/mol-sec; 0.017 L/mol-sec; 1 M

Exercise Four

$t_{1/2}$ = 0.693/k = 0.693/(0.030 min^{-1}) = 23 min
    Because 0.20 M is one-half of 0.40 M, it will take one half life for the concentration to decline from 0.40 M to 0.20 M, namely 23 minutes. It will take 23 min for the concentration to drop from 0.20 M to 0.10 M, and another 23 min to drop from 0.10 M to 0.05 M; therefore to drop from 0.20 M to 0.05 M will require 46 minutes.

Exercise Five

$$\ln \frac{k_1}{k_2} = \frac{E_a}{R} \left(\frac{1}{T_2} - \frac{1}{T_1} \right)$$

For T_1 = 225°C, k_1 = 4.77×10^{-8} L/mol-sec. In Kelvin, T_1 = 498 K.
For T_2 = 616°C, k_1 = 1.66×10^{-6} L/mol-sec. In Kelvin, T_2 = 889 K.
R = 8.314 J/mol-K

$$\ln \frac{4.77 \times 10^{-8}}{1.66 \times 10^{-4}} = \frac{E_a}{8.314 \text{ J/mol-K}} \left(\frac{1}{889 \text{ K}} - \frac{1}{498 \text{ K}} \right)$$

$$-8.155 = (-1.062 \times 10^{-4} \text{ mol/J})\, E_a$$

$$E_a = 7.68 \times 10^4 \text{ J/mol}$$

Exercise Six

E_a = 90300 J/mol
T_1 = 300 K k_1 = 1.5×10^{-5} L/mol-sec
T_2 = 400 K k_2 = ?

$$\ln \frac{1.5 \times 10^{-5} \text{ L/mol-sec}}{k_2} = \frac{90300 \text{ J/mol}}{8.314 \text{ J/mol-K}} \left(\frac{1}{400 \text{ K}} - \frac{1}{300 \text{ K}} \right)$$

$$\ln \frac{1.5 \times 10^{-5} \text{ L/mol-sec}}{k_2} = -9.051$$

$$\frac{1.5 \times 10^{-5} \text{ L/mol-sec}}{k_2} = e^{-9.051}$$

$$k_2 = 0.13 \text{ L/mol-sec}$$

PROBLEMS

1. a. second order   b. $0.080 \text{ M}^{-1} \text{ sec}^{-1}$
2. 0.091 min^{-1}
3. 110 kJ/mol
4. (a) At 300 K, $k = 0.0462 \text{ min}^{-1}$
       At 320 K, $k = 0.231 \text{ min}^{-1}$
    (b) 64.4 kJ/mol
5. $0.0014 \text{ M}^{-1} \text{ sec}^{-1}$
6. The reaction is first order in PCl_5. $k = 0.015 \text{ min}^{-1}$
7. (a) 5.2 h   (b) $1.78 \times 10^{-2} \text{ M}$ (c) 130 min
8. $t_{1/2} = 1 / k [A]_0$; This is less useful because it depends on concentration.

LESSON 24:   MECHANISMS OF CHEMICAL REACTIONS

Exercise One

(a) bimolecular, $\dfrac{d[Cl_2]}{dt} = k [Cl]^2$ (b) unimolecular, $\dfrac{d[O_2]}{dt} = k[O_3]$

(c) termolecular, $\dfrac{d[I_2]}{dt} = k[HI]^2[O]$

Exercise Two

(a) $2 \text{ NO} + Cl_2 \longrightarrow 2 \text{ NOCl}$ (b) $OCl^- + H_2O + I^- \longrightarrow \text{HOI} + OH^- + Cl^-$
(c) $2 \text{ NO} + 2 \text{ H}_2 \longrightarrow 2 \text{ H}_2O + N_2$ (d) $H_2O + CO \longrightarrow H_2 + CO_2$

Exercise Three

The equation for the overall reaction is $OCl^- + I^- \longrightarrow OI^- + Cl^-$.
catalyst: H_2O intermediates: HOCl, OH^-, HOI

Exercise Four

(a) The equation for the overall reaction is $Cl_2 + CHCl_3 \longrightarrow HCl + CCl_4$

For the slow step, the rate equation is: $\dfrac{d[HCl]}{dt} = k[CHCl_3][Cl]$

In this rate equation, Cl is an intermediate.

From step 1, $k_1[Cl_2] = k_{-1}[Cl]^2$, so $[Cl] = \sqrt{\dfrac{k_1}{k_{-1}}[Cl_2]}$

Using this expression for [Cl] in the rate equation yields $\dfrac{d[HCl]}{dt} = k[CHCl_3][Cl_2]^{1/2}$

(b) $\dfrac{d[O_2]}{dt} = k[O_3]$

Exercise Five

For Mechanism A, the rate law is $\dfrac{d[N_2O_5]}{dt} = k[NO_2]^2[O_3]$.

For Mechanism B, the rate law is $\dfrac{d[N_2O_5]}{dt} = k[NO_2][O_3]$.

Therefore, mechanism B is consistent with the experimental rate law, but mechanism A is not.

PROBLEMS

1. (a) bimolecular, $-\dfrac{d[NO]}{dt} = k[NO]^2$ (b) unimolecular, $-\dfrac{d[Cl_2]}{dt} = k[Cl_2]$

    (c) bimolecular, $-\dfrac{d[NO_2]}{dt} = k[NO_2]^2$ (d) termolecular,

    $$-\dfrac{d[OH]}{dt} = k[OH][NO_2][N_2]$$

    (e) bimolecular, $-\dfrac{d[Cl^-]}{dt} = k[Cl^-][H_2O]$

2. (a) $O_3 + O \longrightarrow 2O_2$

    (b) step 1: $-\dfrac{d[O_3]}{dt} = k[O_3][NO]$ step 2: $-\dfrac{d[NO_2]}{dt} = k[NO_2][O]$

    (c) The catalyst is NO.

    (d) The intermediate is NO_2.

3. Mechanism B is consistent with the experimental rate law, but mechanisms A and C are not.

LESSON 25:   COORDINATION COMPOUNDS: COMPOSITION AND STRUCTURE

Exercise One

A. $[Fe(CN)_6]^{3-}$ $K_3[Fe(CN)_6]$
B. $[Ag(OH)_2]^-$ $Na[Ag(OH)_2]$
C. $[Ni(NH_3)_4(H_2O)_2]^{2+}$ $[Ni(NH_3)_4(H_2O)_2]SO_4$
D. $[Pt(H_2O)_4Cl_2]^{2+}$ $[Pt(H_2O)_4Cl_2]Cl_2$

Exercise Two

	central atom	oxidation number	coordination number
A.	Pt	+4	6
B.	Fe	+2	6
C.	Ni	+2	4
D.	Zn	+2	4

Exercise Three

A. $[Pt(NH_3)_4Cl_2]Cl_2$
    $[Pt(NH_3)_2Cl_4]$
    $K[Pt(NH_3)Cl_5]$

B.

formula	counter ions	number of ions
$[Cr(H_2O)_6]^{3+}$	3 Cl$^-$	4
$[Cr(H_2O)_5Cl]^{2+}$	2 Cl$^-$	3
$[Cr(H_2O)_4Cl_2]^+$	Cl$^-$	2
$[Cr(H_2O)_2Cl_4]^-$	NH$_4^+$	2

Exercise Four

A.

coordination number	geometry
6	octahedral
4	tetrahedral
2	linear
6	octahedral
4	square

B.

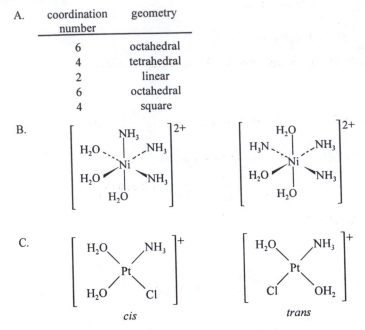

C.

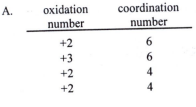

cis   trans

Exercise Five

A.

oxidation number	coordination number
+2	6
+3	6
+2	4
+2	4

B.

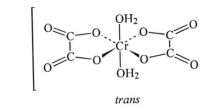

trans

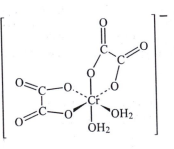

cis

Exercise Six

A. tetraamminecopper(II) sulfate
tetraaquadibromocobalt(III) bromide
potassium diamminetetrachlorochromate(III)
sodium tetrahydroxozincate(II)
potassium trioxalatocobaltate(III)

B. $[Ni(NH_3)_6]SO_4$
$K[Pt(NH_3)Cl_5]$
$[CoCl_2(en)_2]Cl$

PROBLEMS

1. a.   +4   b.   6   c.   2 $(LaCl_3)$

2. a.

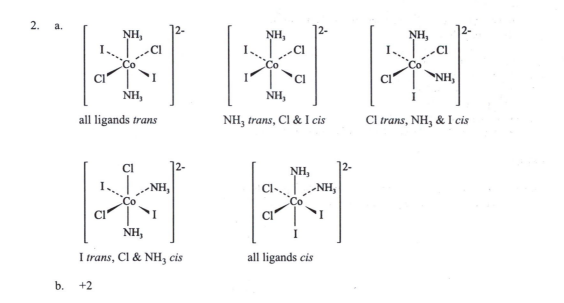

all ligands *trans*   NH₃ *trans*, Cl & I *cis*   Cl *trans*, NH₃ & I *cis*

I *trans*, Cl & NH₃ *cis*   all ligands *cis*

b.   +2

LESSON 26:   COORDINATION COMPOUNDS: COLOR
AND MAGNETIC PROPERTIES

Exercise One

A. violet
B. 650 to 700 nm
C. 1. A   2. C   3. D   4. F

Exercise Two

A.

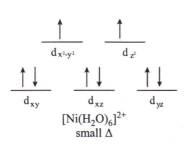

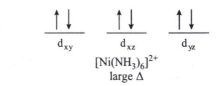

$[Ni(H_2O)_6]^{2+}$
small Δ

$[Ni(NH_3)_6]^{2+}$
large Δ

B. $[Ni(NH_3)_6]^{2+}$ has the larger Δ.
C. 180 kJ/mol

Exercise Three

A. 1.

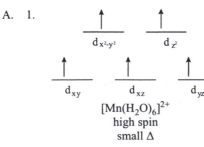

$[Mn(H_2O)_6]^{2+}$
high spin
small Δ

2.

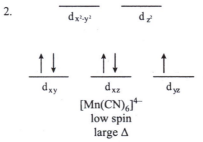

$[Mn(CN)_6]^{4-}$
low spin
large Δ

3.   green
4.   Higher energy than green. Therefore, blue or violet.
5.   Orange or yellow.

B. $[Co(H_2O)_6]^{3+}$ absorbs green
$[Co(CN)_6]^{3-}$ absorbs violet
$[Co(NH_3)_6]^{3+}$ absorbs blue
$[Co(CO_3)_3]^{3-}$ absorbs red
energies of absorbed light: red < green < blue < violet
least splitting = CO_3^{2-} < H_2O < NH_3 < CN^- = most splitting

Exercise Four

A. $[NiCl_4]^{2-}$ has more unpaired electrons than $[Ni(CN)_4]^{2-}$. Therefore, $[NiCl_4]^{2-}$ is high spin.
B.   Tetrahedral complexes are high spin. $[NiCl_4]^{2-}$ is tetrahedral.

C.

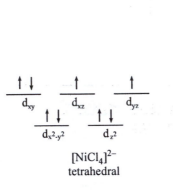

PROBLEMS

1. $[CoF_6]^{3-}$ color observed: violet   high spin
     $[Co(CN)_6]^{3-}$ color observed: orange   low spin
2.   a. $[CoCl_4]^{2-}$ b. $[Co(H_2O)_6]^{2+}$
     c.   seven
     d.   e.

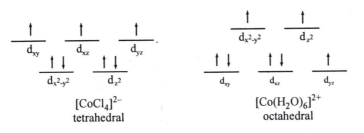

     f. $[CoCl_4]^{2-}$ + 6 H_2O $\rightleftharpoons$ $[Co(H_2O)_6]^{2+}$ + 4 Cl^-

3.

4.   a. 6   b. 4   c. 0   d. H_2O complex   e. high spin

LESSON 27:   COORDINATION COMPOUNDS: COMPLEX ION EQUILIBRIA

Exercise One

	$Fe(CN)_6^{3-}$	$\rightleftharpoons$	Fe^{3+}	+	$6\,CN^-$
initial conc	0.080		0		0
conc change	$-x$		$+x$		$+6x$
equil conc	$0.080-x$		x		$6x$

$$K_{diss} = \frac{[Fe^{3+}][CN^-]^6}{[Fe(CN)_6^-]}$$
$$= 1.0 \times 10^{-31}$$

Assume $0.080 - x = 0.080$

Then, $[Fe(CN)_6^{3-}] = 0.080$
$[Fe^{3+}] = x$
$[CN^-] = 6x$

$$\frac{(x)(6x)^6}{(0.080)} = 1.0 \times 10^{-31}$$

$$46656\,x^7 = (0.080)(1.0 \times 10^{-31}) = 8.0 \times 10^{-33}$$

$$x^7 = 1.7 \times 10^{-37}$$

$$x = (1.7 \times 10^{-37})^{1/7} = 5.6 \times 10^{-6}$$

Then, $[Fe(CN)_6^{3-}] = 0.080$ M
$[Fe^{3+}] = x = 5.6 \times 10^{-6}$ M
$[CN^-] = 6x = 3.4 \times 10^{-5}$ M

Exercise Two

A.

	$Ni(NH_3)_6^{2+}$	$\rightleftharpoons$	Ni^{2+}	+	$6\,NH_3$
initial conc	0		0.020		0.200
conc change	$+x$		$-x$		$-6x$
equil conc	x		$0.020-x$		$0.200-6x$

$$K_{diss} = \frac{[Ni^{2+}][NH_3]^6}{[Ni(NH_3)_6^{2+}]}$$

But $[Ni^{2+}] = 1.5 \times 10^{-4}$

Then, $0.020 - x = 1.5 \times 10^{-4}$
$x = 0.020 - 1.5 \times 10^{-4} = 0.020 = [Ni(NH_3)_6^{2+}]$
$[NH_3] = 0.200 - 6x = 0.200 - 6(0.020) = 0.080$

$$K_{diss} = \frac{(1.5 \times 10^{-4})(0.080)^6}{(0.020)} = 2.0 \times 10^{-9}$$

B. $K_f = \dfrac{1}{K_{diss}} = \dfrac{1}{2.0 \times 10^{-9}} = 5.0 \times 10^8$

Exercise Three

	$Ag(SCN)_2^-$	$\rightleftharpoons$	Ag^+	$+$	$2\,SCN^-$	$K_{diss} = \dfrac{[Ag^+][SCN^-]^2}{[Ag(SCN)_2^-]}$
initial conc	0		0.0500		y	
conc change	$+x$		$-x$		$-2x$	$= 1.0 \times 10^{-10}$
equil conc	x		$0.0500-x$		$y-2x$	

Require 99% of Ag in complex, i.e., $[Ag(SCN)_2^-] = (0.99)(0.0500) = 0.0495 = x$

Then, $[Ag^+] = 0.0500 - x = 0.0500 - 0.0495 = 0.0005$

$[SCN^-] = y - 2x = y - 2(0.0495) = y - 0.0990$

$$\frac{(0.0005)(y-0.0990)^2}{(0.0495)} = 1.0 \times 10^{-10}$$

$$(y-0.0990)^2 = \frac{(0.0495)}{(0.0005)}(1.0 \times 10^{-10}) = 1 \times 10^{-8}$$

$$y-0.0990 = \sqrt{1 \times 10^{-8}} = 1.0 \times 10^{-4}$$

$$y = 0.0990 + 1.0 \times 10^{-4} = 0.0991 \text{ mol/L}$$

Then, 0.0991 mole of NaSCN must be added.

Exercise Four

	$AgBr$	$+$	$2\,S_2O_3^{2-}$	$\rightleftharpoons$	$Ag(S_2O_3)_2^{3-}$	$+$	Br^-	$K = \dfrac{[Ag(S_2O_3)_2^{3-}][Br^-]}{[S_2O_3^{2-}]^2}$
initial conc			0.10		0		0	
conc change			$-2x$		$+x$		$+x$	$= \dfrac{K_{sp}}{K_{diss}} = \dfrac{5.0 \times 10^{-13}}{1.0 \times 10^{-13}}$
equil conc			$0.10-2x$		x		x	$= 5.0$

$$\frac{(x)(x)}{(0.10-2x)^2} = 5.0$$

$$\frac{(x)^2}{(0.10-2x)^2} = 5.0$$

$$\frac{x}{0.10-2x} = \sqrt{5.0} = 2.24$$

$$x = 2.24(0.10-2x) = 0.224 - 4.48x$$

$$5.48\,x = 0.224$$

$$x = (0.224)/(5.48) = 0.041 \text{ mol/L}$$

0.041 mole of AgBr will dissolve in 1.0 L of 0.10 M $Na_2S_2O_3$.

Exercise Five

	ZnS	$+$	$4\,CN^-$	$\rightleftharpoons$	$Zn(CN)_4^{2-}$	$+$	S^{2-}	$K = \dfrac{[Zn(CN)_4^{2-}][S^{2-}]}{[CN^-]^4}$
initial conc	y				0		0	
conc change	$-4x$				$+x$		$+x$	$= \dfrac{K_{sp}}{K_{diss}} = \dfrac{1.0 \times 10^{-21}}{1.0 \times 10^{-17}}$
equil conc	$y-4x$				x		x	$= 1.0 \times 10^{-4}$

If 0.20 mol of ZnS dissolve in 1.0 L, then $[Zn(CN)_4{}^{2-}] = 0.20 = x = [S^{2-}]$.

$$[CN^-] = y - 4(0.20) = y - 0.80$$

$$\frac{(0.20)(0.20)}{(y-0.80)^4} = 1.0 \times 10^{-4}$$

$$(y-0.80)^4 = \frac{(0.20)^2}{1.0 \times 10^{-4}} = 4.0 \times 10^2$$

$$y - 0.80 = (4.0 \times 10^2)^{1/4} = 4.5$$

$$y = 4.5 + 0.80 = 5.3 \text{ mol/L}$$

5.3 moles of NaCN are needed to dissolve 0.20 mole of ZnS in in 1.0 L of water.

PROBLEMS

1. $[Ag^+] = 4.8 \times 10^{-10}$ M $[NH_3] = 1.0$ M $[Ag(NH_3)_2{}^+] = 0.012$ M
2. $[NH_3] = 6.7 \times 10^{-4}$ M
3. 6.7×10^{-12} mol/L
4. 1.5×10^{-8} mol/L
5. 0.014 mole KCN

LESSON 28:   BALANCING OXIDATION-REDUCTION EQUATIONS

Exercise One

A. Redox reaction. C is reduced, H is oxidized.
B. Not a redox reaction.
C. Redox reaction. S is oxidized, Cl is reduced.

Exercise Two

A. oxidation half-reaction: $C_2O_4{}^{2-} \longrightarrow CO_2$
   reduction half-reaction: $MnO_4{}^- \longrightarrow Mn^{2+}$
B. oxidation half-reaction: $H_2S \longrightarrow S$
   reduction half-reaction: $NO_3{}^- \longrightarrow NO$

Exercise Three

A. oxidation half-reaction: $C_2O_4{}^{2-} \longrightarrow 2\ CO_2$
   All others are the same as in Exercise Two.

Exercise Four

A. reduction half-reaction: $MnO_4{}^- \longrightarrow Mn^{2+} + 4\ H_2O$
B. reduction half-reaction: $NO_3{}^- \longrightarrow NO + 2\ H_2O$
   Others are the same as in Exercise Three.

Exercise Five

A. reduction half-reaction: $8\ H^+ + MnO_4{}^- \longrightarrow Mn^{2+} + 4\ H_2O$
   oxidation same as in Exercise Four.
B. oxidation half-reaction: $H_2S \longrightarrow S + 2\ H^+$
   reduction half-reactioin: $4\ H^+ + NO_3{}^- \longrightarrow NO + 2\ H_2O$

Exercise Six

A. oxidation half-reaction: $C_2O_4{}^{2-} \longrightarrow 2\ CO_2 + 2\ e^-$
   reduction half-reaction: $5\ e^- + 8\ H^+ + MnO_4{}^- \longrightarrow Mn^{2+} + 4\ H_2O$

B. oxidation half-reaction: $H_2S \longrightarrow S + 2 H^+ + 2 e^-$
reduction half-reaction: $3 e^- + 4 H^+ + NO_3^- \longrightarrow NO + 2 H_2O$

Exercise Seven

A.

$$5 C_2O_4^{2-} \longrightarrow 10 CO_2 + 10 e^-$$
$$\underline{10 e^- + 16 H^+ + 2 MnO_4^- \longrightarrow 2 Mn^{2+} + 8 H_2O}$$
$$5 C_2O_4^{2-} + 16 H^+ + 2 MnO_4^- \longrightarrow 10 CO_2 + 2 Mn^{2+} + 8 H_2O$$

B.

$$3 H_2S \longrightarrow 3 S + 6 H^+ + 6 e^-$$
$$\underline{6 e^- + 8 H^+ + 2 NO_3^- \longrightarrow 2 NO + 4 H_2O}$$
$$3 H_2S + 2 H^+ + 2 NO_3^- \longrightarrow 3 S + 2 NO + 4 H_2O$$

Exercise Eight

A. Steps

1. & 2.
$$Fe^{2+} \longrightarrow Fe^{3+}$$
$$MnO_4^- \longrightarrow Mn^{2+}$$

3.
$$Fe^{2+} \longrightarrow Fe^{3+}$$
$$MnO_4^- \longrightarrow Mn^{2+}$$

4. same

5.
$$Fe^{2+} \longrightarrow Fe^{3+}$$
$$MnO_4^- \longrightarrow Mn^{2+} + 4 H_2O$$

6.
$$Fe^{2+} \longrightarrow Fe^{3+}$$
$$8 H^+ + MnO_4^- \longrightarrow Mn^{2+} + 4 H_2O$$

7.
$$Fe^{2+} \longrightarrow Fe^{3+} + 1 e^-$$
$$5 e^- + 8 H^+ + MnO_4^- \longrightarrow Mn^{2+} + 4 H_2O$$

8.
$$5 Fe^{2+} \longrightarrow 5 Fe^{3+} + 5 e^-$$
$$5 e^- + 8 H^+ + MnO_4^- \longrightarrow Mn^{2+} + 4 H_2O$$

9.
$$5 Fe^{2+} + 8 H^+ + MnO_4^- \longrightarrow 5 Fe^{3+} + Mn^{2+} + 4 H_2O$$

10. It's balanced.

B. Steps

1. & 2.
$$I_2 \longrightarrow IO_3^-$$
$$NO_3^- \longrightarrow NO_2$$

3.
$$I_2 \longrightarrow 2 IO_3^-$$
$$NO_3^- \longrightarrow NO_2$$

4. same

5.
$$6 H_2O + I_2 \longrightarrow 2 IO_3^-$$
$$NO_3^- \longrightarrow NO_2 + H_2O$$

6.
$$6 H_2O + I_2 \longrightarrow 2 IO_3^- + 12 H^+$$
$$2 H^+ + NO_3^- \longrightarrow NO_2 + H_2O$$

7.
$$6 H_2O + I_2 \longrightarrow 2 IO_3^- + 12 H^+ + 10 e^-$$
$$1 e^- + 2 H^+ + NO_3^- \longrightarrow NO_2 + H_2O$$

8.
$$6 H_2O + I_2 \longrightarrow 2 IO_3^- + 12 H^+ + 10 e^-$$
$$10 e^- + 20 H^+ + 10 NO_3^- \longrightarrow 10 NO_2 + 10 H_2O$$

9.
$$I_2 + 8 H^+ + 10 NO_3^- \longrightarrow 2 IO_3^- + 10 NO_2 + 4 H_2O$$

10. It's balanced.

C. Steps

1. & 2.
$$I^- \longrightarrow I_2$$
$$H_2O_2 \longrightarrow H_2O$$

3.
$$2 I^- \longrightarrow I_2$$
$$H_2O_2 \longrightarrow 2 H_2O$$

4.   same
5. $2 \, I^- \longrightarrow I_2$
                          $H_2O_2 \longrightarrow 2 \, H_2O$
6. $2 \, I^- \longrightarrow I_2$
               $2 \, H^+ + H_2O_2 \longrightarrow 2 \, H_2O$
7. $2 \, I^- \longrightarrow I_2 + 2 \, e^-$
          $2 \, e^- + 2 \, H^+ + H_2O_2 \longrightarrow 2 \, H_2O$
8. $2 \, I^- \longrightarrow I_2 + 2 \, e^-$
          $2 \, e^- + 2 \, H^+ + H_2O_2 \longrightarrow 2 \, H_2O$
9. $2 \, I^- + 2 \, H^+ + H_2O_2 \longrightarrow I_2 + 2 \, H_2O$
10.   It's balanced.

Exercise Nine

A.   Steps

1. & 2. $S^{2-} \longrightarrow S$
                          $NO_3^- \longrightarrow NO_2^-$

3.   same
4.   same
5. $S^{2-} \longrightarrow S$
                          $NO_3^- \longrightarrow NO_2^- + H_2O$
6. $S^{2-} \longrightarrow S$
               $2 \, H^+ + NO_3^- \longrightarrow NO_2^- + H_2O$
7. $S^{2-} \longrightarrow S + 2 \, e^-$
          $2 \, e^- + 2 \, H^+ + NO_3^- \longrightarrow S + NO_2^- + H_2O$
8.   same
9. $S^{2-} + 2 \, H^+ + NO_3^- \longrightarrow S + NO_2^- + H_2O$
         $S^{2-} + H_2O + NO_3^- \longrightarrow S + NO_2^- + 2 \, OH^-$
10.   It's balanced.

B.   Steps

1. & 2. $Fe \longrightarrow Fe(OH)_3$
                          $CrO_4^{2-} \longrightarrow Cr(OH)_3$

3.   same
4.   same
5. $3 \, H_2O + Fe \longrightarrow Fe(OH)_3$
                          $CrO_4^{2-} \longrightarrow Cr(OH)_3 + H_2O$
6. $3 \, H_2O + Fe \longrightarrow Fe(OH)_3 + 3 \, H^+$
               $5 \, H^+ + CrO_4^{2-} \longrightarrow Cr(OH)_3 + H_2O$
7. $3 \, H_2O + Fe \longrightarrow Fe(OH)_3 + 3 \, H^+ + 3 \, e^-$
          $3 \, e^- + 5 \, H^+ + CrO_4^{2-} \longrightarrow Cr(OH)_3 + H_2O$
8.   same
9. $2 \, H_2O + Fe + 2 \, H^+ + CrO_4^{2-} \longrightarrow Fe(OH)_3 + Cr(OH)_3$
     $4 \, H_2O + Fe + CrO_4^{2-} \longrightarrow Fe(OH)_3 + Cr(OH)_3 + 2 \, OH^-$
10.   It's balanced.

C.   Steps

1. & 2. $N_2H_4 \longrightarrow N_2$
                          $ClO^- \longrightarrow Cl^-$

3.   same
4.   same
5. $N_2H_4 \longrightarrow N_2$

$$ClO^- \longrightarrow Cl^- + H_2O$$

6.
$$N_2H_4 \longrightarrow N_2 + 4\,H^+$$
$$2\,H^+ + ClO^- \longrightarrow Cl^- + H_2O$$

7.
$$N_2H_4 \longrightarrow N_2 + 4\,H^+ + 4\,e^-$$
$$2\,e^- + 2\,H^+ + ClO^- \longrightarrow Cl^- + H_2O$$

8.
$$N_2H_4 \longrightarrow N_2 + 4\,H^+ + 4\,e^-$$
$$4\,e^- + 4\,H^+ + 2\,ClO^- \longrightarrow 2\,Cl^- + 2\,H_2O$$

9.
$$N_2H_4 + 2\,ClO^- \longrightarrow N_2 + 2\,Cl^- + 2\,H_2O$$

No H^+ ions remain.

10.   It's balanced.

Exercise Ten

A.   Steps

5.
$$2\,H_2O + UF_6^- \longrightarrow UO_2^{2+} + 6\,HF$$
$$H_2O \longrightarrow 2\,H_2O$$

6.
$$2\,H^+ + 2\,H_2O + UF_6^- \longrightarrow UO_2^{2+} + 6\,HF$$
$$2\,H^+ + H_2O \longrightarrow 2\,H_2O$$

7.
$$2\,H^+ + 2\,H_2O + UF_6^- \longrightarrow UO_2^{2+} + 6\,HF + 1\,e^-$$
$$2\,e^- + 2\,H^+ + H_2O \longrightarrow 2\,H_2O$$

8.
$$4\,H^+ + 4\,H_2O + 2\,UF_6^- \longrightarrow 2\,UO_2^{2+} + 12\,HF + 2\,e^-$$
$$2\,e^- + 2\,H^+ + H_2O_2 \longrightarrow 2\,H_2O$$

9.
$$6\,H^+ + 2\,H_2O + 2\,UF_6^- + H_2O_2 \longrightarrow 2\,UO_2^{2+} + 12\,HF$$

10.   It's balanced.

B.   Steps

1. & 2.
$$SnCl_4^{2-} \longrightarrow SnCl_6^{2-}$$
$$CuCl_4^{2-} \longrightarrow CuCl$$

3.   same

4.
$$2\,Cl^- + SnCl_4^{2-} \longrightarrow SnCl_6^{2-}$$
$$CuCl_4^{2-} \longrightarrow CuCl + 3\,Cl^-$$

5.   same

6.   same

7.
$$2\,Cl^- + SnCl_4^{2-} \longrightarrow SnCl_6^{2-} + 2\,e^-$$
$$e^- + CuCl_4^{2-} \longrightarrow CuCl + 3\,Cl^-$$

8.
$$2\,Cl^- + SnCl_4^{2-} \longrightarrow SnCl_6^{2-} + 2\,e^-$$
$$2\,e^- + 2\,CuCl_4^{2-} \longrightarrow 2\,CuCl + 6\,Cl^-$$

9.
$$SnCl_4^{2-} + 2\,CuCl_4^{2-} \quad\quad SnCl_6^{2-} + 2\,CuCl + 4\,Cl^-$$

10.   It's balanced.

PROBLEMS

1. $5\,C_2H_4 + 36\,H^+ + 12\,MnO_4^- \longrightarrow 10\,CO_2 + 12\,Mn^{2+} + 28\,H_2O$

2. $2\,Cl^- + 4\,H^+ + PbO_2 \longrightarrow Cl_2 + Pb^{2+} + 2\,H_2O$

3. $2\,OH^- + 2\,CrO_2^- + 3\,H_2O_2 \longrightarrow 2\,CrO_4^{2-} + 4\,H_2O$

4. $2\,OH^- + H_2O_2 + 2\,ClO_2 \longrightarrow O_2 + 2\,H_2O + 2\,ClO_2^-$

5. $4\,Cl^- + H_2O_2 + 2\,ClO_2 \longrightarrow O_2 + 2\,H_2O + 2\,ClO_2^-$

6. $8\,NH_3 + 2\,Cu + 2\,H_2O + O_2 \longrightarrow 2\,Cu(NH_3)_4^{2+} + 4\,OH^-$

7. $5\,(NH_4)_3[Co(SCN)_6] + 119\,MnO_4^- + 262\,H^+ \longrightarrow$
$$45\,NO_3^- + 30\,SO_4^{2-} + 30\,CO_2 + 5\,Co^{2+} + 119\,Mn^{2+} + 161\,H_2O$$

LESSON 29:   VOLTAIC CELLS

Exercise One

A. A voltaic cell is a device which uses a spontaneous chemical reaction to produce an electric current.

B. An anode is the electrode where oxidation occurs. A cathode is the electrode where reduction takes place.

C. Standard state conditions are: 1.0 M solute concentration
1.0 atm gas pressure
25°C temperature

D. 1. −0.76 volt   3. −0.77 volt
2. +1.36 volts   4. −1.82 volts

E. 1.36 is greater than −0.76, so reaction 2 occurs more readily than reaction 1.
−0.77 is greater than −1.82, so reaction 3 occurs more readily than reaction 4.

Exercise Two

A. Oxidizing agents become reduced, so we compare *reduction* potentials:

$Au^{3+} + 3\,e^- \longrightarrow Au$ $E° = +1.50$ volts
$Pb^{2+} + 2\,e^- \longrightarrow Pb$ $E° = -0.13$ volt
$Fe^{3+} + e^- \longrightarrow Fe^{2+}$ $E° = +0.77$ volt
$NO_3^- + 4\,H^+ + 3e^- \longrightarrow NO + 2\,H_2O$ $E° = +0.96$ volt

In order of decreasing strength as oxidizing agents:
strongest $= Au^{3+} > NO_3^- > Fe^{3+} > Pb^{2+} =$ weakest

B. Reducing agents become oxidized. So we must find the largest oxidation potential in each set.

1. $Zn \longrightarrow Zn^{2+} + 2e^-$ $E° = +0.76$ volt
2. $Pb \longrightarrow Pb^{2+} + 2e^-$ $E° = +0.13$ volt
3. $2\,I^- \longrightarrow I_2 + 2e^-$ $E° = -0.53$ volt
4. $Cu^+ \longrightarrow Cu^{2+} + e^-$ $E° = -0.15$ volt

Exercise Three

A. $Cu^+ \longrightarrow Cu^{2+} + e^-$ $E° = -0.15$ volt
$Ag^+ + e^- \longrightarrow Ag$ $E° = +0.80$ volt
Reaction potential $= -0.15 + 0.80 = 0.65$ volt

B. $H_2S \longrightarrow S + 2\,H^+ + 2e-$ $E° = -0.14$ V
$NO_3^- + 4\,H^+ + 3\,e^- \longrightarrow NO + 2\,H_2O$ $E° = 0.96$ V
Reaction potential $= -0.14 + 0.96 = 0.82$ volt

C. $Mn^{2+} + 2\,H_2O \longrightarrow MnO_2 + 4\,H^+ + 2e^-$ $E° = -1.23$ V
$Br_2 + 2e^- \longrightarrow 2\,Br^-$ $E° = +1.07$ V
Reaction potential $= -1.23 + 1.07 = -0.16$ volt

Exercise Four

A. $E°$ for reaction $= -0.14 + (-0.14) = -0.28$ volt. Because the reaction potential is less than zero (i.e., $E°$ is negative), the reaction is nonspontaneous.

B. $E°$ for reaction $= -0.53 + 1.07 = 0.54$ volt. Because the reaction potential is positive, the reaction is spontaneous.

C. $E°$ for reaction $= 0.20 + 0.28 = 0.48$ volt. This is positive, so reaction is spontaneous.

Exercise Five

$Cd \longrightarrow Cd^{2+} + 2e^-$ $E° = +0.40$ volt
$Pb^{2+} + 2e^- \longrightarrow Pb$ $E° = -0.13$ volt

A. $E°$ for cell $= 0.40 + (-0.13) = 0.27$ volt.

B. $Cd + Pb^{2+} \longrightarrow Cd^{2+} + Pb$

C. oxidation: $Cd \longrightarrow Cd^{2+} + 2e^-$
reduction: $Pb^{2+} + 2e^- \longrightarrow Pb$

D. Electrons flow from the Cd electrode to the Pb electrode.

E. Cations flow from Cd half–cell toward the Pb half-cell. Anions flow in the opposite direction.

Exercise Six

$$Pb \longrightarrow Pb^{2+} + 2e^- \qquad\qquad E° = 0.13 \text{ volt}$$
$$2\,Ag^+ + 2e^- \longrightarrow 2\,Ag \qquad\qquad E° = 0.80 \text{ volt}$$
$$Pb + 2\,Ag^+ \longrightarrow Pb^{2+} + 2\,Ag \qquad E° = 0.13 + 0.80 = 0.93 \text{ V}$$

$$n = 2$$

$$Q = \frac{[Pb^{2+}]}{[Ag^+]^2} = \frac{(2.0)}{(0.0010)^2} = 2.0 \times 10^6$$

$$E_{cell} = 0.93 - \frac{0.0257}{2} \ln(2.0 \times 10^6)$$

$$= 0.93 - (0.0129)(14.51) = 0.74 \text{ volt}$$

Exercise Seven

From Exercise Six:

$$Pb + 2\,Ag^+ \longrightarrow Pb^{2+} + 2\,Ag \qquad E° = 0.93 \text{ volt}$$
$$n = 2$$

$$Q = \frac{[Pb^{2+}]}{[Ag^+]^2} = \frac{[Pb^{2+}]}{(0.10)^2} = \frac{[Pb^{2+}]}{0.010}$$

$E_{cell} = 1.00$ volt (as given)

$$1.00 = 0.93 - \frac{0.0257}{2} \ln \frac{[Pb^{2+}]}{0.010}$$

$$0.07 = -0.0129 \ln \frac{[Pb^{2+}]}{0.010}$$

$$\ln \frac{[Pb^{2+}]}{0.010} = -5.43$$

$$\frac{[Pb^{2+}]}{0.010} = e^{-5.43} = 4.4 \times 10^{-3}$$

$$[Pb^{2+}] = (0.010)(4.4 \times 10^{-3}) = 4.4 \times 10^{-5} \text{ M}$$

Exercise Eight

$$2\,Fe^{3+} + 2e^- \longrightarrow 2\,Fe^{2+} \qquad E° = 0.77 \text{ volt}$$
$$2\,I^- \longrightarrow I_2 + 2e^- \qquad E° = -0.53 \text{ volt}$$
$$\overline{2\,Fe^{3+} + 2\,I^- \longrightarrow 2\,Fe^{2+} + I_2 \qquad E° = 0.24 \text{ volt}}$$

$$n = 2$$

When $E_{cell} = 0$, then $Q = K_{eq}$

$$0 = 0.24 - \frac{0.0257}{2} \ln K_{eq}$$

$$-0.24 = -0.0129 \ln K_{eq}$$

$$\ln K_{eq} = \frac{-0.24}{-0.0129} = 18.6$$

$$K_{eq} = e^{18.6} = 1.2 \times 10^8$$

PROBLEMS

1. (a) Mg   (b) Mg^{2+} (c) Co^{3+} (d) Co^{2+}
2. (a) +1.24 V   (b) 8.10×10^{41}
3. 1.7×10^{-8}
4. 1.11 volts

5. (a) $Zn + Cu^{2+} \longrightarrow Zn^{2+} + Cu$ (b) 1.10 V   (c) 6.0×10^{-14} M
6. (a) $PbO_2 + Pb(s) + 2\ HSO_4^- + 2\ H^+ \longrightarrow 2\ PbSO_4 + 2\ H_2O$
    (b) $E° = 2.00$ volts   (c) 6
    (d) cathode: $PbO_2(s) + 3\ H^+ + HSO_4^- + 2e^- \longrightarrow PbSO_4(s) + 2\ H_2O$
       anode: $Pb(s) + HSO_4^- \longrightarrow PbSO_4(s) + H^+ + 2e^-$
    (e) 3.9×10^{67}
7. 1.5×10^{-12} This is a solubility product, K_{sp}.

LESSON 30:   ELECTROLYTIC CELLS

Exercise One

A.   cathode: $Na^+ + e^- \longrightarrow Na$
    anode: $2\ F^- \longrightarrow F_2 + 2e^-$
    overall: $2\ Na^+ + 2\ F^- \longrightarrow 2\ Na + F_2$

B.   cathode: $Ca^{2+} + 2e^- \longrightarrow Ca$
    anode: $2\ Cl^- \longrightarrow Cl_2 + 2e^-$
    overall: $Ca^{2+} + 2\ Cl^- \longrightarrow Ca + Cl_2$

C.   cathode: $Li^+ + e^- \longrightarrow Li$
    anode: $2\ O^{2-} \longrightarrow O_2 + 4e^-$
    overall: $4\ Li^+ + 2\ O^{2-} \longrightarrow 4\ Li + O_2$

D.   cathode: $Al^{3+} + 3e^- \longrightarrow Al$
    anode: $S^{2-} \longrightarrow S + 2e^-$
    overall: $2\ Al^{3+} + 3\ S^{2-} \longrightarrow 2\ Al + 3\ S$

Exercise Two

A.   cathode: $Cu^{2+} + 2e^- \longrightarrow Cu$
    anode: $2\ Br^- \longrightarrow Br_2 + 2e^-$
    overall: $Cu^{2+} + 2\ Br^- \longrightarrow Cu + Br_2$

B.   cathode: $Ag^+ + e^- \longrightarrow Ag$
    anode: $2\ H_2O \longrightarrow O_2 + 4\ H^+ + 4e^-$ (note: NO_3^- cannot be oxidized in solution)
    overall: $4\ Ag^+ + 2\ H_2O \longrightarrow 4\ Ag + O_2 + 4\ H^+$

C.   cathode: $2\ H_2O + 2e^- \longrightarrow H_2 + 2\ OH^-$ (note: Ca is a IIA–group metal)
    anode: $2\ Cl^- \longrightarrow Cl_2 + 2e^-$
    overall: $2\ H_2O + 2\ Cl^- \longrightarrow H_2 + Cl_2 + 2\ OH^-$

D.   cathode: $2\ H_2O + 2e^- \longrightarrow H_2 + 2\ OH^-$
    anode: $2\ H_2O \longrightarrow O_2 + 4\ H^+ + 4e^-$
    overall: $4\ H_2O + 2\ H_2O \longrightarrow 2\ H_2 + O_2 + \underbrace{4\ OH^- + 4\ H^+}$

                         combine these
                         to make $4\ H_2O$

    overall reaction becomes: $2\ H_2O \longrightarrow 2\ H_2 + O_2$

Exercise Three

A. $2\ F^- \longrightarrow F_2 + 2e^-$ $38\ g\ F_2 \left(\dfrac{1\ mol\ F_2}{38\ g\ F_2} \right) \left(\dfrac{2\ mol\ e^-}{1\ mol\ F_2} \right) = 2.0\ mol\ e^-$

B. $2\ H_2O \longrightarrow O_2 + 4\ H^+ + 4e^-$ $2\ mol\ e^- \left(\dfrac{1\ mol\ O_2}{4\ mol\ e^-} \right) \left(\dfrac{32\ g\ O_2}{1\ mol\ O_2} \right) = 16\ g\ O_2$

Exercise Four

A. $4.2 \text{ mol } e^- \left(\dfrac{1 \text{ F}}{1 \text{ mol } e^-} \right) \left(\dfrac{9.65 \times 10^4 \text{ C}}{1 \text{ F}} \right) = 4.1 \times 10^5 \text{ C}$

B. $6.0 \text{ hr} \left(\dfrac{60 \text{ min}}{1 \text{ hr}} \right) \left(\dfrac{60 \text{ sec}}{1 \text{ hr}} \right) = 2.2 \times 10^4 \text{ sec}$

   $8.0 \text{ ampere} = 8.0 \text{ C / sec}$

   $2.2 \times 10^4 \text{ sec} \left(\dfrac{8.0 \text{ C}}{1 \text{ sec}} \right) \left(\dfrac{1 \text{ F}}{9.65 \times 10^4 \text{ C}} \right) \left(\dfrac{1 \text{ mol } e^-}{1 \text{ F}} \right) = 1.9 \text{ mol } e^-$

C. $\text{current} = \text{coulomb / sec}$

   $0.3 \text{ mol } e^- \left(\dfrac{1 \text{ F}}{1 \text{ mol } e^-} \right) \left(\dfrac{9.65 \times 10^4 \text{ C}}{1 \text{ F}} \right) = 2.9 \times 10^4 \text{ C}$

   $\text{current} = (2.9 \times 10^4 \text{ C}) / (20 \text{ sec}) = 1.4 \times 10^3 \text{ C/sec} = 1.4 \times 10^3 \text{ ampere}$

D. $0.10 \text{ ampere} = 0.10 \text{ C/sec}$

   $5000 \text{ C} \left(\dfrac{1 \text{ sec}}{0.10 \text{ C}} \right) = 5 \times 10^4 \text{ sec}$

Exercise Five

A. $$Ag^+ + e^- \longrightarrow Ag$$

   $2.0 \text{ hr} \left(\dfrac{3600 \text{ sec}}{1 \text{ hr}} \right) \left(\dfrac{5.0 \text{ C}}{1 \text{ sec}} \right) \left(\dfrac{1 \text{ F}}{9.65 \times 10^5 \text{ C}} \right) \left(\dfrac{1 \text{ mol } e^-}{1 \text{ F}} \right) \left(\dfrac{1 \text{ mol Ag}}{1 \text{ mol } e^-} \right) = 0.37 \text{ mol Ag}$

   $0.37 \text{ mol Ag} \left(\dfrac{108 \text{ g Ag}}{1 \text{ mol Ag}} \right) = 40. \text{ grams Ag}$

B. $$2 \text{ Br}^- \longrightarrow Br_2 + 2e^-$$

   $80.0 \text{ g Br}_2 \left(\dfrac{1 \text{ mol Br}_2}{160 \text{ g Br}_2} \right) \left(\dfrac{2 \text{ mol } e^-}{1 \text{ mol Br}_2} \right) \left(\dfrac{1 \text{ F}}{1 \text{ mol } e^-} \right) \left(\dfrac{9.65 \times 10^5 \text{ C}}{1 \text{ F}} \right) \left(\dfrac{1 \text{ sec}}{5.0 \text{ C}} \right) = 1.9 \times 10^4 \text{ sec}$

Exercise Six

$$\text{cathode: } Ag^+ + e^- \longrightarrow Ag$$
$$\text{anode: } \quad Ag \longrightarrow Ag^+ + e^-$$

$6.00 \text{ hr} \left(\dfrac{3600 \text{ sec}}{1 \text{ hr}} \right) \left(\dfrac{0.46 \text{ C}}{1 \text{ sec}} \right) \left(\dfrac{1 \text{ F}}{9.65 \times 10^5 \text{ C}} \right) \left(\dfrac{1 \text{ mol } e^-}{1 \text{ F}} \right) \left(\dfrac{1 \text{ mol Ag}}{1 \text{ mol } e^-} \right) = 0.103 \text{ mol Ag}$

$0.103 \text{ mol Ag} \left(\dfrac{108 \text{ g Ag}}{1 \text{ mol Ag}} \right) = 11.1 \text{ grams Ag}$

The cathode gains: $20.0 + 11.1 = 31.1 \text{ grams}$
The anode loses: $20.0 - 11.1 = 8.9 \text{ grams}$

PROBLEMS

1.   (a) $Ag^+ + e^- \longrightarrow Ag$ (b) 2 amperes
2. U^{4+} 3.   700 hours

4. 29.8 hours

5. 518 ampere

6. 28.4 kg

7. 6.98 kg

8. 1.0×10^4 ampere

LESSON 31:   NUCLEAR REACTIONS AND RADIOACTIVITY

Exercise One

A. 1. $_{+1}^{0}\beta$ 2. $_0^1 n$ 3. $_2^4\alpha$ or $_2^4 He$ 4. $_{-1}^{0}\beta$ or $_{-1}^{0}e$ 5. $_0^0\gamma$ or γ

B. 1. $_{27}^{60}Co \longrightarrow _{-1}^{0}\beta + _{28}^{60}Ni$

4. $_6^{14}C \longrightarrow _{-1}^{0}\beta + _7^{14}N$

2. $_{87}^{221}Fr \longrightarrow _2^4\alpha + _{85}^{217}At$

5. $_{90}^{230}Th \longrightarrow _2^4 He + _{88}^{226}Ra$

3. $_{15}^{30}P \longrightarrow _{+1}^{0}\beta + _{14}^{30}Si$

6. $_6^{11}C \longrightarrow _{+1}^{0}\beta + _5^{11}B$

C. 1. $_{50}^{121}Sn \longrightarrow _{-1}^{0}\beta + _{51}^{121}Sb$

3. $_{82}^{210}Pb \longrightarrow _{80}^{206}Hg + _2^4\alpha$

2. $_7^{13}N \longrightarrow _{+1}^{0}\beta + _6^{13}C$

Exercise Two

A. 1. $_{26}^{54}Fe + _2^4\alpha \longrightarrow _0^1 n + _{28}^{57}Ni$

3. $_{92}^{238}U + _0^1 n \longrightarrow _{92}^{239}U \longrightarrow _{-1}^{0}\beta + _{93}^{239}Np$

2. $_4^9 Be + _0^0\gamma \longrightarrow _0^1 n + _4^8 Be$

B. 1. $_{99}^{253}Es + _2^4\alpha \longrightarrow _0^1 n + _{101}^{256}Md$

2. $_7^{14}N + _0^1 n \longrightarrow _6^{14}C + _1^1 H$

Exercise Three

A. $_6^{12}C$ contains 6 protons and 6 neutrons having a total mass

$6(1.00728) + 6(1.00867) = 12.09570$ g/mol

The mass of a $_6^{12}C$ nucleus is 11.99671 g/mol

The mass decrement = $12.09570 - 11.99671 = 0.09899$ g/mol

The binding energy in J/mol is

$$\Delta E = 0.09899 \text{ g/mol} \left(\frac{1 \text{ kg}}{1000 \text{ g}} \right) (3.00 \times 10^8 \text{ m/s})^2 = 8.91 \times 10^{12} \text{ J/mol}$$

For one nucleus, the binding energy is

$$\Delta E = 8.91 \times 10^{12} \text{ J/mol} \left(\frac{1 \text{ mol}}{6.022 \times 10^{23} \text{ nuclei}} \right) = 1.48 \times 10^{-11} \text{ J/nucleus}$$

B. $_{92}^{238}U$ contains 92 protons and 146 neutrons which have a total mass

$92(1.00728) + 146(1.00867) = 239.93558$ g/mol

The mass of a $_{92}^{238}U$ nucleus is 238.0003 g/mol

The mass decrement = 239.93558 g/mol $- 238.0003$ g/mol $= 1.9353$ g/mol

The binding energy in J/mol is

$$\Delta E = 1.9353 \text{ g/mol} \left(\frac{1 \text{ kg}}{1000 \text{ g}} \right) (3.00 \times 10^8 \text{ m/s})^2 = 1.74 \times 10^{12} \text{ J/mol}$$

For one nucleus, the binding energy is

$$\Delta E = 1.74 \times 10^{14} \text{ J/mol} \left(\frac{1 \text{ mol}}{6.022 \times 10^{23} \text{ nuclei}} \right) = 2.89 \times 10^{-10} \text{ J/nucleus}$$

Exercise Four

A. $\Delta m = (\text{mass } _{82}^{208}Pb + \text{mass } _2^4 He) - (\text{mass } _{84}^{210}Po + 2 \text{ mass } _0^1 n)$

$= (207.9316 \text{ g/mol} + 4.00150 \text{ g/mol}) - (209.9368 \text{ g/mol} + 2 \times 1.00867 \text{ g/mol}) = -0.0210$ g/mol

$$\Delta E = -0.0210 \text{ g/mol} \left(\frac{1 \text{ kg}}{1000 \text{ g}} \right) (3.0 \times 10^8 \text{ m/s})^2 = -1.89 \times 10^{12} \text{ J/mol}$$

$$\Delta E = -1.89 \times 10^{12} \text{ J/mol} \left(\frac{1 \text{ mol}}{6.022 \times 10^{23} \text{ nuclei}} \right) = -3.14 \times 10^{-12} \text{ J/nucleus}$$

B. For 1 atom, $\Delta E = -3.14 \times 10^{-12}$ J. In 1.0 gram, the number of atoms is

$$1.0 \text{ g Po-210} \left(\frac{1 \text{ mol}}{210 \text{ g}} \right) \left(\frac{6.02 \times 10^{23} \text{ atoms}}{1 \text{ mol}} \right) = 2.87 \times 10^{21} \text{ atoms}$$

$$\Delta E = 2.87 \times 10^{21} \text{ atoms} \left(\frac{-3.14 \times 10^{-12} \text{ J}}{1 \text{ atom}} \right) \left(\frac{1 \text{ kJ}}{1000 \text{ J}} \right) = -9.0 \times 10^6 \text{ kJ}$$

C. $^{241}_{95}\text{Am} \longrightarrow {}^{4}_{2}\alpha + {}^{237}_{93}\text{Np}$ $\Delta m = (\text{mass } {}^{237}_{93}\text{Np} + \text{mass } {}^{4}_{2}\text{He}) - (\text{mass } {}^{241}_{95}\text{Am})$
                                              $= (236.9971 + 4.00150) - (241.0045) = -0.0059 \text{ g/mol}$

$$\Delta E = -0.0059 \text{ g/mol} \left(\frac{1 \text{ kg}}{1000 \text{ g}} \right) (3.00 \times 10^8 \text{ m/s})^2 = -5.3 \times 10^{11} \text{ J/mol}$$

$$\Delta E = -5.3 \times 10^{11} \text{ J/mol} \left(\frac{1 \text{ mol}}{6.022 \times 10^{23} \text{ nuclei}} \right) = -8.8 \times 10^{-13} \text{ J/nucleus}$$

Exercise Five

A.   The $^{56}_{26}\text{Fe}$ nucleus contains 26 protons and 30 neutrons
The mass decrement = 26 (1.00728 g/mol) + 30 (1.00867 g/mol) − 55.92066 g/mol = 0.52872 g/mol
For a mole of nuclei, the binding energy is

$$0.52872 \text{ g/mol} \left(\frac{1 \text{ kg}}{1000 \text{ g}} \right) (3.00 \times 10^8 \text{ m/s})^2 = 4.76 \times 10^{13} \text{ J/mol}$$

For one nucleus, the binding energy is

$$4.76 \times 10^{13} \text{ J/mol} \left(\frac{1 \text{ mol}}{6.022 \times 10^{23} \text{ nuclei}} \right) = 7.90 \times 10^{-11} \text{ J/nucleus}$$

Binding energy per nucleon is

$$7.90 \times 10^{-11} \frac{\text{J}}{\text{nucleus}} \left(\frac{1 \text{ nucleus}}{56 \text{ nucleons}} \right) = 1.41 \times 10^{-12} \text{ J/nucleon}$$

B. $\Delta m = (\text{mass } {}^{87}_{35}\text{Br} + \text{mass } {}^{144}_{59}\text{Pr} + 9 \text{ mass } {}^{1}_{0}\text{n}) \quad - (\text{mass } {}^{239}_{94}\text{Pu} + \text{mass } {}^{1}_{0}\text{n})$
            $= (86.9028 \text{ g/mol} + 143.8807 \text{ g/mol} + 9 \times 1.00867 \text{ g/mol}) - (239.0006 \text{ g/mol} + 1.00867 \text{ g/mol})$
            $= -0.1477 \text{ g/mol}$

For one mole,

$$\Delta E = -0.1477 \text{ g/mol} \left(\frac{1 \text{ kg}}{1000 \text{ g}} \right) (3.00 \times 10^8 \text{ m/s})^2 = -1.33 \times 10^{13} \text{ J/mol}$$

For one nucleus,

$$\Delta E = -1.33 \times 10^{13} \text{ J/mol} \left(\frac{1 \text{ mol}}{6.022 \times 10^{23} \text{ nuclei}} \right) = -2.21 \times 10^{-11} \text{ J/nucleus}$$

C. $\Delta m = (\text{mass of one mole } {}^{4}_{2}\text{He} + \text{mass of one mole } {}^{1}_{0}\text{n}) - (\text{mass of one mole } {}^{2}_{1}\text{H} + \text{mass of one mole } {}^{3}_{1}\text{H})$
            $= (4.00150 \text{ g/mol} + 1.00867 \text{ g/mol}) - (2.01355 \text{ g/mol} + 3.01550 \text{ g/mol}) = -0.01888 \text{ g/mol}$

$$\Delta E = -0.01888 \text{ g/mol} \left(\frac{1 \text{ kg}}{1000 \text{ g}} \right) (3.00 \times 10^8 \text{ m/s})^2 = -1.70 \times 10^{12} \text{ J/mol}$$

$$\Delta E = -1.70 \times 10^{12} \text{ J/mol} \left(\frac{1 \text{ kJ}}{1000 \text{ J}} \right) = -1.70 \times 10^9 \text{ kJ/mol}$$

The negative value indicates that energy is released.

Exercise Six

A. $X_0 = 4.6 \text{ mg}$
    $X = $ amt after 3.0 days
    $t = 3.0 \text{ days}$

$$k = \frac{0.693}{t_{\frac{1}{2}}} = \frac{0.693}{8.0 \text{ days}} = 8.7 \times 10^{-2} \text{ day}^{-1}$$

$$\ln \frac{X}{4.6 \text{ mg}} = -(8.7 \times 10^{-2} \text{ day}^{-1})(3.0 \text{ day})$$

$$\ln \frac{X}{4.6 \text{ mg}} = -0.261$$

$$\frac{X}{4.6 \text{ mg}} = e^{-0.261} = 0.770$$

$$X = (4.6 \text{ mg})(0.770) = 3.5 \text{ mg}$$

B. X_0 = initial amount = 100%
X = amount remaining = 100% - 5% = 95%
t = time for 5% to decompose

$$k = \frac{0.693}{t_{1/2}} = \frac{0.693}{7 \times 10^{10} \text{ yr}} = 1 \times 10^{-11} \text{ yr}^{-1}$$

$$\ln \frac{95\%}{100\%} = -(1 \times 10^{-11} \text{ yr}^{-1})t$$

$$t = -\frac{\ln 0.95}{1 \times 10^{-11} \text{ yr}^{-1}} = -\frac{-0.051}{1 \times 10^{-11} \text{ yr}^{-1}}$$

$$t = 5 \times 10^9 \text{ years}$$

Exercise Seven

$X_0 = 15.3$ $X = 8.6$ t = age of charcoal

$$k = \frac{0.693}{t_{1/2}} = \frac{0.693}{5720 \text{ yr}} = 1.21 \times 10^{-4} \text{ yr}^{-1}$$

$$\ln \frac{15.3 \text{ counts}}{8.6 \text{ counts}} = (1.21 \times 10^{-4} \text{ yr}^{-1})t$$

$$t = \frac{\ln 1.78}{1.21 \times 10^{-4} \text{ yr}^{-1}} = 4800 \text{ years}$$

PROBLEMS

1. (a) ^1_1H (b) ^1_1H (c) $^{257}_{103}\text{Lr}$ (d) ^1_1H

2. (a) $^{14}_7\text{N} + ^4_2\text{He} \longrightarrow ^{17}_8\text{O} + ^1_1\text{H}$

    (b) $^{246}_{96}\text{Cm} + ^{12}_6\text{C} \longrightarrow ^{254}_{102}\text{No} + 4\,^1_0\text{n}$

    (c) $^{10}_5\text{B} + ^4_2\text{He} \longrightarrow ^{13}_7\text{N} + ^1_0\text{n}$

    $^{13}_7\text{N} \longrightarrow ^0_{+1}\beta + ^{13}_6\text{C}$

3. 8.50×10^8 kJ   4. 29.0 years   5. around 3060 B.C.E.

6. (a) 9.04×10^{-13} J   (b) 4.5×10^5 J

LESSON 32: STRUCTURES AND FORMULAS OF ORGANIC COMPOUNDS

Exercise One

A. 1. $CH_3CH_2CH_2OCH_3$ 2. $CH_2{=}CHC{=}CH_2$ 3. $CHCl_2CH_2CH_2OH$
                                              $|$
                                              CH_3

4.
$\underset{CH_2ClCOCH_2CH_3}{\overset{O}{\overset{||}{}}}$ 5. $\underset{CH_3CNH_2}{\overset{O}{\overset{||}{}}}$ 6. $\underset{CF_3CF_2CCH_3}{\overset{O}{\overset{||}{}}}$

B. 1.   2.

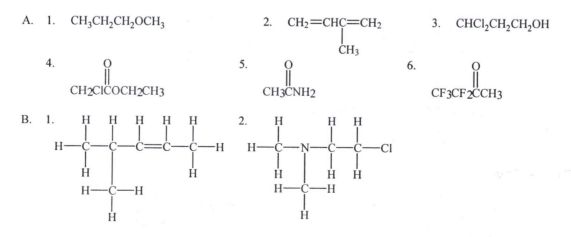

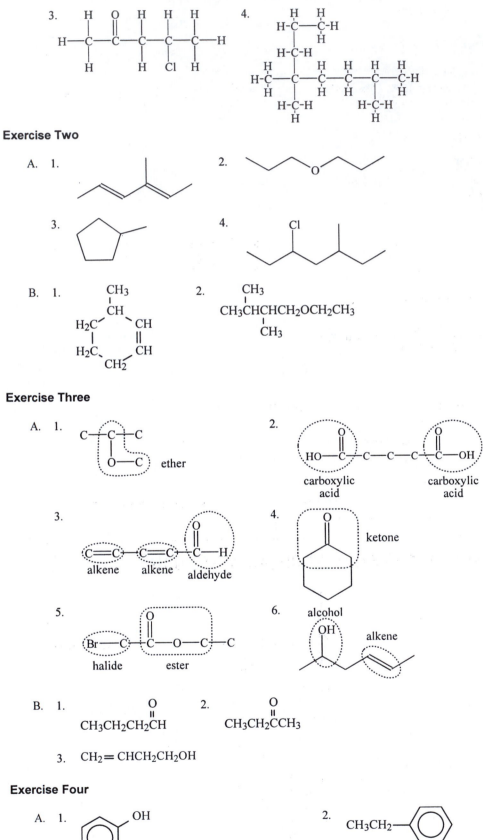

Exercise Two

A. 1.   2.

3.   4.

B. 1.   2.
   CH₃
   CH₃CHCHCH₂OCH₂CH₃
        CH₃

Exercise Three

A. 1. ether   2. carboxylic acid   carboxylic acid

3. alkene alkene aldehyde   4. ketone

5. halide   ester   6. alcohol   alkene

B. 1. CH₃CH₂CH₂CH   2. CH₃CH₂CCH₃

3. CH₂=CHCH₂CH₂OH

Exercise Four

A. 1.   OH   2. CH₃CH₂—

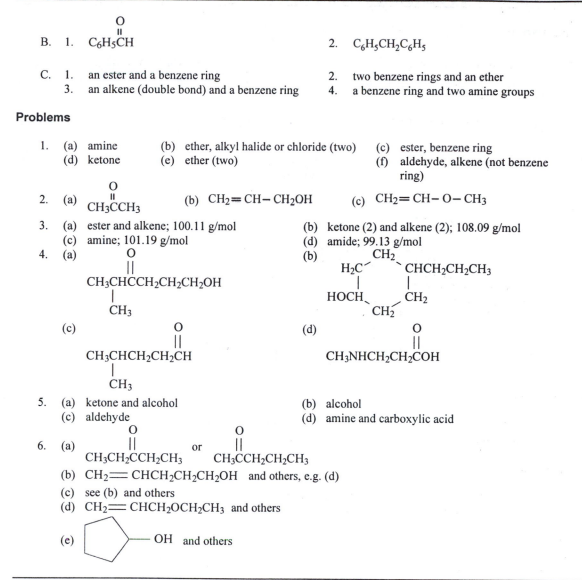

B. 1. $\overset{\overset{\displaystyle O}{\|}}{C_6H_5CH}$ 2. $C_6H_5CH_2C_6H_5$

C. 1. an ester and a benzene ring   2. two benzene rings and an ether
    3. an alkene (double bond) and a benzene ring   4. a benzene ring and two amine groups

Problems

1. (a) amine   (b) ether, alkyl halide or chloride (two)   (c) ester, benzene ring
    (d) ketone   (e) ether (two)   (f) aldehyde, alkene (not benzene
                                                                        ring)

2. (a) $\overset{\overset{\displaystyle O}{\|}}{CH_3CCH_3}$ (b) $CH_2\!=\!CH-CH_2OH$ (c) $CH_2\!=\!CH-O-CH_3$

3. (a) ester and alkene; 100.11 g/mol   (b) ketone (2) and alkene (2); 108.09 g/mol
    (c) amine; 101.19 g/mol   (d) amide; 99.13 g/mol

4. (a) $\overset{\overset{\displaystyle O}{\|}}{CH_3CHCCH_2CH_2CH_2OH}$ (b)
           $\underset{\displaystyle CH_3}{|}$

        (c) $\overset{\overset{\displaystyle O}{\|}}{CH_3CHCH_2CH_2CH}$ (d) $\overset{\overset{\displaystyle O}{\|}}{CH_3NHCH_2CH_2COH}$
             $\underset{\displaystyle CH_3}{|}$

5. (a) ketone and alcohol   (b) alcohol
    (c) aldehyde   (d) amine and carboxylic acid

6. (a) $\overset{\overset{\displaystyle O}{\|}}{CH_3CH_2CCH_2CH_3}$ or $\overset{\overset{\displaystyle O}{\|}}{CH_3CCH_2CH_2CH_3}$
    (b) $CH_2\!=\!CHCH_2CH_2CH_2OH$ and others, e.g. (d)
    (c) see (b) and others
    (d) $CH_2\!=\!CHCH_2OCH_2CH_3$ and others
    (e) 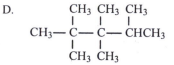 $-\,OH$ and others

---

LESSON 33:   NAMING SIMPLE ORGANIC COMPOUNDS

Exercise One

A. CH_2CH_3
            $|$
    $CH_3CH_2CHCH_2CH_2CH_3$

B. $CH_3\ \ CH_3$
         $|\quad\ \ |$
    $CH_3CCH_2CHCH_3$
         $|$
         CH_3

C. $CH_3\ \ CH_2CH_2CH_3$
        $|\quad\ \ |$
    $CH_3CHCHCHCH_2CH_2CH_2CH_3$
           $|$
           CH_3

D. $CH_3\ CH_3\ CH_3$
            $|\quad\ |\quad\ |$
    $CH_3-C-C-CHCH_3$
            $|\quad\ |$
            $CH_3\ CH_3$

Exercise Two

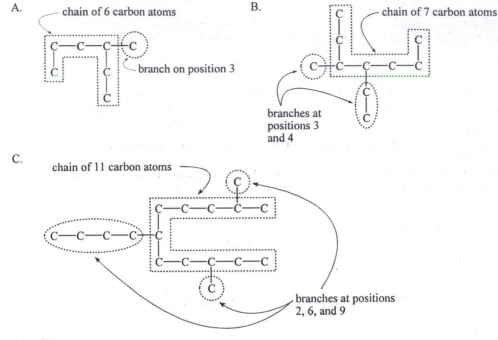

A. — chain of 6 carbon atoms — branch on position 3

B. — chain of 7 carbon atoms — branches at positions 3 and 4

C. — chain of 11 carbon atoms — branches at positions 2, 6, and 9

Exercise Three

A. 3-ethyl-2-methylpentane
C. 4-methyl-5-propylnonane

B. 1-methyl-4-propylcyclohexane
D. 6-ethyl-3-methylnonane

Exercise Four

A. 2-ethyl-4-methyl-1-pentene
C. 4-ethyl-3-methyl-1-hexyne

B. 2,3-dimethyl-2-heptene
D. cyclohexene

Exercise Five

A. 2,4-dichlorohexane
C. 3,4-dichloro-1,1,1-trifluorohexane

B. 4,6-dibromo-2,3,5-trimethyl-2-hexene
D. 2,4-difluoro-1-methylcyclohexane

Exercise Six

A. 4-methyl-2-pentanol
C. 4-ethylhexanoic acid

B. 5,6-dichloroheptanal
D. cyclohexanone

Exercise Seven

A. ethyl trifluoromethyl ether
C. methyl 2-propyl ether

B. ethyl methyl amine

Exercise Eight

A. 2-methylpropyl propanoate

B. cyclohexyl ethanoate (also called cyclohexyl acetate)

Exercise Nine

A. 1,2,4,5-tetrafluorobenzene
C. 1,3-dimethyl -5-propylbenzene

B. diphenyl ether

Exercise Ten

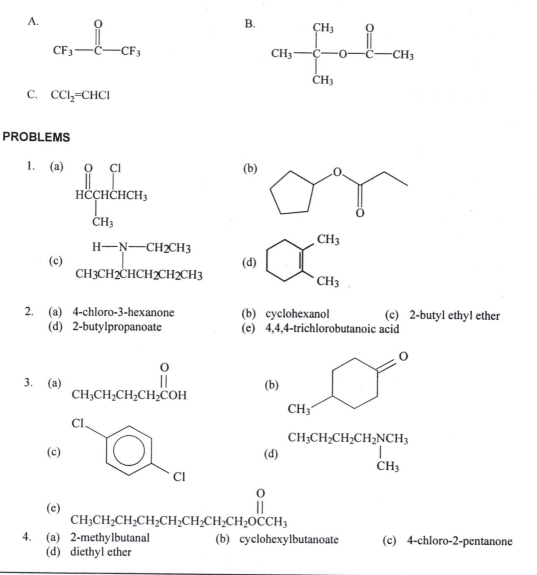

A.

CF_3—C—CF_3 (with O double bonded to C)

B.

CH_3—C(CH_3)(CH_3)—O—C(=O)—CH_3

C. CCl_2=$CHCl$

PROBLEMS

1. (a) HCCHCHCH$_3$ with O double bond, Cl and CH$_3$ substituents

   (b) cyclopentyl propanoate structure

   (c) H—N—CH$_2$CH$_3$; CH$_3$CH$_2$CHCH$_2$CH$_2$CH$_3$

   (d) cyclohexene with two CH$_3$ groups

2. (a) 4-chloro-3-hexanone   (b) cyclohexanol   (c) 2-butyl ethyl ether
   (d) 2-butylpropanoate   (e) 4,4,4-trichlorobutanoic acid

3. (a) $CH_3CH_2CH_2CH_2COH$ (with O double bond)   (b) 4-methylcyclohexanone structure

   (c) 1,4-dichlorobenzene structure

   (d) $CH_3CH_2CH_2CH_2NCH_3$ with CH$_3$ branch

   (e) $CH_3CH_2CH_2CH_2CH_2CH_2CH_2CH_2OCCH_3$ (with O double bond)

4. (a) 2-methylbutanal   (b) cyclohexylbutanoate   (c) 4-chloro-2-pentanone
   (d) diethyl ether

LESSON 34:   REACTIONS OF ORGANIC COMPOUNDS

Exercise One

A.

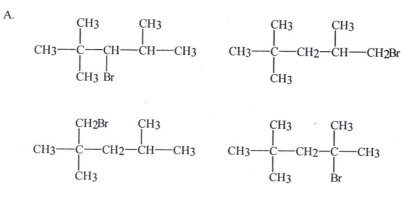

CH_3—C(CH_3)(CH_3Br)—CH(Br)—CH(CH_3)—CH_3

CH_3—C(CH_3)(CH_3)—CH_2—CH(CH_3)—CH_2Br

CH_3—C(CH_2Br)(CH_3)—CH_2—CH(CH_3)—CH_3

CH_3—C(CH_3)(CH_3)—CH_2—C(CH_3)(Br)—CH_3

B.

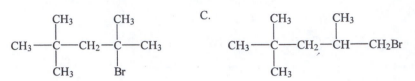

C.

D.   2-chloro-2,3-dimethylbutane

Exercise Two

A.

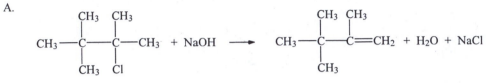

B.   HOCH₂CH₂CH₂CH₂OH

C.   1.   butane   2.   2-butanol   3.   2-butanol

Exercise Three

A.

CH₃—CH₂—CH—CH₃   and   CH₃—CH₂—CH—CH₃

B.

C.

Exercise Four

A.   cyclohexanol

B.

Exercise Five

A.

CH₃CCH₂CH₂—OH   HO—C—H
     alcohol   acid

B.

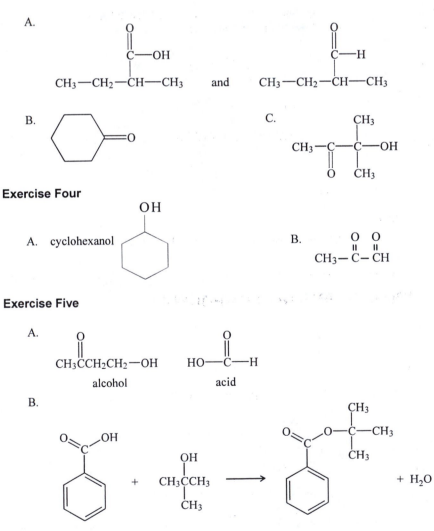

+ H₂O

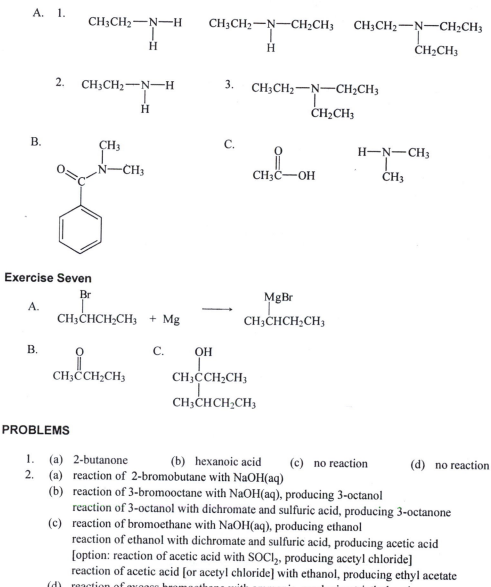

C.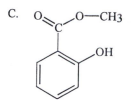

Exercise Six

A. 1. CH₃CH₂—N—H   CH₃CH₂—N—CH₂CH₃   CH₃CH₂—N—CH₂CH₃
            | | |
            H   H   CH₂CH₃

2. CH₃CH₂—N—H   3. CH₃CH₂—N—CH₂CH₃
            | |
            H   CH₂CH₃

B.   CH₃
       |
O=C—N—CH₃

C.   O
       ‖
   CH₃C—OH   H—N—CH₃
                       |
                      CH₃

Exercise Seven

A.   Br   MgBr
       | |
  CH₃CHCH₂CH₃ + Mg ⟶ CH₃CHCH₂CH₃

B.   O   C.   OH
       ‖ |
  CH₃CCH₂CH₃   CH₃CCH₂CH₃
                                  |
                             CH₃CHCH₂CH₃

PROBLEMS

1. (a) 2-butanone   (b) hexanoic acid   (c) no reaction   (d) no reaction
2. (a) reaction of 2-bromobutane with NaOH(aq)
   (b) reaction of 3-bromooctane with NaOH(aq), producing 3-octanol
       reaction of 3-octanol with dichromate and sulfuric acid, producing 3-octanone
   (c) reaction of bromoethane with NaOH(aq), producing ethanol
       reaction of ethanol with dichromate and sulfuric acid, producing acetic acid
       [option: reaction of acetic acid with SOCl₂, producing acetyl chloride]
       reaction of acetic acid [or acetyl chloride] with ethanol, producing ethyl acetate
   (d) reaction of excess bromoethane with ammonia, producing triethyl amine

3. (a)   O
               ‖
       CH₃CH₂CH₂COCHCH₃
                    |
                   CH₃

   (b)   O
                    ‖
         CH₃CH₂CH₂COH

   (c) CH₃CH₂CH₂CH₂OH

LESSON 35:   MOLECULAR STRUCTURES AND ISOMERS

Exercise One

    A.   isomers   B.   same   C.   nonisomeric   D.   isomers

Exercise Two

    A.   geometric isomers   B.   structural isomers
    C.   same   D.   nonisomeric
    E.   geometric isomers

Exercise Three

    A.   chiral   B.   achiral   C.   chiral   D.   achiral

Exercise Four

A.   B.   no stereogenic center

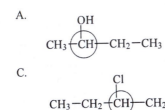

C.   D.

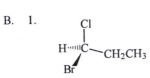   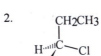

Exercise Five

    A. 1. R   2. R

    B. 1.   2.

Exercise Six

$l = 5$ cm $= 0.5$ dm

$c = (0.28 \text{ g}) / (75 \text{ mL}) = 0.0037$ g/mL

$\alpha_D^{25} = -13°$

$$[\alpha]_D^{25} = \frac{-13°}{(0.5 \text{ dm})(0.0037 \text{ g/mL})} = -7.0 \times 10^3 \text{ °mL g}^{-1} \text{ dm}^{-1}$$

Exercise Seven

    A.   diastereomers   B.   enantiomers   C.   diastereomers

PROBLEMS

    1.   optical   2.   geometric   3.   both

    4.   neither   5.   geometric   6.   neither

LESSON 36:   POLYMERS

Exercise One

A.

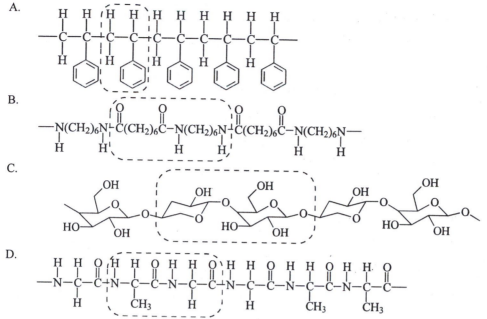

B.

C.

D.

Exercise Two

A.

B.

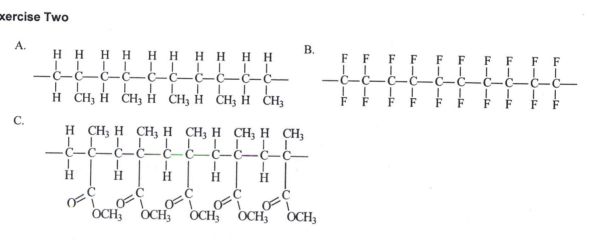

C.

Exercise Three

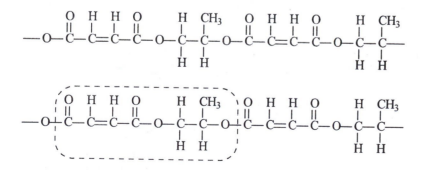

Exercise Four

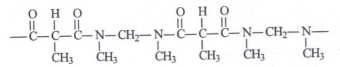

Exercise Five

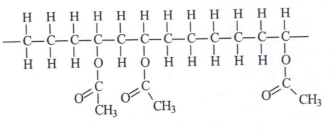

Exercise Six

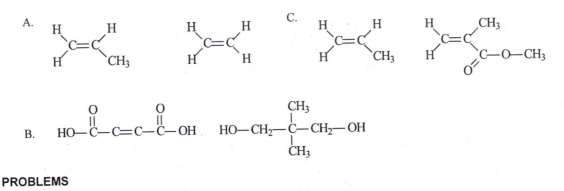

A.

B.

C.

PROBLEMS

1. (a)

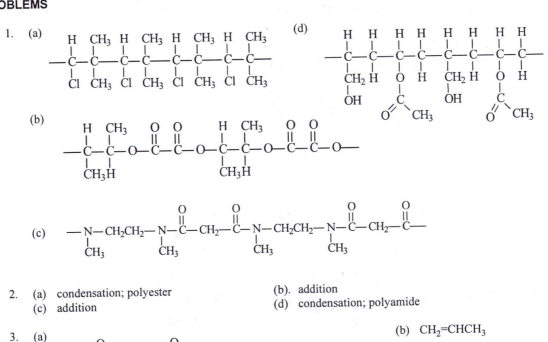

(b)

(d)

(c)

2. (a) condensation; polyester   (b). addition
    (c) addition   (d) condensation; polyamide

3. (a)   (b) $CH_2=CHCH_3$

HO—C(=O)⬡C(=O)—OH $HOCH_2CH_2OH$

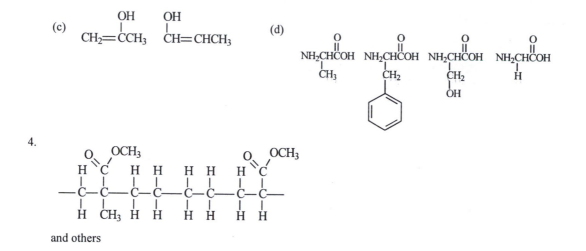

(c) $CH_2 = \overset{\overset{\displaystyle OH}{|}}{C}CH_3$ $\overset{\overset{\displaystyle OH}{|}}{CH} = CHCH_3$

4.

and others

LESSON 37:   SIGNIFICANT FIGURES AND EXPONENTIAL NOTATION

Exercise One

A.   three
D.   two

B.   four
E.   one

C.   three
F.   five

Exercise Two

A. 3.42×10^4
D. 6.024×10^{-9}

B. 7.29×10^{-4}
E. 2×10^5

C. 4.020×10^1
F. 3.00×10^{-3}

Exercise Three

A. 0.00327×10^7

B. 633×10^{-4}

C. 106.4×10^{-5}

Exercise Four

A.   23.6
D.   1.30

B.   24
E.   7.2

C.   0.36
F.   0.36

Exercise Five

A. 1.58×10^5
D. 4×10^{-18}

B. 2.9×10^{-7}

C. 1.5×10^{-3}

Exercise Six

A. 3.72×10^{-3}
D. 1.8×10^5

B. 2.63×10^{-2}

C. 6.92×10^7

INDEX